COMPOSITION AND FUNCTION OF THE EXTRACELLULAR MATRIX IN THE HUMAN BODY

Edited by **Francesco Travascio**

Composition and Function of the Extracellular Matrix in the Human Body
http://dx.doi.org/10.5772/61601
Edited by Francesco Travascio

Contributors

Cevdet Kaya, Max Happel, Oliver Lieleg, Fabienna Arends, Zongjin Li, Yan Nie, Shuaiqiang Zhang, Na Liu, Katarzyna Grzelkowska-Kowalczyk, Stephen Crocker, Cory Willis, Kay Ohlendieck, Sandra Murphy, Zhonglin Tang, Zishuai Wang, Eva-Maria Riso, Elena Vecino, Jessica Kwok, Takao Sakai, Irene Tadeo, Tomás Álvaro, Samuel Navarro, Rosa Noguera, Stéphane Chabaud, Tamas Fulop, Sylvain Vigier, Sonia Lobo-Planey

Published by InTech

Janeza Trdine 9, 51000 Rijeka, Croatia

Publishing Process Manager Edi Lipovic

Technical Editor InTech DTP team

Cover InTech Design team

First published June, 2016

Printed in Croatia

Additional hard copies can be obtained from orders@intechopen.com

Composition and Function of the Extracellular Matrix in the Human Body, Edited by Francesco Travascio

p. cm.

Print ISBN 978-953-51-2415-3

Online ISBN 978-953-51-2416-0

For EU product safety concerns:
IN TECH d.o.o., Prolaz Marije Krucifikse Kozulić 3, 51000 Rijeka, Croatia,
info@intechopen.com or visit our website at intechopen.com.

Meet the editor

Dr. Francesco Travascio received a doctoral degree in Biomedical Engineering at the University of Miami. He is currently Assistant Professor at the College of Engineering of the University of Miami, where he conducts research in the field of orthopedics at the Biomechanics Research Laboratory. In particular, Dr. Travascio investigates the mechanisms regulating the homeostasis of the extracellular matrix of cartilaginous tissues and the etiology of its degeneration with the ultimate goal of developing therapeutic approaches for tissue regeneration.

Contents

Preface

The extracellular matrix (ECM) is an ensemble of non-cellular components present within all tissues and organs of the human body. The ECM provides structural support for scaffolding cellular constituents and biochemical and biomechanical support for those events leading to tissue morphogenesis, differentiation and homeostasis. Essential components of all ECMs are water, proteins and polysaccharides. However, their composition, architecture and bioactivity greatly vary from tissue to tissue in relation to the specific role the ECM is required to assume. This book overviews the role of the ECM in different tissues and organs of the human body.

The composition and the function of the ECM in skeletal muscle are discussed with respect to the role of ECM in the motor function, as well as on mechanisms of remodeling during adaptation to physical activity. Special attention is paid to changes in skeletal muscle ECM assembly and function during muscle aging and in diseases such as fibrosis and myopathies.

Rather than static and passive, the ECM is an active and dynamic environment for cells. The importance of ECM remodeling in the central nervous system is hereby discussed. For instance, ECM proteins may modulate glial functions, such as response to injury or inflammation. Also, it is reported how the study of ECM plasticity in the juvenile and adult brain may promote complex forms of learning and may have further implications of regenerative and therapeutic potential.

The importance of the ECM is vividly illustrated by the wide range of pathologies and syndromes that arise from genetic abnormalities in its proteins. For instance, alterations of the basal lamina contribute to several physiological malfunctions. Also, the analysis of ECM composition can provide prognosis and progress of pathologies of the urinary tract, wound healing, mechanisms of collagen network organization following tissue damage, pathological processes of chronic fibrotic diseases as well as bone and cartilage repair.

Stem cell therapy and tissue engineering are two strategic approaches for repair and regeneration of damaged tissue, which are often combined together. The success of a cell therapy is greatly based on the role the ECM has in providing a supportive microenvironment for cell viability and biosynthesis. For several applications, the ECM must be synthetically reproduced. At its conclusion, this book reports ample reviews on the available biomaterials and on the current techniques in tissue engineering for the production of near-to-native synthetic ECMs.

Francesco Travascio, Ph.D.
University of Miami
Coral Gables, FL, USA

Section 1

ECM of Skeletal Muscles

Chapter 1

The Importance of Extracellular Matrix in Skeletal Muscle Development and Function

Katarzyna Grzelkowska-Kowalczyk

Additional information is available at the end of the chapter

http://dx.doi.org/10.5772/62230

Abstract

Skeletal muscle tissue makes up approximately 40% of the total body mass in adult mammals. Contractile muscle fibers building skeletal muscle tissue are coated by an extracellular matrix material (ECM), accounting for 1–10% of the muscle mass. The ECM in skeletal muscle was initially considered as a structure, providing mechanical support for bearing force transmission. Now it is evident that muscle cells adhere to and connect with the ECM, also for signaling, and the ECM provides an appropriate and permissive environment for muscle development and functioning. This chapter summarizes current knowledge on the role of ECM components in skeletal muscle growth and regeneration, which is of great importance for potential therapeutic interventions. It also focuses on the contribution of ECM in the motor function of skeletal muscle as well as on mechanisms mediating muscle ECM remodeling during adaptation to physical activity. The role of the ECM in the metabolic function of skeletal muscle tissue and the ECM disturbances associated with insulin resistance are described. Finally, the attention is paid on potential implications of changes in skeletal muscle ECM assembly and function in health and disease.

Keywords: myogenesis, satellite cell niche, exercise, insulin signaling, myopathies

1. Introduction

Skeletal muscle tissue, making up approximately 40% of the total body mass in adult mammals, is composed of multinucleated contractile muscle cells, myofibers. Intramuscular connective tissue accounts for 1–10% of the skeletal muscle mass and varies substantially between muscles [1]. Muscle fibers are coated by an extracellular matrix material (ECM), called the basement membrane, and composed of two layers: an internal, basal lamina, directly linked to the plasma membrane of myofibers (sarcolemma), and an external, reticular lamina. Extracellular matrix surrounding muscle fibers is composed of collagens (dominated by collagen IV), laminins, fibronectin, and proteoglycans, formed by glycosaminoglycans bound to a protein core. Initially, it was considered as a structure that provides mechanical support for bearing force transmission [2]. The ECM gives mechanical structure to myofibers during contractions, provides the tissue with elastic properties, and participates in the transmission of force from the myofiber to tendon. It also serves as a basic mechanical support for nerves and vessels present in skeletal muscle tissue, and determines the spatial barrier between endothelium and muscle cell surface. A great progress in cell biology, molecular biology and genetics, gives new insight into skeletal muscle biology, and now it becomes evident that cells adhere to and connect with the ECM not only for structural stability but also for signaling. The integrins, heterodimeric transmembrane receptors comprising unrelated alpha and beta subunits, play critical roles in converting extracellular signals to intracellular responses (outside-in signaling) as well as in extracellular matrix interactions based upon intracellular changes (inside-out signaling) [3]. They bind to ECM or cell surface ligands and link the actin microfilament system with ECM, providing a connection between the ECM, the cytoskeleton, and signaling molecules. Integrins are considered as sensors of tensile strain at the cell surface, and together with the cytoskeleton form a mechanically sensitive organelle. Despite the large overall number of integrin receptor complexes, skeletal muscle integrin receptors are limited to seven alpha subunits, all associated with the beta1 integrin subunit. Integrin signal transmission depends on the activation of focal adhesion kinase (FAK), a nonreceptor tyrosine kinase, localized at focal adhesions. Integrin engagement causes the formation of transient signaling complex, initiated by the recruitment of Src-family protein SH2 to the FAK Tyr-397 autophosphorylation site, and by serving as a signaling element in cytoskeleton-associated networks [4]. Integrin-linked kinase (ILK), initially considered as a kinase, but, in fact, incapable to perform phosphorylation due to pseudoactive domain, mediates interactions of integrins with numerous cellular proteins and regulates focal adhesion assembly, cytoskeleton organization, and signaling [5]. The major enzymes responsible for the ECM breakdown under physiological conditions are matrix metalloproteinases (MMPs, or matrixins), which belong to a family of zinc-dependent and calcium-activated neutral endopeptidases, comprising secreted and membrane-associated members. MMPs are involved in degradation of the ECM and basement membrane; however, they also cleave a variety of other ECM-related proteins, including cytokines, chemokines, and growth factors [6]. There is some specificity of certain MMPs toward collagen types, that is, MMP-2 and 9 (gelatinases) primarily degrade type IV collagen and other compounds of the ECM in muscle, whereas MMP-1 and 8 (collagenases) traditionally are thought to break down types I and III collagen, being more relevant for tendon. MMP activities are regulated by tissue inhibitors of matrix metalloproteinases (TIMPs). Four TIMPs, responsible for the inhibition of over 20 MMPs, are identified; of these, TIMP-1 and TIMP-2 are capable of inhibiting, of all MMPs, preferably MMP-2 and 9, respectively [1]. In addition to MMP-dependent mechanisms, TIMPs can alter cell growth and survival in an MMP-independent manner, mediated by integrins. A good example is TIMP-2, which regulates beta1 integrin expression and the size of myotubes formed during myoblast differentiation [7]. MMPs play an important role in skeletal muscle cell growth and differentiation, as they are engaged in release and activation of cytokines and growth factors. The main contributors to ECM assembly in skeletal muscle are resident fibroblasts; however, muscle cells also synthesize and secrete numerous ECM components and ECM-related molecules, suggesting their active and direct participation in ECM remodeling. Thus, the composition of the ECM exerts mechanical, metabolic, hemodynamic, and angiogenic effects in skeletal muscle tissue. The extracellular matrix and its receptors also provide an appropriate and permissive environment for muscle development and some ECM components, in addition to muscle-specific factors, can serve as good indicators of skeletal muscle functioning. This chapter summarizes current knowledge on the role of ECM components related to skeletal muscle development and regeneration, which is of great importance for potential therapeutic interventions. It also focuses on the contribution of ECM in motor and metabolic functions of skeletal muscle tissue. Finally, the attention is paid on potential implications of changes in ECM assembly and function in health and disease.

2. Extracellular matrix in regulation of muscle stem cell niche

Fetal stage is crucial for skeletal muscle development, when muscle fibers are formed by fusion of mesodermal progenitor cells, myoblasts. During postnatal period, the number of myofibers remains constant; however, the size of each myofiber can increase by fusion with muscle stem cells, called satellite cells. Skeletal muscle is one of the most adaptive tissues in the body, and the adult regenerative myogenesis after muscle injury depends on satellite cells. These cells are normally quiescent, but in response to overloading or muscle damage, they become activated; that is, they begin to proliferate, and their progeny myoblasts terminally differentiate and fuse with one another or with existing myofibers to restore the contractile muscle apparatus and normal tissue architecture [8].

Proper muscle regeneration depends on the cross-talk between the satellite cells and their microenvironment (cell niche). According to the stem cell niche concept, the structural and biochemical stimuli emanating from surrounding environment determine the fate of stem cells present in tissues. Muscle satellite cells exist in highly specific niches, consisting of the basement membrane of myofibers, different types of resident cells (i.e., fibroblasts, adipocytes, etc.), vascular and neural systems, and extracellular matrix [9]. Each of these niche elements exerts profound effects on satellite cell functioning. Satellite cells reside between the basal lamina and the apical sarcolemma of myofibers, covered in laminin. They bind to collagen type IV and laminin through integrins, which also connect with collagen type VI and several proteoglycans, that is, perlecan and decorin. The ECM protein, nidogen (or entactin), supports cross-links between laminin and collagens. Basal lamina directly contacts satellite cells and separates them from muscle interstitium. It also acts as a mechanical barrier to prevent migration of satellite cells and their loss from normal muscle, and could be involved in repressing satellite cell mitosis and differentiation in the absence of muscle injury [2]. On the other site of the satellite cell niche, the myofiber sarcolemma links to the basal lamina, more particularly to laminin, through the dystroglycan complex [10]. Myofibers influence satellite cell behavior as a result of the physical interactions and by the secretion of paracrine factors. Nerves and associated neuromuscular apparatus exert their effects through the control of myofiber activity. Fibroblasts primarily contribute to matrix formation and, as adipocytes, secrete paracrine factors. Circulating blood transports hormones and other systemic factors; endothelial cells lining blood vessels serve as a source of growth factors, whereas immune cells, infiltrating muscle tissue upon injury, transiently affect satellite cells through the secretion of cytokines [11].

An important function of muscle progenitor cell niche is maintaining the balance between quiescence and activation. The quiescent satellite cells sense the stiffness of their niche through integrins and express various matrix proteins to maintain the stable ECM structure. Within the ECM, growth factors and other bioactive molecules are sequestered, supporting the "quiet" state [10]. Communication between the ECM and satellite cells is essential in the regulation of cellular events crucial for muscle growth and repair, such as gene expression, cell proliferation, adhesion, and differentiation of activated satellite cells. In response to muscle injury, components of the basal lamina are degraded by matrix metalloproteinases, and growth factors and signaling molecules are liberated, which is essential for regulation of processes ongoing in activated satellite cells. Presence of the ECM is required for muscle stem cells to respond to growth factors [12]. Proteoglycans expressed on the surface of satellite cells function as low-affinity receptors and bind to the secreted, inactive growth factor precursors, including hepatocyte growth factor (HGF), basic fibroblast growth factor (bFGF), epidermal growth factor (EGF), insulin-like growth factor isoforms (IGF-1, IGF-2), originating from myofibers, satellite cells, muscle-residing cells, or serum. All these growth factors play crucial roles in myogenesis, and in vivo exist in matrix-associated form. Some ECM molecules, that is, decorin [13], fibronectin [14], and laminin [15], can bind to and suppress the activity of myostatin, a negative regulator of muscle cell proliferation and differentiation. Through interactions with these growth factors, the extracellular matrix regulates the ability of skeletal muscle satellite cells to proliferate or differentiate. Differences in the expression of proteoglycans alter satellite cell responsiveness to the growth factor, that is, overexpression of glypican-1 (heparin sulfate proteoglycan) in satellite cells increases their responsiveness to FGF-2, whereas underexpression diminishes cell proliferation and differentiation [12]. Taken together, the major components of basal lamina orchestrate muscle satellite cell development by presentation of mitogenic and myogenic factors. Muscle cells play an active role in creating their own microenvironment via ECM remodeling. Supporting this idea, numerous studies prove changes in expression and/or secretion of proteoglycans, metalloproteinases, adhesion molecules, and growth factors in regenerating muscle tissue and differentiating myoblasts [8, 16–18]. Activated satellite cells dynamically remodel their niche via transient high expression of fibronectin, and knockdown of this protein expression in satellite cells markedly impaired the ability to repopulate the niche [19].

When satellite cells move to the injured site, the surrounding ECM should be degraded for allowing cell migration. Matrix metalloproteinases degrade extracellular matrix components such as collagens, elastin, fibronectin, laminin, and proteoglycans. MMPs play an important role in creating cell niche in regenerating muscle and are essential for satellite cell activation, migration, and differentiation. Expression of matrix metalloproteases is up-regulated upon satellite cell activation, whereas transcripts for proteinase inhibitors are high in quiescent cells

[20]. Migration of satellite cells underneath the basement membrane requires the expression of MMP-2, 3, 7, 9, and 10. The specific inhibition of these MMPs decreases the migration velocity and increases the sustainability of moving direction of myoblasts in vitro [21]. Among the MMPs expressed in skeletal muscle, MMP-2 and 9 appear particularly critical. MMP-2 is secreted by satellite cells and regenerating myofibers, whereas MMP-9 is expressed by leukocytes and macrophages. Upon injury, the release of the nitric oxide synthase (NOS) from damaged basal lamina leads to nitric oxide (NO) production, which in turn up-regulates protein level and activity of MMP-2 and 9. Activated proteases degrade collagen IV, facilitating satellite cell migration across the basement membrane to injured regions [22]. The most important details concerning the ECM structure and cues emanating from cellular elements of muscle satellite cell niche are summarized in **Figure 1**.

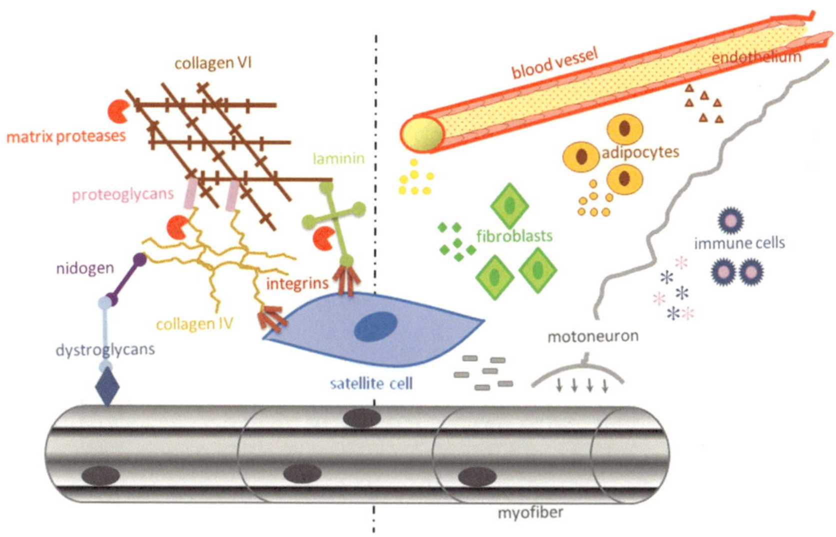

Figure 1. Schematic representation of the complex microenvironment (niche) of satellite cells in skeletal muscle. Left part illustrates the networks and cross-linkings of major ECM proteins in the immediate environment of muscle satellite cells. Right part presents contributions of cellular components in creating the satellite cell niche. The small symbols represent humoral factors released by different types of cells (the colors used correspond with the source of appropriate bioactive factors).

The role of specific niche for muscle stem cell's self-renewal and differentiation is supported by observations, that after removal from the microenvironment, the satellite cells quickly withdraw from quiescence, begin to proliferate, and lose their myogenic properties. On the other hand, myogenic cells cultured on the ECM extracted from large thigh adult muscles manifest enhanced proliferation and differentiation in comparison to standard growth surfaces [23]. In order to study the role of specific ECM components in creating the niche of muscle stem cells, in vitro cell culture models are employed, where the environmental conditions can be easily controlled. In such experiments, primary muscle stem cells derived from muscle tissue are cultured in vitro on surfaces coated with the ECM components (i.e. collagen, laminin, fibronectin, gelatin, or Matrigel—a balanced mixture of different ECM proteins) to mimic the muscle extracellular environment. Usually, the primary muscle stem cells show distinct proliferation and differentiation pattern, as well as different muscle-specific and ECM-related gene expressions, dependent on the coating type used [9]. These experiments reveal that the loss of mitogenic and/or myogenic potential of muscle stem cells, due to their transfer from the specific niche to an ex vivo situation, could be reduced by using some ECM components/mixture coating. For example, fibronectin and laminin could be used for sorting myoblasts from fibroblasts. Such observations are of great interest and importance in tissue engineering and stem cell therapies.

3. Changes in ECM assembly and function during myogenesis

Skeletal muscle growth and development is a complex process controlled by interactions between muscle cells and surrounding microenvironment. Several cellular events take place during skeletal myogenesis, that is, migration of muscle precursor cells, proliferation of myoblasts, cell cycle arrest, and myoblast terminal differentiation, followed by transcription of muscle-specific genes and myoblast fusion. Muscle cell differentiation is governed by an ordered sequence of the expression of muscle regulatory factors (MRFs) such as MyoD (Myoblast determination protein), Myf-5 (Myogenic factor-5), myogenin, and MRF-4 [24]. The commitment of muscle precursor cells requires MyoD expression, whereas the proliferation arrest and terminal myoblast differentiation are driven by myogenin, a key transcription factor, which activates skeletal muscle-specific genes encoding creatine kinase, myosin heavy chain, and acetylcholine receptor. The formation of myotubes expressing muscle-specific genes is essential for the specialization of myofiber function.

The importance of extracellular matrix molecules as a part of myogenesis signaling mechanism has also been demonstrated. An inhibition of cell-surface transmembrane proteoglycan sulfation results in delayed proliferation and altered MyoD expression, indicating that heparan sulfate is required for proper progression of the early myogenic program [25]. Neither the expression of myogenin nor its localization to myoblast nuclei was sufficient to drive skeletal muscle differentiation, if the cell–ECM interactions were inhibited [26]. Inhibition of proteoglycan sulfation in myoblast cultures strongly affects ECM synthesis and deposition, and induces the expression of the osteogenic markers (alkaline phosphatase and osteocalcin), without alterations in expression of specific muscle transcription factors, such as MyoD and Myf-5 [27]. The above observations support the idea that extracellular matrix provides stimuli for muscle cell development, which are independent of muscle-specific factor expression.

Myogenesis is accompanied by remodeling of ECM proteins as well as by changes in integrin receptor expression pattern [28]. Fibronectin and laminins display an opposite pattern of changes in time during myogenesis, that is, myoblasts secrete a large amount of fibronectin, which is replaced by laminins in myotubes. As a consequence, the location of these proteins in muscle is different, that is, fibronectin is absent in regions manifesting active myogenesis, whereas laminin adjoins myotubes. In myoblasts subjected to differentiation in vitro, fibronectin is detected primarily in the extracellular environment as a thick mesh. At the same time, laminin appears ultimately in the cytosolic fraction, which confirms delayed synthesis of this protein during myogenesis, in comparison to fibronectin [29]. During myogenic differentiation, the laminin synthesis increases, and laminin begins to accumulate in the medium in soluble form, followed by the formation of insoluble cell-associated fraction [30]. Both fibronectin and laminin per se can affect myogenesis. Fibronectin promotes myoblast adhesion and proliferation; however, it inhibits differentiation and participates in collagen fibrillogenesis, thus providing the ECM assembly [1]. Fibronectin also stimulates adhesion of fibroblasts and may facilitate dedifferentiation of myoblasts. This protein is required for somitogenesis, and it may function to regulate fiber organization and limit fast-twitch muscle fiber length [31]. Laminin is crucial for several processes involved in myogenesis, as it enhances myoblast proliferation, migration, and alignment preceding the fusion. Myotube formation is markedly impaired in the absence of laminin [2]. Changes in integrin receptor expression pattern reflect the ECM remodeling during myogenesis. Proliferating and migrating myoblasts express high amounts of the fibronectin-binding alpha5beta1 integrin, while during myotube formation they switch to the laminin-binding alpha7beta1 integrin, which is the major integrin receptor in adult muscles [32]. Moreover, there is a negative cooperativity between alpha7 and alpha5 integrin subunits. Transfection with integrin alpha7 resulted in the marked reduction of alpha5beta1 surface complex expression and its decreased affinity to fibronectin in myoblasts. Such a relationship may play an important role in determining functional regulation of integrins during myogenesis. A critical phase of myogenesis is the fusion of mononucleated myoblasts and the formation of long multinucleated myotubes. Myoblast fusion and myotube formation are associated with increased expression of integrin alpha3, particularly abundant in myotube membrane [29]. Overexpression of the full-length integrin alpha3 subunit induces myoblast fusion, whereas the inhibition of integrin alpha3 extracellular domain impairs this process [33]. Myogenesis is largely normal in the absence of alpha4, alpha5, alpha6, and alpha7 integrin subunits, indicating the redundancy in integrin functions. In contrast, disruption of the integrin beta1 in vivo and in vitro profoundly influences myogenesis. Lack of integrin beta1 had no apparent effect on the migration and proliferation of myoblasts; however, clear alterations occur at the later stages of myogenesis and are manifested by impaired fusion [34]. According to an early study, muscle-specific integrin beta1, appearing in a doublet form, was used as a marker of differentiation [35]. Integrin beta1 subunit is also involved in muscle cell survival. In response to the activation of integrin beta1, focal adhesion kinase phosphorylates tyrosine at residue 397, leading to the activation of cell survival signal transduction and inhibition of apoptosis [36]. Moreover, FAK appears as a mediator by which integrins may regulate myoblast fusion. Specific disruption of gene encoding FAK suppresses the transcription of caveolin 3 and integrin subunit beta1D isoform, both considered as essential for morphological muscle differentiation. As a consequence, the cell fusion and myotube formation are defective, while the expression of muscle terminal differentiation genes, such as sarcomeric alpha-actin, alpha-actinin, and vinculin, remain unaltered [37]. It suggests a specific role of FAK in the regulation of cell fusion, as a part of the myogenic differentiation program.

A characteristic feature of proliferating and quiescent undifferentiated myoblasts is the high expression of a disintegrin and metalloprotease, ADAM12, which combines features of adhesion molecules and proteinases [38]. ADAM12 cleaves insulin-like growth factor binding proteins IGFBP3 and IGFBP5, and heparin binding-EGF. The cysteine-rich domain of

ADAM12 supports cell adhesion by binding to syndecan-4, whereas the cytoplasmic domain interacts with signaling proteins, that is, tyrosine kinase Src phosphatidylinositol-3-kinase, and cytoskeletal alpha-actinin 1 and 2 [38]. ADAM12 in transiently upregulated at the onset of differentiation, whereas other ADAMs, such as ADAM9, 10, 15, 17, and 19 are expressed at all stages of myogenesis [35]. Inhibition of ADAM12 by siRNA approach in myogenic cell cultures was accompanied by lower expression of both quiescent markers (p130 and p27 proteins) and differentiation markers (cell cycle inhibitor p21 and myogenin). Overexpression of ADAM12 induces a quiescent-like phenotype and does not stimulate differentiation. Possible role of ADAM12 in myogenesis is associated with the preservation of "reserve pool" of myoblasts, which do not trigger the myogenic differentiation program and maintain regeneration potential. A 100 kDa long isoform of ADAM12 is increased in myoblasts differentiating for 3 days in the presence of IL-1beta [39] and IGF-I [29], suggesting similar effects of proinflammatory cytokines and anabolic growth factors on ECM regulation at early stages of myogenesis. On the other hand, there are studies that implicate the involvement of ADAM12 in the fusion of muscle cells. The expression of ADAM12 and integrin alpha9 subunit parallels and culminates at the time of myoblast fusion, and inhibition of ADAM12/alpha9beta1 integrin interaction dramatically impairs this process [40]. ADAM12 is linked to the cytoskeleton via alpha-actinin [35], and thus the cytoskeleton may regulate the distribution of ADAM12 on the cell surface, where localized proteolysis and/or cell–cell contacts occur [41]. The most important modifications of the ECM structure and function associated with skeletal myogenesis are depicted in **Figure 2**.

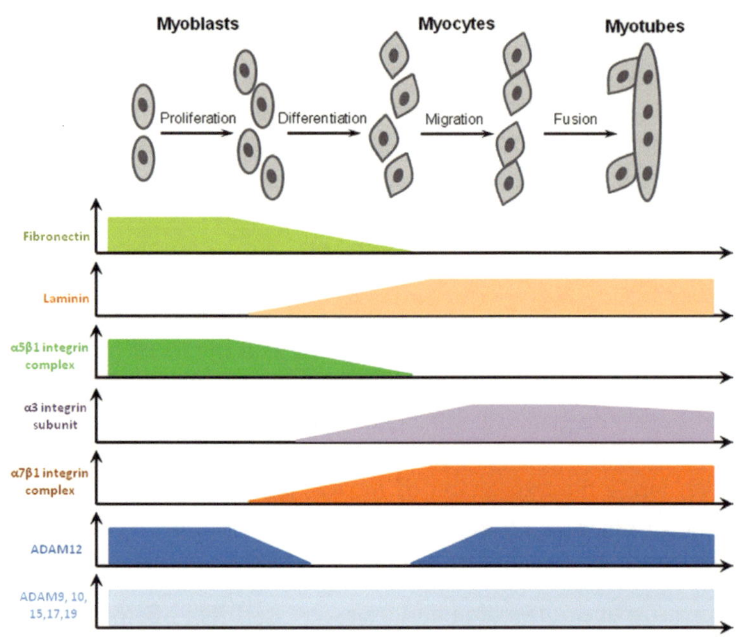

Figure 2. Schematic illustration of ECM remodeling and ECM-related proteins level/activity during skeletal myogenesis. The most important events during myogenic development are presented in the upper panel.

http://dx.doi.org/10.5772/62230

4. ECM and the motor function of skeletal muscle

Skeletal muscle provides structural support, enables the body to maintain posture, and controls motor movements. Muscle tissue is strong, flexible, stress-resistant, and in view of its mechanical properties, it consists of contractile elements (i.e. sarcomeres) and elastic components, supported by extracellular matrix. Majority of ECM elements, which account for muscle strength and elasticity, reside in the basement membrane, especially in basal lamina. The basic structure of basal lamina consists of different networks of triple-helical collagen IV, composed of alpha chains, and the major noncollagenous protein, laminin, which is a heterodimer of alpha, beta and gamma chains. The collagen network contains covalent cross-links; moreover, distinct networks are linked by another noncollagenous protein, nidogen (entactin). These major elements display several further functions: (i) they possess multiple sites binding other protein of basal lamina, (ii) they anchor components of reticular lamina to basal lamina, and (iii) they serve as ligands for membrane-associated receptors (i.e., integrins, dystroglycans, etc.), which interact with cytoskeleton [2]. Taken together, in the context of the mechanical function of skeletal muscle, the extracellular matrix may be considered as a series of networks that connect reticular lamina, basal lamina, sarcolemma, and cytoskeletal structure.

Overload of healthy skeletal muscle leads to myofiber hypertrophy and ECM remodeling, the processes that are thought to contribute to muscle growth. Several ECM components are controlled by the level of mechanical loading, and multiple intracellular proteins involved in mechanotransduction signaling are suggested, including focal adhesion kinase (FAK), paxillin, integrin-linked kinase (ILK), and mitogen-activated protein kinase (MAPK) [1]. The latter is crucial for the conversion of mechanical load to tissue adaptation, transmitting signaling from the cytosol to the nucleus. Laminin, integrin alpha7, and integrin-linked kinase (ILK) are all critical for mechanical stability of skeletal muscle [42]. ILK is recruited to the myotendinous junction, which requires the presence of laminin in the ECM and integrin alpha7 in sarcolemma. Moreover, ILK is essential for strengthening the adhesion of the muscle fibers with the ECM and acts with the dystrophin/dystroglycan adhesion complex in maintaining mechanical stability of skeletal muscles.

Endurance and resistance exercises accelerate the turnover of ECM components in skeletal muscle. Several studies reveal an increase in collagen synthesis and accumulation induced by exercise (summarized in [43]). Transcription of genes encoding types I, III, and IV collagen increases after endurance training. In another study, endurance exercise augments concentration of type IV collagen in slow (soleus), but not in fast (rectus femoris) muscle. Matrix metalloproteinases are activated in human skeletal muscle in response to voluntary exercise, and the expression and time pattern indicate differences between the MMPs in regards of production sites as well as in the regulating mechanism. TIMPs are often activated together with MMPs in response to physical activity, indicating the simultaneous stimulation and inhibition of the ECM degradation. Probably, MMPs' activation precede TIMPs' activation, and the latter serve as "guardians" of degradation termination, providing limits in the ECM breakdown [1]. Levels of MMP-2, 14, and TIMP-1 mRNA in muscle tissue increase after 10 days of training. MMP-2 and 9 proteins were both present in the ECM, around myofibers and capillaries, but MMP-2 was also visible within the skeletal muscle fibers [44].

Mechanical loading induces the secretion of TGF-beta, PDGF, and bFGF in tendon fibroblasts; moreover, it increases the expression of collagen and other ECM components, such as proteoglycans. TGF-beta stimulates collagen formation and reduces its degradation, also via activation of the TIMPs, together with a suppression of MMPs, leading to the ECM accumulation. TGF-beta is known to function as a modulator of ECM proteins and to induce both collagen gene activation and protein formation. In a human model of microdialysis of the Achilles tendon, both local and systemic levels of TGF-beta increase in response to 1 h of running, proving a release of this cytokine from tissues that are mechanically activated during exercise and suggesting a role in the response to mechanical loading in vivo [1]. Mechanical loading induces FGF release from skeletal muscle cells in vitro. Several isoforms of FGFs exist; of these, basic FGF (or FGF2) and, to a lesser extent, the acidic FGF (FGF1) stimulate fibroblast proliferation and collagen synthesis. Interleukin-6 (IL-6) is considered as a physical activity-associated myokine released from working muscles [45]. It can stimulate fibroblasts to increase the synthesis of collagens, glycosaminoglycans, hyaluronic acid, and chondroitin sulfates. Increased expression of IL-6 is necessary for the regulation of ECM remodeling during the hypertrophic response of skeletal muscle to overload [46]. Mechanical activity increases expression of IL-1beta in human and rabbit tendon cells, leading to increased MMPs activity, diminished collagen synthesis, and initiating tissue degradation and remodeling in response to loading. IGF-I is directly involved in skeletal muscle ECM synthesis after mechanical loading. This growth factor increases the expression of types I and III collagen in intramuscular fibroblasts. Bioavailability of IGF-I is controlled by IGF-binding proteins, and increased proteolysis of IGFBPs occurs in response to prolonged training in humans. Interestingly, MMPs can degrade IGFBPs, which provides a possible mechanism of regulation of the free IGF-I in skeletal muscle tissue and circulation.

The blood flow in skeletal muscle is tightly coupled with the metabolic demands of contracting myofibers. During exercise, local mechanisms cause rapid dilation of muscle arterioles to increase the flow of blood to the working muscle. It appears that fibronectin fibrils in the extracellular matrix transduce signals from actively shortening skeletal muscle fibers to local blood vessels to increase blood flow. Skeletal muscle contraction alters the conformation of ECM fibronectin, which results in transient exposure of specific matricryptic sites. These sequences are not exposed in the soluble form of ECM molecules, but may be expressed due to structural or conformational changes, providing "a reserve" of signaling sites activated during ECM remodeling. Matricryptic fibronectin sites (FNIII-1) interact with FNIII-1H receptors on smooth muscle cells and/or skeletal muscle fibers. This activates the neuronal nitrogen oxide (NO) synthase to release NO, which leads to smooth muscle relaxation, vasodilation, and increased blood flow. Thus, FNIII-1 sites in ECM fibronectin serve as important mechanical coupling between skeletal muscle contraction and arteriolar dilation [47]. **Figure 3** summarizes the cellular mechanisms activated during exercise leading to skeletal muscle ECM remodeling.

http://dx.doi.org/10.5772/62230

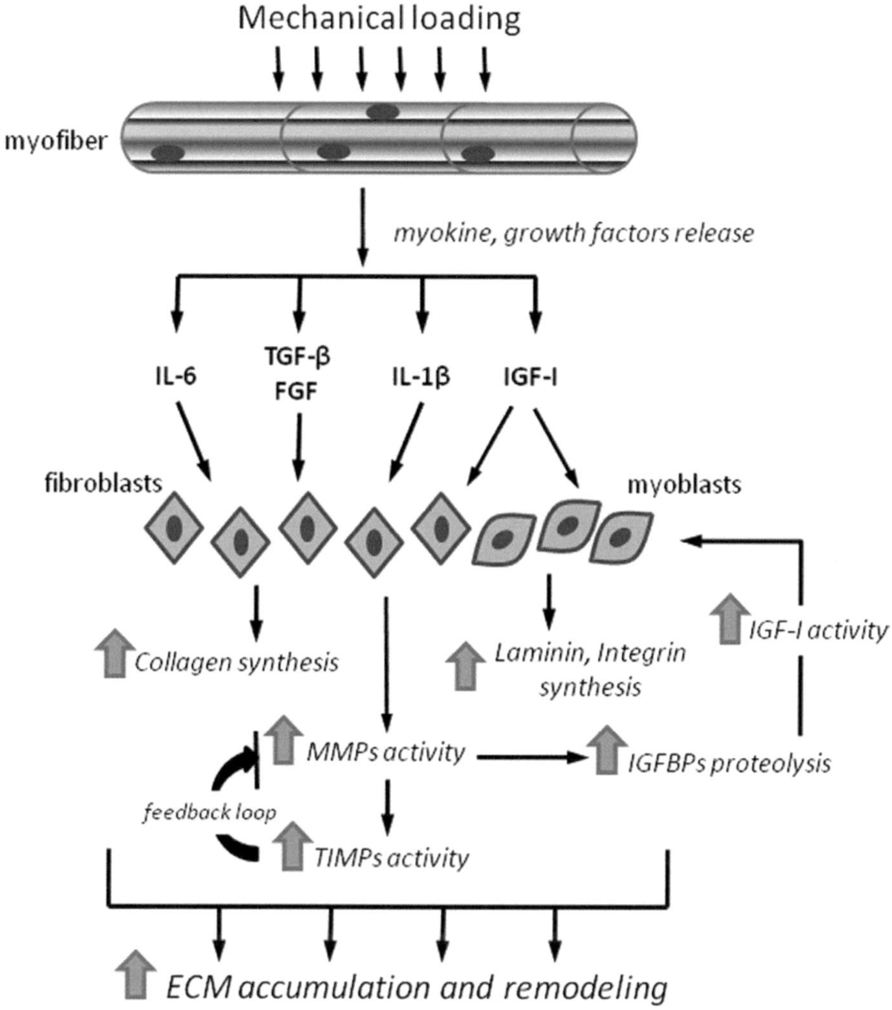

Figure 3. Proposed schema illustrating the mechanisms of alterations in the ECM in skeletal muscle induced by mechanical loading. → means activation/stimulation, ⊣ means inhibition. Gray block arrows indicate total stimulation of particular processes resulting from the regulation of upstream pathways.

5. ECM and the metabolic function of skeletal muscle

Skeletal muscle is a key insulin-sensitive tissue, important in maintaining homeostasis, due to its relatively large mass and energy needs [48,49]. Postprandial, insulin-stimulated glucose disposal in skeletal muscle results from the activation of a complex signaling network with multiple alternative and complementary pathways. Insulin binding to the insulin receptor causes tyrosine autophosphorylation of the receptor beta-subunit, activation of its intrinsic tyrosine kinase, and subsequent phosphorylation of several intracellular proteins, including insulin receptor substrate (IRS) proteins [50]. This leads to the recruitment of further signaling components such as phosphatidylinositol-3 kinase (PI-3 kinase), the tyrosine phosphatase SHPTP2, the growth factor receptor-binding protein-2 (GRB-2), as well as protein serine/threonine kinases: phosphoinositide-dependent protein kinase (PDK1), protein kinase B (PKB), atypical isoforms of protein kinase C (PKC) lambda and zeta, mitogen-activated protein kinase (MAPK), and others, which support the signal divergency and function as messengers for various biological effects of insulin. Regarding postprandial glucose uptake in skeletal muscle, the activation of insulin signaling leads to the translocation of the insulin-responsive glucose transporter, Glut4, from intracellular storage sites to cell surface membrane, which is a critical step in cellular glucose utilization. Dysregulation of any step of this process in skeletal muscle results in insulin resistance, predisposing for diabetes.

There is an important cross-talk between extracellular matrix and insulin signaling in skeletal muscle. Integrin engagement stimulates both IRS-1-associated PI-3 kinase activity and PKB/Akt pathway. Integrin receptor beta1 subunit increases insulin-stimulated IRS phosphorylation, IRS-associated PI-3 kinase, and activation of PKB (summarized in [51]). Regulation of focal adhesion kinase (FAK) by integrin receptors modulates insulin-dependent cytoskeleton organization, glucose transport, and glycogen synthesis in myoblasts [4]. FAK can interact with IRS-1, PI-3 kinase, PKC, and glycogen synthase kinase-3beta, leading to translocation of Glut4. A decrease in tyrosine phosphorylation and activation of FAK was reported in skeletal muscle of insulin-resistant Spraque-Dowley rats fed with a high-fat diet, as well as in insulin-resistant C2C12 myoblasts [52]. The expression of IRS-1 mRNA is abolished in FAK knockout mouse fibroblasts. Apart from the regulation of skeletal muscle insulin signaling and action by FAK, the reciprocal interaction is documented. It appears that FAK tyrosine phosphorylation, essential for skeletal muscle differentiation, is modulated by insulin. Insulin causes an increase in FAK phosphorylation in proliferating myoblasts, while in differentiating cells, there is an inhibition of FAK phosphorylation [53]. Under insulin resistance, the phosphatase PTEN and SHIP2, usually recognized as negative regulators of insulin signaling, are up-regulated, and they impair insulin action through FAK dephosphorylation [54]. The integrin-linked kinase (ILK) can phosphorylate and activate PKB, and function as its potential upstream regulator. Integrin beta1 knockout mice manifest an impairment of insulin-stimulated skeletal muscle glucose uptake and glycogen synthesis in skeletal muscle, resulting from marked reduction in ILK expression and concomitant decrease in PKB phosphorylation.

Insulin resistance is tightly associated with the ECM remodeling in muscle, and the ECM defects predisposing to diabetes-related symptoms are known. The deposition of collagens, the most abundant structural ECM components, is increased in insulin-resistant muscles, both in humans and rodent experimental models [55]. Synthesis of fibronectin, laminin, and collagen IV is up-regulated by high glucose and diabetes [56], which may lead to basement membrane thickening and the development of diabetes-associated microangiopathy. Similarly, a high-fat diet causes an increase in collagen IV in skeletal muscle [57]. As MMPs are responsible for the degradation of all components of the ECM, their dysregulation is also implicated in the pathology of diabetes and obesity. MMP-9 activity in skeletal muscle is decreased in high fat-fed mice, and it is related inversely to muscle collagen deposition and

http://dx.doi.org/10.5772/62230

directly to muscle insulin resistance [58]. The genetic deletion of MMP-9 worsens diet-induced muscle insulin resistance, indicating that this metalloproteinase is necessary to protect against more serious metabolic disturbances associated with high fat feeding. Collagen V, widely expressed and a less abundant fibrillar protein, which regulates collagen fibril geometry and strength, is important for skeletal muscle glucose homeostasis. Mutant mice lacking *col5a3* gene manifest hyperglycemia, glucose intolerance, and insulin resistance [59]. Skeletal muscle of these mutants is defective in glucose uptake and mobilization of intracellular Glut4 glucose transporter to the plasma membrane in response to insulin.

High-ambient glucose markedly elevates the level of fibronectin in myogenic cells in vitro and causes a decrease in cellular content of the full length 100 kDa form of ADAM12, without affecting integrin alpha5 and integrin beta1 subunit expressions [60]. Such alterations could result in the disturbances in ECM remodeling and accumulation, which in turn contribute to the impairment of the myogenic differentiation, manifested by decrease in MyoD, myogenin,

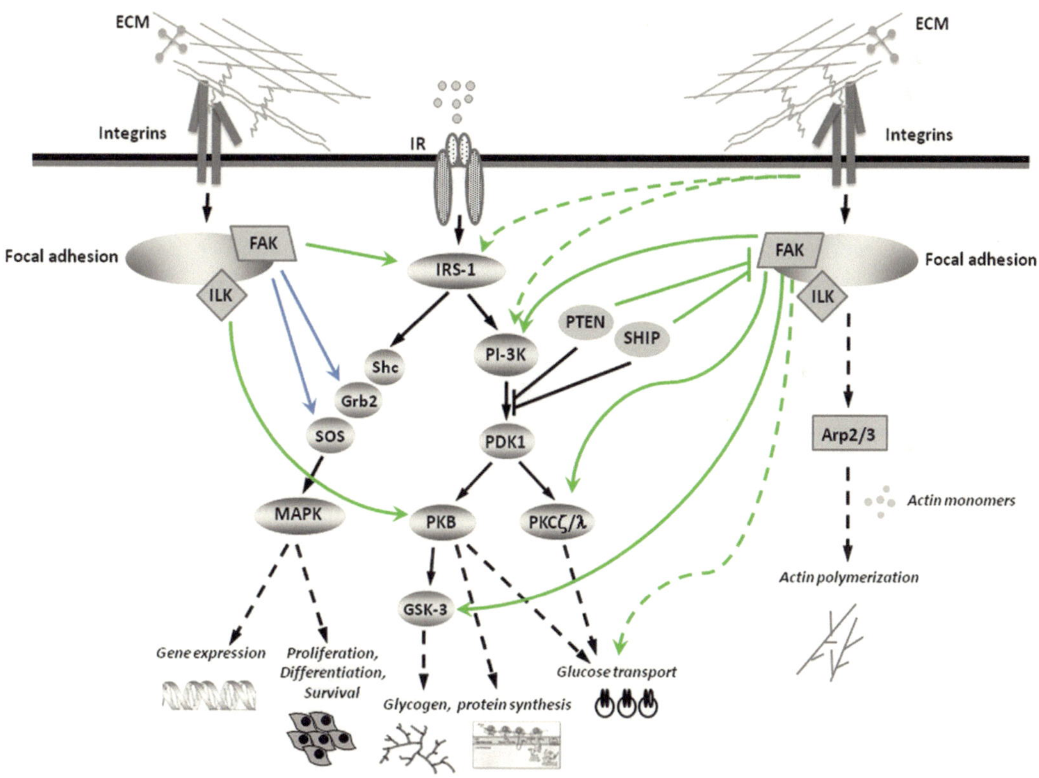

Figure 4. Proposed schema illustrating the cross-talk between insulin signaling and ECM signaling in skeletal muscle. To clarify the picture, both the insulin signaling pathway and ECM signaling are markedly simplified, as they present only the most important linkings and biological effects. Solid lines mean direct connections, dashed lines mean indirect effects. → means activation/stimulation, ⊣ means inhibition. Green lines indicate interactions between insulin- and integrin-activated pathways (direct or indirect) reported in skeletal muscle. Blue lines indicate interactions described in other cell types [61], and only potentially functioning in skeletal muscle tissue.

myosin heavy chain levels, and fusion index. In view of an important cross-talk between ECM and insulin signaling [4,51], the high-glucose-induced alterations in ECM can, at least partly, contribute to the attenuated insulin and growth factors' action in skeletal muscle under hyperglycemia and diabetes. The ECM turnover also plays a role in the metabolic regulation of skeletal muscle in the pathology of diet-induced insulin resistance. **Figure 4** illustrates the most important points of the cross-talk between insulin signaling and the ECM-related signaling cascades in skeletal muscle.

6. Alterations of muscle ECM components in health and disease

Skeletal muscles have a great ability to adapt and regenerate, and usually injured areas of muscle tissue are replaced with healthy contractile fibers, which results in a full recovery and mechanical function, or even gains in muscle mass and strength. The regenerative potential of skeletal muscle is markedly impaired in aging and several diseases, and is associated with disturbances of muscle ECM.

The efficiency of skeletal muscle regeneration decreases with age, and this phenomenon is primarily associated with the changes in satellite cell functions, that is, the reduction of cell number and/or proliferative capacity. The basal lamina of aged muscle is thicker, and its structure is irregular and amorphous. During aging, type IV collagen abundance increases in slow muscles, whereas laminin increases in fast muscles, which can affect the ability of the basal lamina to store and release growth factors and other bioactive compounds creating the satellite cell microenvironment. Another alteration in the basal lamina during aging is increase in osteopontin, the cytokine, which negatively regulates myogenesis in vitro and muscle regeneration in vivo. Satellite cell niche during aging also contains other extracellular matrix-associated negative regulators of muscle differentiation, such as transforming growth factor-beta and Wnt signaling [10]. The composition of local milieu in aged muscles changes also due to the remodeling of the neuromuscular junction, the functional alterations in endothelial cells (i.e., apoptosis) and in immune cells (i.e. impaired chemotaxis). Taken together, the satellite cell niche during aging shifts toward an increasingly inhibitory influence on satellite cell activity and muscle regeneration potential [11]. Age-related changes in content and structure of ECM in skeletal muscle can also lead to decrease in the local expression or limited access to matricriptic sites in fibronectin [47]. As a consequence, the disturbances of vascular dilation in working muscles can occur and contribute to the impairment of skeletal muscle function in aging.

Muscle atrophy can be divided into primary muscular disease and secondary muscular disorders [62], both of them characterized by pathological changes in muscle ECM. Genetic studies of several primary muscle diseases show that the basement membrane is critical for the maintenance of muscle integrity. In all of these diseases, skeletal muscle tissue development is normal, but they are characterized by progressive muscle weakness, fibrosis, and fatty infiltration [2]. Muscle dystrophy can result from the loss or impairment of any of the elements in the reticular lamina–basal lamina–sarcolemma–cytoskeleton linkage. The examples include laminin alpha2 and its transmembrane receptors, that is, integrin alpha7 and dystroglycan (congenital muscular dystrophy), dystrophin (Duchenne muscular dystrophy), and the dystroglycan- and dystrophin-associated sarcoglycans (limb girdle muscular dystrophy), collagen IV (Walker-Warburg syndrome), and the alpha chains of collagen VI, which connect reticular lamina to basal lamina (Ulrich congenital muscular dystrophy and Bethlem myopathy). For muscle maintenance, both structural and signaling properties of the basement membrane are required. Signaling from laminin alpha2 provides survival stimuli for myofibers; thus, its absence in congenital muscle dystrophy is associated with high level of apoptosis.

The best known primary muscular disease is Duchenne muscular dystrophy (DMD) resulted from the mutation in the gene encoding dystrophin, which leads to the lack of dystrophin protein at the sarcolemma of muscle fibers. It is characterized by progressive muscle weakness associated with continuous degeneration and regeneration of skeletal myofibers [63]. The loss of satellite cell regenerative capacity due to continuous needs for regeneration may contribute to disease progression in DMD [64]. The absence of dystrophin per se can exert a direct influence on the homeostasis of the ECM by allowing leakage of cellular components to the extracellular space or by abnormal cellular uptake of growth factors, cytokines, and enzymes. This in turn can affect muscle fibroblasts, either directly by altering their adhesion properties or indirectly by interacting with molecules released by muscle or inflammatory cells. Apart from disturbances in dystrophin complex, muscles from DMD patients manifest decreased accumulation of laminin alpha2 and beta1, increased accumulation of collagen IV, higher expression of integrin alpha 7, and profibrotic cytokines, which inhibit myogenesis, that is, TGF-beta and osteopontin [10]. An up-regulation of decorin, myostatin, and MMP-7 transcripts and proteins, as well as a down-regulation of MMP-1 and TIMP-3 expression are reported in DMD fibroblasts [65]; the latter may result in increased ECM deposition leading to tissue fibrosis.

Diabetic muscles are more vulnerable to exercise-induced myofiber damage than healthy muscles. Diabetes-induced changes in skeletal muscle concern the structure of the basement membrane and the activities of the enzymes of collagen synthesis. Microarray analysis of skeletal muscle transcriptom in streptozotocin-diabetic mice show reduced gene expression of types I, III, IV, V, VI, and XV collagen. Moreover, mRNA expressions for some noncollagenous proteins and proteoglycans, that is, elastin, thrombospondin-1, laminin-2, and decorin, as well as connective tissue growth factor (CTGF) increase in diabetic muscles [43]. This can alter the structure of the basement membrane in a less collagenous direction and affect its properties. Patients with congestive heart failure (CHF) experience increased skeletal muscle fatigue. The mechanism underlying this phenomenon involves increased MMPs' activity and collagen content, accompanied by a drop in VEGF expression, which may disturb the normal contractile function of skeletal muscle [66].

Apart from the alteration, loss or impairment of some specific ECM components in physiological and pathological states, the stiffness of the ECM per se, seems to be an important factor regulating muscle cell growth and function. Resting skeletal muscle and myotubes in culture display a similar elastic stiffness (elastic modulus approximately 12 kPa), whereas aged and dystrophic muscles are several-fold stiffer (summarized in [22]). The reason for such alterations is increased extracellular matrix accumulation, especially collagen deposition by fibroblasts, resulted from repeated muscle degeneration–regeneration events. Another mechanism could be the accumulation of advanced glycation end products (AGEs), nonspecific cross-linkings mediated by condensation of reducing sugars with amino groups, observed in aging and pathological states with elevated glucose levels. Glycated intramuscular ECM has stiffer and more load-resistant structure; however, it also manifests a reduced ability to adapt to altered loading, probably due to decreased collagen turnover. Moreover, AGEs up-regulate the expression of CTGF in fibroblasts, which can promote fibrosis in old and diabetic individuals [1]. Numerous studies using in vitro model reveal that proper myogenesis requires an optimal ECM stiffness and that both softer and stiffer coatings markedly diminish the myoblast's ability to proliferate and differentiate. These results confirm the importance of mechanical and biophysical stimuli in skeletal muscle maintenance and remodeling.

Acknowledgements

This work was supported through funding from the Department of Physiological Sciences, Faculty of Veterinary Medicine, Warsaw University of Life Sciences (SGGW).

Author details

Katarzyna Grzelkowska-Kowalczyk

Address all correspondence to: k_grzel_kow@poczta.fm

Faculty of Veterinary Medicine, Department of Physiological Sciences, Warsaw University of Life Sciences (SGGW), Warsaw, Poland

References

[1] Kjaer M: Role of extracellular matrix in adaptation of tendon and skeletal muscle to mechanical loading. Physiol Rev. 2004;84:649–698. DOI: 10.1152/physrev.00031.2003

[2] Sanes JR: The basement membrane/basal lamina of skeletal muscle. J Biol Chem. 2003;278:12601–12604. DOI: 10.1074/jbc.R200027200

[3] Askari JA, Buckley PA, Mould PA, Humphries JM: Linking integrin conformation to function. J Cell Sci. 2009;122:165–170. DOI: 10.1242/jcs.018556

[4] Huang D, Khoe M, Ilic D, Bryer-Ash M: Reduced expression of focal adhesion kinase disrupts insulin action in skeletal muscle cells. Endocrinology 2006;147:3333–3343. DOI: 10.1210/en.2005-0382

[5] Qin J, Wu C: ILK: a pseudokinase in the center stage of cell-matrix adhesion and signaling. Curr Opin Cell Biol. 2012;24:607–613. DOI: 10.1016/j.ceb.2012.06.003

[6] Hrabec E, Naduk J, Stręk M, Hrabec Z: Type IV collagenases (MMP-2 and MMP-9) and their substrates—intracellular proteins, hormones, cytokines, chemokines and their receptors. (pol). Adv Biochem. 2007;53:37–45.

[7] Lluri G, Langlois GD, Soloway PD, Jaworski DM: Tissue inhibitor of metalloproteinase-2 (TIMP-2) regulates myogenesis and beta1 integrin expression in vitro. Exp Cell Res. 2008;314:11–24. DOI: 10.1016/j.excr.2007.06.007

[8] Griffin CA, Apponi LH, Long KK, Pavlath GK: Chemokine expression and control of muscle cell migration during myogenesis. J Cell Sci. 2010;123:3052–3060. DOI:10.1242/jcs.066241

[9] Wilschut KJ, Haagsman HP, Roelen BAJ: Extracellular matrix components direct porcine muscle stem cell behavior. Exp Cell Res. 2010;316:341–352. DOI: 10.1016/j.yexcr.2009.10.014

[10] Thomas K, Engler AJ, Meyer GA: Extracellular matrix regulation in the muscle satellite cell niche. Connect Tissue Res. 2015;56:1–8. DOI: 10.3109/03008207.2014.947369

[11] Gopinath SD, Rando TA: Stem cell review series: aging of the skeletal muscle stem cell niche. Aging Cell. 2008;7:590–598. DOI: 10.1111/j.1474-9726.2008.00399.x

[12] Dodson MV, Hausman GJ, Guan L, Du M, Rasmussen TP, Poulos SP, Mir P, Bergen WG, Fernyhough ME, McFarland DC, Rhoads RP, Soret B, Reecy JM, Velleman SG, Jiang Z. Skeletal muscle stem cells from animals. I. Basic cell biology. Int J Biol Sci. 2010;6:465–474. DOI: 10.7150/ijbs.6.465

[13] Kishioka Y, Thomas M, Wakamatsu J, Hattori A, Sharma M, Kambadur R, Nishimura T: Decorin enhances the proliferation and differentiation of myogenic cells through suppressing myostatin activity. J Cell Physiol. 2008;215:856–867. DOI: 10.1002/jcp.21371

[14] Miura T, Kishioka Y, Wakamatsu J, Hattori A, Nishimura T: Interaction between myostatin and extracellular matrix components. Anim Sci. 2010;81:102–107. DOI: 10.1111/j.1740-0929.2009.00700.x

[15] Yasaka N, Suzuki K, Kishioka Y, Wakamatsu J, Nishimura T: Laminin binds to myostatin and attenuates its signaling. Anim Sci. 2013;84:663–668. DOI: 10.1111/asj.12052

[16] Goetsch SC, Hawke TJ, Galladro TD, Richardson JA, Garry DJ: Transcriptional profiling and regulation of the extracellular matrix during muscle regeneration. Physiol Genomics. 2003;15:261–271

[17] Henningsen J, Rigbolt KT, Blagoev B, Pedersen BK, Kratchmarova I: Dynamics of the skeletal muscle secretome during myoblast differentiation. Mol Cell Proteom. 2010;9:2482–2496. DOI: 10.1074/mcp.M110.002113

[18] Grzelkowska-Kowalczyk K, Wicik Z, Majewska A, Tokarska J, Grabiec K, Kozłowski M, Milewska M, Błaszczyk M: Transcriptional regulation of important cellular processes in skeletal myogenesis through interferon-γ. J Interferon Cytokine Res. 2015;35:89–99. DOI 10.1089/jir.2014.0018

[19] Betzinger CF, Wang YX, von Maltzahn J, Soleimani VD, Yin H, Rudnicki MA: Fibronectin regulates Wnt7a signaling and satellite cell expansion. Cell Stem Cell. 2013;12:75–87. DOI: 10.1016/j.stem.2012.09.015

[20] Pallafacchina G, François S, Regnault B, Czarny B, Dive V, Cumano A, Montarras D, Buckingham M: An adult tissue-specific stem cell in its niche: a gene profiling analysis of in vivo quiescent and activated muscle satellite cells. Stem Cell Res. 2010;4:77–91. DOI: 10.1016/j.scr.2009.10.003

[21] Nishimura T, Nakamura K, Kisioka Y, Kato-Mori Y, Wakamatsu J, Hattori A: Inhibition of matrix metalloproteinases suppresses the migration of skeletal muscle cells. J Muscle Res Cell Motil. 2008;29:37–44. DOI: 10.1007/s10974-008-9140-2

[22] Yin H, Price F, Rudnicki MA: Satellite cells and the muscle stem cell niche. Physiol Rev. 2013;93:23–67. DOI: 10.1152/physrev.00043.2011

[23] Stern MM, Myers RL, Hammam N, Stern KA, Eberli D, Kritchevsky SB, Soker S, Van Dyke M: The influence of extracellular matrix derived from skeletal muscle tissue on the proliferation and differentiation of myogenic progenitor cells ex vivo. Biomaterials 2009;30:2393–2399. DOI: 10.1016/j.biomaterials.2008.12.069

[24] Zammit PS, Patridge TA, Yablonka-Reuveni Z: The skeletal muscle satellite cells: the stem cell that came in from the cold. J Histochem Cytochem. 2006;54:1177–1191. DOI: 10.1369/jhc.6R6995.2006

[25] Cornelison DD, Filla MS, Stanley HM, Rapraeger AC, Olwin BB: Syndecan-3 and syndecan-4 specifically mark skeletal muscle satellite cells and are implicated in satellite cell maintenance and muscle regeneration. Dev Biol. 2001;239:79–94. DOI: 10.1006/dbio.2001.0416

[26] Osses N, Brandan E: ECM is required for skeletal muscle differentiation independently of muscle regulatory factor expression. Am J Physiol Cell Physiol. 2002;282:C383–C394. DOI: 10.1152/ajpcell.00322.2001

[27] Osses N, Casar JC, Brandan E: Inhibition of extracellular matrix assembly induces the expression of osteogenic markers in skeletal muscle cells by a BMP-2 independent mechanism. BMC Cell Biol. 2009;10:73. DOI: 10.1186/1471-2121-10-73

[28] Knoblauch A, Will C, Goncharenko G, Ludwig S, Wixler V: The binding of Mss4 to alpha-integrin subunits regulates matrix metalloproteinase activation and fibronectin remodeling. FASEB J. 2007;21:497–510. DOI: 10.1096/fj.06-7022com

[29] Grzelkowska-Kowalczyk K, Grabiec K, Tokarska J, Gajewska M, Błaszczyk M, Milewska M: Insulin-like growth factor-I increases laminin, integrin subunits and metallo-

protease ADAM12 in mouse myoblasts. Folia Biol. (Krakow) 2015;63:241–247. DOI: 10.3409/fb63_4.241

[30] Olwin BB, Hall ZW: Developmental regulation of laminin accumulation in the extracellular matrix of a mouse muscle cell line. Dev Biol. 1985;112:359–367. DOI: 10.1016/0012-1606(85)90407-5

[31] Snow CJ, Peterson MT, Khalil A, Henry CA: Muscle development is disrupted in zebrafish embryos deficient for fibronectin. Dev Dyn. 2008;237:2542–2553. DOI: 10.1002/dvdy.21670

[32] Mayer U: Integrins: redundant or important players in skeletal muscle? J Biol Chem. 2003;278:14587–14590. DOI: 10.1074/jbc.R200022200

[33] Brzóska E, Bello V, Darribere T, Moraczewski J: Integrin *a*3 subunit participates in myoblast adhesion and fusion in vitro. Differentiation 2006;74:105–118. DOI: 10.1111/j.1432-0436.2005.00059.x

[34] Schwander M, Leu M, Stumm M, Dorchies OM, Ruegg UT, Schittny J, Muller U: β1 integrins regulate myoblast fusion and sarcomere assembly. Dev Cell. 2003;4:673–685.

[35] Galliano MF, Huet C, Frygelius J, Polgren A, Wewer UM, Engvall E: Binding of ADAM12, a marker of skeletal muscle regeneration, to the muscle-specific actin-binding protein, alpha -actinin-2, is required for myoblast fusion. J Biol Chem. 2000;275:13933–13939. DOI: 10.1074/jbc.275.18.13933

[36] Li X, McFarland DC, Velleman SG: Transforming growth factor-beta1-induced satellite cell apoptosis in chickens is associated with beta1 integrin-mediated focal adhesion kinase activation. Poult Sci. 2009;88:1725–1734. DOI: 10.3382/ps.2008-00534

[37] Quach NL, Biressi S, Reichardt LF, Keller C, Rando TA: Focal adhesion kinase signaling regulates the expression of caveolin 3 and beta1 integrin, genes essential for normal myoblast fusion. Mol Biol Cell. 2009;20:3422–3435. DOI: 10.1091/mbc.E09-02-0175

[38] Cao Y, Zhao Z, Gruszczynska-Biegala J, Zolkiewska A: Role of metalloprotease disintegrin ADAM12 in determination of quiescent reserve cells during myogenic differentiation in vitro. Mol Cell Biol. 2003;23:6725–6738. DOI: 10.1128/MCB.23.19.6725-6738.2003

[39] Grabiec K, Tokarska J, Milewska M, Błaszczyk M, Gajewska M, Grzelkowska-Kowalczyk K: Interleukin-1ß stimulates early myogenesis of mouse C2C12 myoblasts: the impact on myogenic regulatory factors, extracellular matrix components, IGF binding proteins and protein kinases. Pol J Vet Sci. 2013;16:255–264. DOI: 10.2478/pjvs-2013-0036

[40] Lafuste P, Sonnet C, Chazaud B, Dreyfus PA, Gherardi RK, Wewer UM, Authier FJ: ADAM12 and α9β1 integrin are instrumental in human myogenic cell differentiation. Mol Biol Cell. 2005;16:861–870. DOI: 10.1091/mbc.E04-03-0226

[41] Wewer UM, Albrechtsen R, Engvall E. ADAM12. The long and the short of it. In: Hooper NM, Lendeckel U, editors. The ADAM Family of Proteases. Springer; 2005. pp. 123–146.

[42] Postel R, Vakeel P, Topczewski J, Knöll R, Bakkers J: Zebrafish integrin-linked kinase is required in skeletal muscles for strengthening the integrin-ECM adhesion complex. Dev Biol. 2008;318:92–101. DOI: 10.1016/j.ydbio.2008.03.024

[43] Lehti TM, Silvennoinen M, Kivelä R, Kainulainen H, Komulainen J: Effects of streptozotocin-induced diabetes and physical training on gene expression of extracellular matrix proteins in mouse skeletal muscle. Am J Physiol Endocrinol Metab. 2006;290:E900–E907. DOI: 10.1152/ajpendo.00444.2005

[44] Rullman E, Norrbom J, Strömberg A, Wågsäter D, Rundqvist H, Haas T, Gustafsson T: Endurance exercise activates matrix metalloproteinases in human skeletal muscle. J Appl Physiol. 2009;103:804–812. DOI: 10.1152/japplphysiol.90872.2008

[45] Pedersen BK, Febbraio MA: Muscle as an endocrine organ: focus on muscle-derived interleukin-6. Physiol Rev. 2008;88:1379–1406. DOI: 10.1152/physrev.90100.2007

[46] White JP, Reecy JM, Washington TA, Sato S, Le ME, Davis JM, Wilson LB, Carson JA: Overload-induced skeletal muscle extracellular matrix remodeling and myofibre growth in mice lacking IL-6. Acta Physiol. (Oxf) 2009;197:321–332. DOI: 10.1111/j. 1748-1716.2009.02029.x

[47] Hocking DC. Titus PA, Sumagin R, Sarelius IH: Extracellular matrix fibronectin mechanically couples skeletal muscle contraction with local vasodilation. Circ Res. 2008;102:372–379. DOI: 10.1161/CIRCRESAHA.107.158501

[48] Lowell BB, Shulman GI: Mitochondrial dysfunction and type 2 diabetes. Science 2005;307:384–387. DOI: 10.1126/science.1104343

[49] Houmard JA: Intramuscular lipid oxidation and obesity. Am J Physiol Regul Integr Comp Physiol. 2008;294:R1111–R1116. DOI: 10.1152/ajpregu.00396.2007

[50] Schinner S, Scherbaum WA, Bornstein SR, Barthel A: Molecular mechanisms of insulin resistance. Diabet Med. 2005;22:674–682. DOI: 10.1111/j.1464-5491.2005.01566.x

[51] Zong H, Bastie CC, Xu J, Fassler R, Campbell KP, Kurland IJ, Pessin JE: Insulin resistance in striated muscle-specific integrin receptor beta1-deficient mice. J Biol Chem. 2009;284:4679–4688. DOI: 10.1074/jbc.M807408200

[52] Bisht B, Goel HL, Dey CS: Focal adhesion kinase regulates insulin resistance in skeletal muscle. Diabetologia 2007;50:1058–1069. DOI: 10.1007/s00125-007-0591-6

[53] Goel HL, Dey CS: Focal adhesion kinase tyrosine phosphorylation is associated with myogenesis and modulated by insulin. Cell Prolif. 2002;35:131–142. DOI: 10.1046/j. 1365-2184.2002.00232.x

[54] Gupta A, Dey CS: PTEN and SHIP2 regulates PI3K/Akt pathway through focal adhesion kinase. Mol Cell Endocrinol. 2009;309:55–62. DOI: 10.1016/j.mce.2009.05.018

[55] Berria R, Wang L, Richardson DK, Richardson DK, Finlayson J, Belfort R, Pratipanawatr T, De Filippis EA, Kashyap S, Mandarino LJ: Increased collagen content in insulin-resistant skeletal muscle. Am J Physiol Endocrinol Metab. 2006;290:E560–E565. DOI: 10.1152/ajpendo.00202.2005

[56] Cherian S, Roy S, Pinheiro A, Roy S: Tight glycemic control regulates fibronectin expression and basement membrane thickening in retinal and glomerular capillaries of diabetic rats. Invest Ophthalmol Vis Sci. 2009;50:943–949. DOI: 10.1167/iovs.08-2377

[57] Kang L, Mayes WH, James FD, Bracy DP, Wasserman DH: Matrix metalloproteinase 9 opposes diet-induced muscle insulin resistance in mice. Diabetologia 2014;57:603–613. DOI: 10.1007/s00125-013-3128-1

[58] Kang L, Ayala JE, Lee-Young RS, Zhang Z, James FD, Neufer PD, Pozzi A, Zutter MM, Wasserman DH: Diet-induced muscle insulin resistance is associated with extracellular matrix remodeling and interaction with integrin alpha2beta1 in mice. Diabetes 2011;60:416–426. DOI: 10.2337/db10-1116

[59] Huang G, Ge G, Wang D, Gopalakrishnan B, Butz DH, Colman RJ, Nagy A, Greenspan DS: Alpha3(V) collagen is critical for glucose homeostasis in mice due to effects in pancreatic islets and peripheral tissue. J Clin Invest. 2011;121:769–783. DOI: 10.1172/JCI45096

[60] Grzelkowska-Kowalczyk K, Wieteska-Skrzeczyńska W, Grabiec K, Tokarska J: High glucose-mediated alterations of mechanisms important in myogenesis of mouse C2C12 myoblasts. Cell Biol Int. 2013;37:29–35. DOI: 10.1002/cbin.10004

[61] Wary KK, Kohler EE, Chatterjee I: Focal adhesion kinase regulation of neovascularization. Microvasc Res. 2012;83:64–70. DOI: 10.1016/j.mvr.2011.05.002

[62] Wang XH: Micro RNA in myogenesis and muscle atrophy. Curr Opin Clin Nutr Metab Care 2013;16:258–266. DOI: 10.1097/MCO.0b013e32835f81b9

[63] Pichavant C, Aartsma-Rus A, Clemens PR, Davies KE, Dickson G, Takeda S, Wilton SD, Wolff JA, Wooddell CI, Xiao X, Tremblay JP: Current status of pharmaceutical and genetic therapeutic approaches to treat DMD. Mol Ther. 2011;19:830–840. DOI: 10.1038/mt.2011.59

[64] Mouly V, Aamiri A, Périé S, Mamchaoui K, Barani A, Bigot A, Bouazza B, François V, Furling D, Jacquemin V, Negroni E, Riederer I, Vignaud A, St Guily JL, Butler-Browne GS: Myoblast transfer therapy: is there any light at the end of the tunnel? Acta Myol. 2005;24:128–133.

[65] Zanotti S, Gibertini S, Mora M: Altered production of extracellular matrix components by muscle-derived Duchenne muscular dystrophy fibroblasts before and after TGF-beta1 treatment. Cell Tissue Res. 2010;339:397–410. DOI: 10.1007/s00441-009-0889-4

[66] Rehn TA, Borge BA, Lunde PK, Munkvik M, Sneve ML, Grøndahl F, Aronsen JM, Sjaastad I, Prydz K, Kolset SO, Wiig H, Sejersted OM, Iversen PO: Temporary fatigue and altered extracellular matrix in skeletal muscle during progression of heart failure in rats. Am J Physiol Regul Integr Comp Physiol. 2009;297:R26–R33. DOI: 10.1152/ajpregu.90617.2008

Chapter 2

Composition and Function of Extracellular Matrix in Development of Skeletal Muscle

Zishuai Wang and Zhonglin Tang

Additional information is available at the end of the chapter

http://dx.doi.org/10.5772/62645

Abstract

Skeletal muscle extracellular matrix (ECM), surrender of muscle fibers, the amount of which is just <5%, appeals less attention in the field of skeletal muscle physiology. Thus, at one time, the function of skeletal muscle ECM was arbitrarily considered as general structural support that is typical in other tissues. However, an increasing number of recent evidences have indicated that the ECM plays a critical role in muscle fiber force transmission, proliferation, differentiation, migration, and polarization of cells. Alterations of molecules within the ECM are involved in fibrosis, muscle aging, regeneration, and myopathies. In this chapter, we review the composition and functions of ECM in skeletal muscle development.

Keywords: extracellular matrix, skeletal muscle, myogenesis, regeneration, fibrosis, myopathies

1. Introduction

The process of skeletal muscle formation in vertebrates begins from myogenic progenitors originating in the somites. However, somitic cells are the source of several cell lineages and only a subset are committed to a muscle fate [1]. Those cells destined for a muscle fate then undergo the process of myogenesis, during which the progenitors become specified and determined as myoblasts, which will proliferate, migrate, and fuse to one another to form multinucleated myofibers [2]. Thus, myogenesis seem to be critical in myoblast alignment and fusion into multinucleated myotubes. And the formation of myotubes is central to skeletal muscle development.

Extracellular matrix (ECM) has been considered as a structural scaffold between cells. It has been clear for many years that the ECM is a dynamic structure that influences cell behavior through the interaction of ECM molecules with each other, interaction with growth factors, and through cell– ECM signal transduction pathways [3]. Although the compositions of the ECM differ between tissues, all ECMs share the common function of structural support, cell adhesion, cell-to-cell communication, and differentiation [4]. Since the discovery that skeletal muscle ECM participate in the conversion of myoblasts to myotubes [5], the field of skeletal muscle physiology begins to focus on the relationship between muscle cells and ECM. In this review, we will give more details about the compositions of skeletal muscle ECM and how they affects muscle's normal functions.

2. Composition of skeletal muscle ECM

Anatomic studies indicate that vertebrate skeletal muscle can be typically classified into three layers: skeletal muscle fibers, enclosed by endomysium; muscle fasciculus, enclosed by perimysium; and entire muscle enclosed by epimysium. Thus, skeletal muscle ECM can also be organized into hierarchical structure: endomysial, perimysial, and epimysial connective tissues. According to the structure topology studies, the ECM can be classified into two layers: the interstitial matrix and the basement membrane. Interstitial matrix appears in the intercellular spaces, while basement membrane is a static structure on which cells rest. The interstitial matrix is filled by fibrous proteins and fibroblasts which is responsible for producing collagen, fibronectin, proteoglycans (PGs) and glycosidase, and matrix metalloproteinase (MMPs) [6–8]; while basement membrane is composed of basal lamina and fibrillar reticular lamina [9]. Muscle ECM is made up of numerous macromolecules including collagens, glycoprotein and matricellular proteins, PGs, and matrix remodeling enzymes [10].

In common with other tissues, the major protein of skeletal muscle ECM is collagen [11], synthesized and excreted by fibroblasts, including types I, III, IV, VI, XI, XII, XIII, XIV, XV, and XVIII [12–15]. According to their structure and functions, these types can be divided into several groups. Fibrillar collagens: collagens that have the ability to self assemble into fibrils including types I, III, XI. Network-forming collagens: collagens that have the ability to form a network including types IV and VI. Association collagens: collagens that have the ability to associate with fibrils including types XII and XIV. Transmembrane collagens including type XIII. Multiplexin: multiple triple helix domains with interruptions including types XV and XVIII [16]. Among these isoforms, the predominant distributors in ECM are types I and III as type I appears in perimysium, whereas type III prefers to distribute between endomysium and epimysium [17]. Types IV, VI, XV/XVIII, and XIII collagen are ingredients of the basement membrane [12, 18, 19]. Types XII and XIV collagen are perimysial fibril-associated collagens with interrupted triple helices [14].

Basement membrane, the specific region of ECM, is a reticular lamina knitted by collagen IV and glycoproteins including laminins, fibronectins, and entactin/nidogen [20]. Specifically, laminins bind to integrins and α-dystroglycan, while fibronectins bind to integrins and

http://dx.doi.org/10.5772/62645

laminins. Laminins and collagen type IV are linked to each other by entactin/nidogen [21–25]. Besides, there are other functional matricellular proteins appear in skeletal muscle ECM including tenascin-C, tenascin-Y, osteopontin, thrombospondin. Particularly, only during muscle regeneration can osteopontin be detected. And Tenascin-C appear to be located to the neuromuscular junction [26–31].

PG is heavily glycosylated proteins that is composed of a central core protein with one or more covalently attached glycosaminoglycan (GAG) chain(s) [32, 33]. Typically, the GAG is a polymer of disaccharide repeats including hyaluronan (HA), chondroitin sulfate (CS), dermatan sulfate (DS), heparan sulfate (HS), and keratan sulfate (KS). Most of the PGs appeared in skeletal muscle ECM belongs to the small leucine-rich proteoglycan (SLRP) family. And the majority of SLRP family present in muscle ECM is decorin that is covalently attached by CS/DS and biglycan [34]. Decorin can associate with fibrillar collagen, types I and III collagens [3]. Moreover, heparan sulfate proteoglycans (HSPGs) including types XV, VIII collagen, perlecan, and agrin are intrinsic constituents of basement membranes that are famous for its interaction with growth factors [35, 36]. Matrilins are a novel family of oligomeric ECM proteins. The matrilin family has four members, which are named matrilin 1, 2, 3, and 4 that all share a structure made up of von willebrand factor A (VWA) domains [37, 38]. In skeletal muscle ECM, matrilin-2 is widely distributed while other members are rarely present. Matrilin-2 has two VWA domains that are connected by ten epidermal growth factor (EGF)-like modules and is believed to be involved in the development and homeostasis of the ECM network by participating in filamentous network forming [38–41].

Dynamic equilibrium of skeletal muscle ECM is maintained by degradation enzyme and cells that can secrete ECM productions. It is well known that the majority of ECM components are secreted the fibroblast. Besides, myogenic cells can also secrete collagens, MMP-2 and decorin [42–44], and embryonic myoblasts secrete collagens [45]. There are at least six categories of enzymes that can digest ECM compositions: prolinase, serine protease, cysteine protease, asparagine proteinase, glycosidase, and matrix metalloproteinase (MMP). Since MMP can widely degrade collages and PGs, it is regarded as the most important regulator in keeping the integrity and homeostasis of ECM [43, 46–48].

Briefly, ECM is a complicated supermolecular network composed by collagen, glycoprotein, and PGs. Each component contains different isoforms and form complicated complexes by connecting with each other. Thus, it is hard to characterize skeletal muscle ECM constructors fully, and for much more details about these components, new techniques are needed.

3. Role of ECM in skeletal muscle development

As a fundamental component of the microenvironment of muscle fibers, the functions of ECM are traditionally considered as force transmission and structure integrity maintenance. However, an increasing number of evidence demonstrating ECM also plays an important role in myogenesis, cell proliferation, differentiation, migration, and muscle regeneration [49].

As mentioned above, providing structural and biochemical support to the surrounding cells is a common function of ECM in all cells. However, the transmission of force from contractile elements in the muscle fiber to the resultant movement of a joint seems to be the primary function of skeletal muscle ECM [50]. In order to achieve this function, ECM was linked to cytoskeleton by integrins, dystroglycan, and PGs at the cell surface [51–53]. Specifically, integrins can convert mechanical signals to adaptive responses in the cell [54–56] and dystrophin–glycoprotein complex is critical in mechanotransduction of muscle and tendon tissue [56]. In this way, adhesion complexes composed by ECM and transmembrane proteins establish a mechanical continuum along which forces can be transmitted from inside of the cell to outside, and vice versa.

One generally held idea is that many growth factors bind to their signaling receptors using GAG chains attached to ECM and membrane proteins as cofactors. For example, the binding of fibroblast growth factor (FGF) to FGF receptor depends on a HS chain binding at the same time [57]. Fibronectin and vitronectin bind to hepatocyte growth factor (HGF) and form the HGF receptor complexes to enhance cell migration [58]. And vascular endothelial growth factor (VEGF) binds to fibronectin type III (FN3) domains to promote cell proliferation [59]. Together, these evidences suggested that ECM proteins bind and present growth factors as organized solid-phase ligands. And considering growth factors including HGF, IGF, FGF, and the TGF-β superfamily are involved in controlling the proliferation and differentiation of myoblasts. Thus, it seems to be clear that ECM proteins can participate in skeletal muscle development by connecting with growth factors.

In vitro studies have shown that collagen fibrils are necessary during orientation and alignment of muscle fibers [60], and the inhibition of collagen synthesis suppresses the differentiation of myoblasts [49]. The functional importance of collagen network can be further proved through studies of mutant knockout models. Defection of types IV, IX, XIII, XV collagen [61–64] and mutations of collagen type VI will cause myopathy symptomatology [65]. Furthermore, lacking collagen types IX or XI will lead to abnormal collagen fibrils [66, 67], while lacking collagen type X chondrodysplasia will present [68].

PGs can also affect skeletal muscle development by modulating the activation of growth factors. For instance, perlecan can activate basic FGF (bFGF) tyrosine kinase receptors, which is a strong inhibitor of myogenic differentiation [69]. Syndecan-4 and glypican-1 participate in muscle cell proliferation and differentiation by regulating FGF2 [70]. Furthermore, syndecan-1 and -3 can also modulate the biological activity of FGF-2 [71, 72].

Decorin obstructs muscle cell proliferation [73, 74], by inhibiting the activity of transforming growth factor-β1 (TGF-β1). Myostatin, belonging to TGF-β superfamily, is a negative factor in muscle development. And decorin can also enhance myoblasts differentiation by restraining myostatin [75]. Moreover, fibromodulin, lumican, and biglycan can stimulate myostatin, insulin-like growth factor (IGF), or HGF [76–78].

Laminin is another critical matrix component that affects myogenesis. Specifically, evidences indicate that laminin can promote myoblast adhesion, proliferation, and myotube formation by regulating myostatin activity [79–81]. And lacking laminin mice characterize growth retardation and muscle dystrophy. On the other hand, laminin and collagen IV provide binding sites for PGs that can regulate growth factors activity. However, fibronectin, another glycoprotein, prevents myoblast differentiation by selectively promoting adhesion of fibroblasts [81, 82].

TGF-β1 signal pathway is reported to prevent myogenic differentiation partly by inhibiting matrilin-2 expression. In return, the matrilin-2 promotes cell differentiation and regeneration processes in myogenic by binding to other ECM proteins and integrins to regulate the TGF-β/BMP-7/Smad and other signaling pathways [83].

Skeletal muscle is a regenerative tissues and such regeneration requires the activity of a population of tissue-specific adult stem cells referred to as satellite cells. The satellite cell reside in mature skeletal muscle and is normally quiescent; however, when injury occurs, these muscle progenitor populations will proliferate, migrate, and fuse into new muscle fibers [84]. These special cells are wedged in basal lamina, of which the most abundant proteins are collagen type IV and laminin-2. In vitro studies showed that when satellite cells will rapidly enter cell cycle and proliferate after leaving basal lamina [85]. What is more, satellite cells cultured on matrigel with collagen VI are more inclined to be quiescence compare to these without collagen VI [86]. Thus, it seems that the basal lamina can prevent satellite cell proliferation and differentiation in the absence of damage [20]. When it comes to muscle regeneration, ECM components will positively participate in cell mitosis and differentiation as we mentioned before. Syndecan-3, one member of HSPGs, can regulate homeostasis of the satellite cell population and myofiber size by cooperating with Notch [87]. Together, these evidences show that ECM compositions play an important role in keeping satellite cells quiescent under normal circumstances and proliferation, differentiation during regeneration process.

4. ECM and myopathies

Abnormal accumulation of ECM is clinically termed “fibrosis”, which is characterized by increased endomysium and perimysium in skeletal muscle. Skeletal muscle fibrosis can be detected in nearly all muscular dystrophies, aging, and muscle injury [88–92]. However, it is hard to precisely quantify skeletal muscle fibrosis as the components are complicated and dynamically changed. Furthermore, in normal muscle, the amount of ECM area fraction is 5%, but this value can dramatically increase in muscle fibrosis cases. This is because the muscle fibers will become atrophic in diseased, such as severe atrophy, chronic inflammation, and dystrophies or injured states even ECM structure remains the same [93]. Whether muscle fibrosis is characterized by excessive production of ECM components remains unclear, but the participation of these components in muscle fibrosis has been proved.

TGF-β has long been believed to be a central mediator of the fibrotic response as it can induces fibroblasts to synthesize type-I collagen and fibronectin [94]. Moreover, TGF-β can induces the expression of connective tissue growth factor (CTGF), a downstream mediator of the effects of TGF-β on fibroblasts [95, 96], and the matrix protein fibronectin, a critical factor in enhancing the expression of collagen type I [97].

In skeletal muscular dystrophies, the expressions of decorin and biglycan are increased [98, 99], which will cause alteration of TGF-β signaling and eventual fibrosis [100]. Besides, treatments using decorin and TGF-β inhibitors in injured muscle enhance regeneration and prevent fibrosis [101–103].

Fibrin, a structural component of the ECM, accumulates in areas of degeneration and inflammation in dystrophic muscle, whereas knockout fibrinogen was shown to reduce fibrosis development in mdx mice. Fibrin can induce the expression of TGF-β to promote fibrosis [104]. Fibrin can activate fibroblasts to synthesize and secrete collagens by binding to αVβ3 integrin receptor [105]. Considering the synthesis and degradation of collagens is controlled by MMPs, the importance of proteases in muscle fibrosis is absolutely obvious [106].

On the other hand, defects in or deficiencies of ECM molecules will cause myopathies and inherited connective tissue disorders. As we mentioned before, ECM and cytoskeleton are connected by transmembrane proteins named dystroglycan, sarcoglycan, integrin. Dystroglycan has two subunits α and β, β-dystroglycan intracellularly binds to dystrophin and extracellularly to α-dystroglycan, which is associated with the ECM proteins laminin α2, biglycan, and perlecan [16, 107]. Defects in α-dystroglycan can lead to congenital muscular dystrophy (MDC) and limb–girdle muscular dystrophy (LGMD) that can also be caused by deficiency of laminin α2 [108]. Sarcoglycans can extracellularly binds to biglycan and is closely associated with the dystroglycan complex [109–111]. Mutations in sarcoglycans result in autosomal-recessive limb–girdle muscular dystrophies. In integrin knockouted mice, mild form of muscular dystrophy appears [112]. Furthermore, clinical studies show that collagen VI deficiency lead to Bethlem myopathy and Ullrich congenital muscular dystrophy [61, 113, 114].

Extracellular fat is another pathological response of skeletal muscle to disease or injury that is accompanied by pathological diseases include Duchenne muscular dystrophy, obesity, type-2 diabetes, and aged muscle [115–117]. Recent studies have identified a PDGFRα+ progenitor cell population that is responsible for intracellular fat deposition as the cell can differentiate into adipose tissue under nonregenerating conditions [118]. Moreover, these cells were found to distribute more in perimysium than endomysium [119].

5. MMPs and skeletal muscle

MMPs are famous for its irreplaceable role in degrading ECM compositions. In skeletal muscle, MMP-2 and MMP-9 [43] can degrade type-IV collagen, fibronectin, PGs, and laminin, while MMP-1 [48] and MMP-13 [120] degrade types I and III collagen. The activities of MMPs are controlled by tissue inhibitors of matrix metalloproteinases (TIMPs). TIMP-1 binds to active forms of MMPs forming noncovaent complexes, whereas TIMP-2 stabilizes the inactive form of the enzyme, and thus inhibits the formation of active proteolyticenzyme [47, 48]. In normal muscle tissues, the expression of MMPs are very low but increased in injured muscles mainly because they are secreted by inflammatory cells [121]. Although studies rarely show the functions of MMPs in skeletal muscle, they have been implicated in many pathological processes including myogenesis, muscle growth, development, aging, and regeneration [122, 123]. MMP-2 knockout mice developed significantly less hypertrophy and ECM remodeling in response to overload compared to a significant increase in MMP-2 activity and upregulation of ECM components and remodeling enzymes in wild-type mice [124]. In vivo study shows that MMP-2 is essential for myoblast migration [125], while in vitro study indicates MMP-2 is secreted at all stages from cell to myotubes [126]. Acute muscle ischemia results in remodeling of the basal lamina which is accompanied by increased MMP gelatinases [127]. And increased MMPs (MMP-2 and MMP-9) are also responsible to the degradation of ECM in skeletal muscle atrophy [128]. Furthermore, satellite cells are reported to synthesize and secrete MMP-2 and induce MMP-9 activity in human skeletal muscles [129]. During regeneration, MMP-2 activation appears go along with the formation of new myofibers, whereas MMP-9 expression is related to the inflammatory response [43]. Expression changes of MMPs have been involved in different myopathies. Distinctly increased MMP-9 appears in inflammatory myopathies [130], MMP-7 upregulation is prominent in case of polymyositis, whereas MMP-2 is only slightly elevated in inflamed muscle [131].

6. Conclusion

Skeletal muscle fibers are surrounded by ECM, and the ECM is an important part of the cellular microenvironment consists of a complex mixture of structural and functional proteins including glycoproteins, collagen, and PGs. These molecules interact with each other and form a super molecular network in order to maintain skeletal muscle integrity and participate in the development of skeletal muscle. Additionally, skeletal muscle fibrosis, characterized by abnormal accumulation of ECM, is an obvious clinical characteristic of myopathies such as age-related sarcopenia, muscular dystrophy, and Duchenne muscular dystrophy. Genetic diseases, dysregulation of TGF-β signaling and physical activity can cause defects in or deficiencies of molecules within the skeletal muscle ECM.

Author details

Zishuai Wang and Zhonglin Tang*

*Address all correspondence to: zhonglinqy_99@sina.com

Institute of Animal Science, Chinese Academy of Agricultural Sciences, Beijing, PR China

References

[1] Christ B, Jacob M, Jacob HJ. On the origin and development of the ventrolateral abdominal muscles in the avian embryo. An experimental and ultrastructural study. Anatomy and Embryology (Berl). 1983; 166: 87–101.

[2] Buckingham M, Vincent SD. Distinct and dynamic myogenic populations in the vertebrate embryo. Current Opinion in Genetics & Development. 2009; 19: 444–453. doi:10.1016/j.gde.2009.08.001.

[3] Nishimura T. Role of extracellular matrix in development of skeletal muscle and postmortem aging of meat. Meat Science. 2015; 109: 48–55. doi:10.1016/j.meatsci.2015.05.015.

[4] Watt FM, Fujiwara H. Cell-extracellular matrix interactions in normal and diseased skin. Cold Spring Harbor Perspectives in Biology. 2011; 3: a005124. doi:10.1101/cshperspect.a005124.

[5] Hauschka SD, Konigsberg IR. The influence of collagen on the development of muscle clones. Proceedings of the National Academy of Sciences USA. 1966; 55: 119–126.

[6] Gatchalian CL, Schachner M, Sanes JR. Fibroblasts that proliferate near denervated synaptic sites in skeletal muscle synthesize the adhesive molecules tenascin(J1), N-CAM, fibronectin, and a heparan sulfate proteoglycan. Journal of Cell Biology. 1989; 108: 1873–1890.

[7] Kuhl U, Ocalan M, Timpl R, Mayne R, Hay E, von der Mark K. Role of muscle fibroblasts in the deposition of type-IV collagen in the basal lamina of myotubes. Differentiation. 1984; 28: 164–172.

[8] Archile-Contreras AC, Mandell IB, Purslow PP. Phenotypic differences in matrix metalloproteinase 2 activity between fibroblasts from 3 bovine muscles. Journal of Animal Science. 2010; 88: 4006–4015. doi:10.2527/jas.2010-3060.

[9] Bosman FT, Stamenkovic I. Functional structure and composition of the extracellular matrix. The Journal of Pathology. 2003; 200: 423–428. doi:10.1002/path.1437.

[10] Gillies AR, Lieber RL. Structure and function of the skeletal muscle extracellular matrix. Muscle & Nerve. 2011; 44: 318–331. doi:10.1002/mus.22094.

[11] Dransfield E. Intramuscular composition and texture of beef muscles. Journal of the Science of Food and Agriculture. 1977; 28: 833–842.

[12] Myers JC, Dion AS, Abraham V, Amenta PS. Type XV collagen exhibits a widespread distribution in human tissues but a distinct localization in basement membrane zones. Cell and Tissue Research. 1996; 286: 493–505.

[13] Marvulli D, Volpin D, Bressan GM. Spatial and temporal changes of type VI collagen expression during mouse development. Developmental Dynamics. 1996; 206: 447–454. doi:10.1002/(SICI)1097-0177(199608)206:4<447::AID-AJA10>3.0.CO;2-U.

[14] Listrat A, Lethias C, Hocquette JF, Renand G, Menissier F, Geay Y, Picard B. Age-related changes and location of types I, III, XII and XIV collagen during development of skeletal muscles from genetically different animals. Journal of Histochemistry & Cytochemistry. 2000; 32: 349–356.

[15] Brandan E, Gutierrez J. Role of skeletal muscle proteoglycans during myogenesis. Matrix Biology. 2013; 32: 289–297. doi:10.1016/j.matbio.2013.03.007.

[16] Voermans NC, Bonnemann CG, Huijing PA, Hamel BC, van Kuppevelt TH, de Haan A, Schalkwijk J, van Engelen BG, Jenniskens GJ. Clinical and molecular overlap between myopathies and inherited connective tissue diseases. Neuromuscular Disorders. 2008; 18: 843–856. doi:10.1016/j.nmd.2008.05.017.

[17] Light N, Champion AE. Characterization of muscle epimysium, perimysium and endomysium collagens. Biochemical Journal. 1984; 219: 1017–1026.

[18] Whittaker CA, Bergeron KF, Whittle J, Brandhorst BP, Burke RD, Hynes RO. The echinoderm adhesome. Developmental Biology. 2006; 300: 252–266. doi:10.1016/j.ydbio.2006.07.044.

[19] Sanes JR. Laminin, fibronectin, and collagen in synaptic and extrasynaptic portions of muscle fiber basement membrane. Journal of Cell Biology. 1982; 93: 442–451.

[20] Sanes JR. The basement membrane/basal lamina of skeletal muscle. Journal of Biological Chemistry. 2003; 278: 12601–12604. doi:10.1074/jbc.R200027200.

[21] Ervasti JM, Campbell KP. A role for the dystrophin-glycoprotein complex as a transmembrane linker between laminin and actin. Journal of Cell Biology. 1993; 122: 809–823.

[22] Wu C, Keivens VM, O'Toole TE, McDonald JA, Ginsberg MH. Integrin activation and cytoskeletal interaction are essential for the assembly of a fibronectin matrix. Cell. 1995; 83: 715–724.

[23] Fox JW, Mayer U, Nischt R, Aumailley M, Reinhardt D, Wiedemann H, Mann K, Timpl R, Krieg T, Engel J, et al. Recombinant nitrogen consists of three globular domains and mediates binding of laminin to collagen type IV. EMBO Journal. 1991; 10: 3137–3146.

[24] Mayer U, Nischt R, Poschl E, Mann K, Fukuda K, Gerl M, Yamada Y, Timpl R. A single EGF-like motif of laminin is responsible for high affinity nidogen binding. EMBO Journal. 1993; 12: 1879–1885.

[25] Rao CN, Margulies IM, Liotta LA. Binding domain for laminin on type IV collagen. Biochemical and Biophysical Research Communicate. 1985; 128: 45–52.

[26] Cotman SL, Halfter W, Cole GJ. Identification of extracellular matrix ligands for the heparan sulfate proteoglycan agrin. Experimental Cell Research. 1999; 249: 54–64. doi: 10.1006/excr.1999.4463.

[27] Chung CY, Erickson HP. Glycosaminoglycans modulate fibronectin matrix assembly and are essential for matrix incorporation of tenascin-C. Journal of Cell Science. 1997; 110(Pt 12): 1413–1419.

[28] Uaesoontrachoon K, Yoo HJ, Tudor EM, Pike RN, Mackie EJ, Pagel CN. Osteopontin and skeletal muscle myoblasts: association with muscle regeneration and regulation of

myoblast function in vitro. International Journal Biochemistry & Cell Biology. 2008; 40: 2303–2314. doi:10.1016/j.biocel.2008.03.020.

[29] Malek MH, Olfert IM. Global deletion of thrombospondin-1 increases cardiac and skeletal muscle capillarity and exercise capacity in mice. Experimental Physiology. 2009; 94: 749–760. doi:10.1113/expphysiol.2008.045989.

[30] Chiquet M, Fambrough DM. Chick myotendinous antigen. I. A monoclonal antibody as a marker for tendon and muscle morphogenesis. Journal of Cell Biology. 1984; 98: 1926–1936.

[31] Chiquet M, Fambrough DM. Chick myotendinous antigen. II. A novel extracellular glycoprotein complex consisting of large disulfide-linked subunits. Journal of Cell Biology. 1984; 98: 1937–1946.

[32] Fransson LA, Havsmark B, Chiarugi VP. Co-polymeric glycosaminoglycans in transformed cells. Transformation-dependent changes in the co-polymeric structure of heparan sulphate. Biochemical Journal. 1982; 201: 233–240.

[33] Kjellen L, Lindahl U. Proteoglycans: structures and interactions. Annual Review of Biochemistry, Vol 81. 1991; 60: 443–475. doi:10.1146/annurev.bi.60.070191.002303.

[34] Brandan E, Inestrosa NC. Isolation of the heparan sulfate proteoglycans from the extracellular matrix of rat skeletal muscle. Journal of Neurobiology. 1987; 18: 271–282. doi:10.1002/neu.480180303.

[35] Mundhenke C, Meyer K, Drew S, Friedl A. Heparan sulfate proteoglycans as regulators of fibroblast growth factor-2 receptor binding in breast carcinomas. American Journal of Pathology. 2002; 160: 185–194. doi:10.1016/S0002-9440(10)64362-3.

[36] Iozzo RV, San Antonio JD. Heparan sulfate proteoglycans: heavy hitters in the angiogenesis arena. Journal of Clinical Investigation. 2001; 108: 349–355. doi:10.1172/JCI13738.

[37] Deak F, Wagener R, Kiss I, Paulsson M. The matrilins: a novel family of oligomeric extracellular matrix proteins. Matrix Biology. 1999; 18: 55–64.

[38] Deak F, Piecha D, Bachrati C, Paulsson M, Kiss I. Primary structure and expression of matrilin-2, the closest relative of cartilage matrix protein within the von Willebrand factor type A-like module superfamily. Journal of Biological Chemistry. 1997; 272: 9268–9274.

[39] Piecha D, Muratoglu S, Morgelin M, Hauser N, Studer D, Kiss I, Paulsson M, Deak F. Matrilin-2, a large, oligomeric matrix protein, is expressed by a great variety of cells and forms fibrillar networks. Journal of Biological Chemistry. 1999; 274: 13353–13361.

[40] Klatt AR, Becker AK, Neacsu CD, Paulsson M, Wagener R. The matrilins: modulators of extracellular matrix assembly. International Journal of Biochemistry & Cell Biology. 2011; 43: 320–330. doi:10.1016/j.biocel.2010.12.010.

[41] Li L, Zhang L, Shao Y, Wang G, Gong R, Wang Z, Peng J, Wang S, Genochio D, Zhao B, Luo J. Distinct roles of two alternative splice variants of matrilin-2 in protein oligomerization and proteolysis. Molecular Medicine Reports. 2012; 6: 1204–1210. doi: 10.3892/mmr.2012.1056.

[42] Beach RL, Rao JS, Festoff BW. Extracellular-matrix synthesis by skeletal muscle in culture. Major secreted collagenous proteins of clonal myoblasts. Biochemical Journal. 1985; 225: 619–627.

[43] Kherif S, Lafuma C, Dehaupas M, Lachkar S, Fournier JG, Verdiere-Sahuque M, Fardeau M, Alameddine HS. Expression of matrix metalloproteinases 2 and 9 in regenerating skeletal muscle: a study in experimentally injured and mdx muscles. Developmental Biology. 1999; 205: 158–170. doi:10.1006/dbio.1998.9107.

[44] Brandan E, Fuentes ME, Andrade W. The proteoglycan decorin is synthesized and secreted by differentiated myotubes. European Journal of Cell Biology. 1991; 55: 209–216.

[45] Sasse J, von der Mark H, Kuhl U, Dessau W, von der Mark K. Origin of collagen types I, III, and V in cultures of avian skeletal muscle. Developmental Biology. 1981; 83: 79–89.

[46] Ohuchi E, Imai K, Fujii Y, Sato H, Seiki M, Okada Y. Membrane type 1 matrix metalloproteinase digests interstitial collagens and other extracellular matrix macromolecules. Journal of Biological Chemistry. 1997; 272: 2446–2451.

[47] Chin JR, Werb Z. Matrix metalloproteinases regulate morphogenesis, migration and remodeling of epithelium, tongue skeletal muscle and cartilage in the mandibular arch. Development. 1997; 124: 1519–1530.

[48] Singh A, Nelson-Moon ZL, Thomas GJ, Hunt NP, Lewis MP. Identification of matrix metalloproteinases and their tissue inhibitors type 1 and 2 in human masseter muscle. Archives of Oral Biology. 2000; 45: 431–440.

[49] Kjaer M. Role of extracellular matrix in adaptation of tendon and skeletal muscle to mechanical loading. Physiological Reviews. 2004; 84: 649–698. doi:10.1152/physrev.00031.2003.

[50] Patel TJ, Lieber RL. Force transmission in skeletal muscle: from actomyosin to external tendons. Exercise and Sport Sciences Reviews. 1997; 25: 321–363.

[51] Mayer U. Integrins: redundant or important players in skeletal muscle? Journal of Biological Chemistry. 2003; 278: 14587–14590. doi:10.1074/jbc.R200022200.

[52] Michele DE, Campbell KP. Dystrophin-glycoprotein complex: post-translational processing and dystroglycan function. Journal of Biological Chemistry. 2003; 278: 15457–15460. doi:10.1074/jbc.R200031200.

[53] Ingber DE, Dike L, Hansen L, Karp S, Liley H, Maniotis A, McNamee H, Mooney D, Plopper G, Sims J, et al. Cellular tensegrity: exploring how mechanical changes in the

cytoskeleton regulate cell growth, migration, and tissue pattern during morphogenesis. International Review of Cytology. 1994; 150: 173–224.

[54] Carson JA, Wei L. Integrin signaling's potential for mediating gene expression in hypertrophying skeletal muscle. Journal of Applied Physiology (1985). 2000; 88: 337–343.

[55] Shyy JY, Chien S. Role of integrins in cellular responses to mechanical stress and adhesion. Current Opinion in Cell Biology. 1997; 9: 707–713.

[56] Ilsley JL, Sudol M, Winder SJ. The interaction of dystrophin with beta-dystroglycan is regulated by tyrosine phosphorylation. Cell Signal. 2001; 13: 625–632.

[57] Mohammadi M, Olsen SK, Goetz R. A protein canyon in the FGF-FGF receptor dimer selects from an a la carte menu of heparan sulfate motifs. Current Opinion Structural Biology. 2005; 15: 506–516. doi:10.1016/j.sbi.2005.09.002.

[58] Rahman S, Patel Y, Murray J, Patel KV, Sumathipala R, Sobel M, Wijelath ES. Novel hepatocyte growth factor (HGF) binding domains on fibronectin and vitronectin coordinate a distinct and amplified Met-integrin induced signalling pathway in endothelial cells. BMC Cell Biology. 2005; 6: 8. doi:10.1186/1471-2121-6-8.

[59] Wijelath ES, Rahman S, Namekata M, Murray J, Nishimura T, Mostafavi-Pour Z, Patel Y, Suda Y, Humphries MJ, Sobel M. Heparin-II domain of fibronectin is a vascular endothelial growth factor-binding domain: enhancement of VEGF biological activity by a singular growth factor/matrix protein synergism. Circulation Research. 2006; 99: 853–860. doi:10.1161/01.RES.0000246849.17887.66.

[60] Lawson MA, Purslow PP. Differentiation of myoblasts in serum-free media: effects of modified media are cell line-specific. Cells Tissues Organs. 2000; 167: 130–137. doi: 10.1159/000016776.

[61] Bonaldo P, Braghetta P, Zanetti M, Piccolo S, Volpin D, Bressan GM. Collagen VI deficiency induces early onset myopathy in the mouse: an animal model for Bethlem myopathy. Human Molecular Genetics. 1998; 7: 2135–2140.

[62] Bonnemann CG, Cox GF, Shapiro F, Wu JJ, Feener CA, Thompson TG, Anthony DC, Eyre DR, Darras BT, Kunkel LM. A mutation in the alpha 3 chain of type IX collagen causes autosomal dominant multiple epiphyseal dysplasia with mild myopathy. Proceedings of the National Academy of Sciences USA. 2000; 97: 1212–1217.

[63] Kvist AP, Latvanlehto A, Sund M, Eklund L, Vaisanen T, Hagg P, Sormunen R, Komulainen J, Fassler R, Pihlajaniemi T. Lack of cytosolic and transmembrane domains of type XIII collagen results in progressive myopathy. American Journal of Pathology. 2001; 159: 1581–1592. doi:10.1016/S0002-9440(10)62542-4.

[64] Eklund L, Piuhola J, Komulainen J, Sormunen R, Ongvarrasopone C, Fassler R, Muona A, Ilves M, Ruskoaho H, Takala TE, Pihlajaniemi T. Lack of type XV collagen causes a

skeletal myopathy and cardiovascular defects in mice. Proceedings of the National Academy of Sciences USA. 2001; 98: 1194–1199. doi:10.1073/pnas.031444798.

[65] Jobsis GJ, Keizers H, Vreijling JP, de Visser M, Speer MC, Wolterman RA, Baas F, Bolhuis PA. Type VI collagen mutations in Bethlem myopathy, an autosomal dominant myopathy with contractures. Nature Genetics. 1996; 14: 113–115. doi:10.1038/ng0996-113.

[66] Li SW, Prockop DJ, Helminen H, Fassler R, Lapvetelainen T, Kiraly K, Peltarri A, Arokoski J, Lui H, Arita M, et al. Transgenic mice with targeted inactivation of the Col2 alpha 1 gene for collagen II develop a skeleton with membranous and periosteal bone but no endochondral bone. Genes & Development. 1995; 9: 2821–2830.

[67] Fassler R, Schnegelsberg PN, Dausman J, Shinya T, Muragaki Y, Mc Carthy MT, Olsen BR, Jaenisch R. Mice lacking alpha 1 (IX) collagen develop noninflammatory degenerative joint disease. Proceedings of the National Academy of Sciences USA. 1994; 91: 5070–5074.

[68] Wallis GA, Rash B, Sykes B, Bonaventure J, Maroteaux P, Zabel B, Wynne-Davies R, Grant ME, Boot-Handford RP. Mutations within the gene encoding the alpha 1 (X) chain of type X collagen (COL10A1) cause metaphyseal chondrodysplasia type Schmid but not several other forms of metaphyseal chondrodysplasia. Journal of Medical Genetics. 1996; 33: 450–457.

[69] Larrain J, Alvarez J, Hassell JR, Brandan E. Expression of perlecan, a proteoglycan that binds myogenic inhibitory basic fibroblast growth factor, is down regulated during skeletal muscle differentiation. Experimental Cell Research. 1997; 234: 405–412. doi:10.1006/excr.1997.3648.

[70] Velleman SG. Meat science and muscle biology symposium: extracellular matrix regulation of skeletal muscle formation. Journal of Animal Science. 2012; 90: 936–941. doi:10.2527/jas.2011-4497.

[71] Larrain J, Carey DJ, Brandan E. Syndecan-1 expression inhibits myoblast differentiation through a basic fibroblast growth factor-dependent mechanism. Journal of Biological Chemistry. 1998; 273: 32288–32296.

[72] Fuentealba L, Carey DJ, Brandan E. Antisense inhibition of syndecan-3 expression during skeletal muscle differentiation accelerates myogenesis through a basic fibroblast growth factor-dependent mechanism. Journal of Biological Chemistry. 1999; 274: 37876–37884.

[73] Riquelme C, Larrain J, Schonherr E, Henriquez JP, Kresse H, Brandan E. Antisense inhibition of decorin expression in myoblasts decreases cell responsiveness to transforming growth factor beta and accelerates skeletal muscle differentiation. Journal of Biological Chemistry. 2001; 276: 3589–3596. doi:10.1074/jbc.M004602200.

[74] Yamaguchi Y, Ruoslahti E. Expression of human proteoglycan in Chinese hamster ovary cells inhibits cell proliferation. Nature. 1988; 336: 244–246. doi:10.1038/336244a0.

[75] Kishioka Y, Thomas M, Wakamatsu J, Hattori A, Sharma M, Kambadur R, Nishimura T. Decorin enhances the proliferation and differentiation of myogenic cells through suppressing myostatin activity. Journal of Cellular Physiology. 2008; 215: 856–867. doi: 10.1002/jcp.21371.

[76] Bolton SJ, Barry ST, Mosley H, Patel B, Jockusch BM, Wilkinson JM, Critchley DR. Monoclonal antibodies recognizing the N- and C-terminal regions of talin disrupt actin stress fibers when microinjected into human fibroblasts. Cell Motility and the Cytoskeleton. 1997; 36: 363–376. doi:10.1002/(SICI)1097-0169(1997)36:4<363::AID-CM6>3.0.CO;2–6.

[77] Collinsworth AM, Torgan CE, Nagda SN, Rajalingam RJ, Kraus WE, Truskey GA. Orientation and length of mammalian skeletal myocytes in response to a unidirectional stretch. Cell Tissue Research. 2000; 302: 243–251.

[78] Schmalbruch H, Lewis DM. Dynamics of nuclei of muscle fibers and connective tissue cells in normal and denervated rat muscles. Muscle & Nerve. 2000; 23: 617–626.

[79] Kroll TG, Peters BP, Hustad CM, Jones PA, Killen PD, Ruddon RW. Expression of laminin chains during myogenic differentiation. Journal of Biological Chemistry. 1994; 269: 9270–9277.

[80] Kubo K, Kanehisa H, Miyatani M, Tachi M, Fukunaga T. Effect of low-load resistance training on the tendon properties in middle-aged and elderly women. Acta Physiologica Scandinavica. 2003; 178: 25–32. doi:10.1046/j.1365-201X.2003.01097.x.

[81] Foster RF, Thompson JM, Kaufman SJ. A laminin substrate promotes myogenesis in rat skeletal muscle cultures: analysis of replication and development using antidesmin and anti-BrdUrd monoclonal antibodies. Developmental Biology. 1987; 122: 11–20.

[82] von der Mark K, Ocalan M. Antagonistic effects of laminin and fibronectin on the expression of the myogenic phenotype. Differentiation. 1989; 40: 150–157.

[83] Korpos E, Deak F, Kiss I. Matrilin-2, an extracellular adaptor protein, is needed for the regeneration of muscle, nerve and other tissues. Neural Regeneration Research. 2015; 10: 866–869. doi:10.4103/1673-5374.158332.

[84] Seale P, Rudnicki MA. A new look at the origin, function, and "stem-cell" status of muscle satellite cells. Developmental Biology. 2000; 218: 115–124. doi:10.1006/dbio.1999.9565.

[85] Gilbert PM, Havenstrite KL, Magnusson KE, Sacco A, Leonardi NA, Kraft P, Nguyen NK, Thrun S, Lutolf MP, Blau HM. Substrate elasticity regulates skeletal muscle stem cell self-renewal in culture. Science. 2010; 329: 1078–1081. doi:10.1126/science.1191035.

[86] Urciuolo A, Quarta M, Morbidoni V, Gattazzo F, Molon S, Grumati P, Montemurro F, Tedesco FS, Blaauw B, Cossu G, Vozzi G, Rando TA, Bonaldo P. Collagen VI regulates

satellite cell self-renewal and muscle regeneration. Nature Communications. 2013; 4: 1964. doi:10.1038/ncomms2964.

[87] Pisconti A, Cornelison DD, Olguin HC, Antwine TL, Olwin BB. Syndecan-3 and Notch cooperate in regulating adult myogenesis. Journal of Cell Biology. 2010; 190: 427–441. doi:10.1083/jcb.201003081.

[88] Duance VC, Stephens HR, Dunn M, Bailey AJ, Dubowitz V. A role for collagen in the pathogenesis of muscular dystrophy? Nature. 1980; 284: 470–472.

[89] Williams PE, Goldspink G. Connective tissue changes in immobilised muscle. Journal of Anatomy. 1984; 138(Pt 2): 343–350.

[90] Alnaqeeb MA, Al Zaid NS, Goldspink G. Connective tissue changes and physical properties of developing and ageing skeletal muscle. Journal of Anatomy. 1984; 139(Pt 4): 677–689.

[91] [91] Berria R, Wang L, Richardson DK, Finlayson J, Belfort R, Pratipanawatr T, De Filippis EA, Kashyap S, Mandarino LJ. Increased collagen content in insulin-resistant skeletal muscle. American Journal of Physiology-Endocrinology and Metabolism. 2006; 290: E560–565. doi:10.1152/ajpendo.00202.2005.

[92] Alexakis C, Partridge T, Bou-Gharios G. Implication of the satellite cell in dystrophic muscle fibrosis: a self-perpetuating mechanism of collagen overproduction. American Journal of Physiology-Cell Physiology. 2007; 293: C661–C669. doi:10.1152/ajpcell.00061.2007.

[93] Lieber RL, Ward SR. Cellular mechanisms of tissue fibrosis. 4. Structural and functional consequences of skeletal muscle fibrosis. American Journal of Physiology-Cell Physiology. 2013; 305: C241–C252. doi:10.1152/ajpcell.00173.2013.

[94] LeRoy EC, Trojanowska MI, Smith EA. Cytokines and human fibrosis. European Cytokine Network. 1990; 1: 215–219.

[95] Grotendorst GR. Connective tissue growth factor: a mediator of TGF-beta action on fibroblasts. Cytokine Growth Factor Review. 1997; 8: 171–179.

[96] Leask A, Denton CP, Abraham DJ. Insights into the molecular mechanism of chronic fibrosis: the role of connective tissue growth factor in scleroderma. Journal of Investigative Dermatology. 2004; 122: 1–6. doi:10.1046/j.0022-202X.2003.22133.x.

[97] Serini G, Bochaton-Piallat ML, Ropraz P, Geinoz A, Borsi L, Zardi L, Gabbiani G. The fibronectin domain ED-A is crucial for myofibroblastic phenotype induction by transforming growth factor-beta1. Journal of Cell Biology. 1998; 142: 873–881.

[98] Fadic R, Mezzano V, Alvarez K, Cabrera D, Holmgren J, Brandan E. Increase in decorin and biglycan in Duchenne muscular dystrophy: role of fibroblasts as cell source of these proteoglycans in the disease. Journal of Cellular and Molecular Medicine. 2006; 10: 758–769.

[99] Zanotti S, Negri T, Cappelletti C, Bernasconi P, Canioni E, Di Blasi C, Pegoraro E, Angelini C, Ciscato P, Prelle A, Mantegazza R, Morandi L, Mora M. Decorin and biglycan expression is differentially altered in several muscular dystrophies. Brain. 2005; 128: 2546–2555. doi:10.1093/brain/awh635.

[100] Zanotti S, Saredi S, Ruggieri A, Fabbri M, Blasevich F, Romaggi S, Morandi L, Mora M. Altered extracellular matrix transcript expression and protein modulation in primary Duchenne muscular dystrophy myotubes. Matrix Biology. 2007; 26: 615–624. doi: 10.1016/j.matbio.2007.06.004.

[101] Fukushima K, Badlani N, Usas A, Riano F, Fu F, Huard J. The use of an antifibrosis agent to improve muscle recovery after laceration. American Journal of Sports Medicine. 2001; 29: 394–402.

[102] Chan YS, Li Y, Foster W, Fu FH, Huard J. The use of suramin, an antifibrotic agent, to improve muscle recovery after strain injury. American Journal of Sports Medicine. 2005; 33: 43–51.

[103] Sato K, Li Y, Foster W, Fukushima K, Badlani N, Adachi N, Usas A, Fu FH, Huard J. Improvement of muscle healing through enhancement of muscle regeneration and prevention of fibrosis. Muscle & Nerve. 2003; 28: 365–372. doi:10.1002/mus.10436.

[104] Wynn TA. Cellular and molecular mechanisms of fibrosis. Journal of Pathology. 2008; 214: 199–210. doi:10.1002/path.2277.

[105] Vidal B, Serrano AL, Tjwa M, Suelves M, Ardite E, De Mori R, Baeza-Raja B, Martinez de Lagran M, Lafuste P, Ruiz-Bonilla V, Jardi M, Gherardi R, Christov C, Dierssen M, Carmeliet P, Degen JL, Dewerchin M, Munoz-Canoves P. Fibrinogen drives dystrophic muscle fibrosis via a TGFbeta/alternative macrophage activation pathway. Genes & Development. 2008; 22: 1747–1752. doi:10.1101/gad.465908.

[106] Chen X, Li Y. Role of matrix metalloproteinases in skeletal muscle: migration, differentiation, regeneration and fibrosis. Cell Adhesion & Migration. 2009; 3: 337–341.

[107] Michele DE, Barresi R, Kanagawa M, Saito F, Cohn RD, Satz JS, Dollar J, Nishino I, Kelley RI, Somer H, Straub V, Mathews KD, Moore SA, Campbell KP. Post-translational disruption of dystroglycan-ligand interactions in congenital muscular dystrophies. Nature. 2002; 418: 417–422. doi:10.1038/nature00837.

[108] Brockington M, Blake DJ, Prandini P, Brown SC, Torelli S, Benson MA, Ponting CP, Estournet B, Romero NB, Mercuri E, Voit T, Sewry CA, Guicheney P, Muntoni F. Mutations in the fukutin-related protein gene (FKRP) cause a form of congenital muscular dystrophy with secondary laminin alpha2 deficiency and abnormal glycosylation of alpha-dystroglycan. American Journal of Human Genetics. 2001; 69: 1198–1209. doi:10.1086/324412.

[109] Watkins TC, Zelinka LM, Kesic M, Ansevin CF, Walker GR. Identification of skeletal muscle autoantigens by expression library screening using sera from autoimmune

rippling muscle disease (ARMD) patients. Journal of Cellular Biochemistry. 2006; 99: 79–87. doi:10.1002/jcb.20857.

[110] Matalon R, Surendran S, Campbell GA, Michals-Matalon K, Tyring SK, Grady J, Cheng S, Kaye E. Hyaluronidase increases the biodistribution of acid alpha-1,4 glucosidase in the muscle of Pompe disease mice: an approach to enhance the efficacy of enzyme replacement therapy. Biochemical and Biophysical Research Communications. 2006; 350: 783–787. doi:10.1016/j.bbrc.2006.09.133.

[111] Taniguchi M, Kurahashi H, Noguchi S, Sese J, Okinaga T, Tsukahara T, Guicheney P, Ozono K, Nishino I, Morishita S, Toda T. Expression profiling of muscles from Fukuyama-type congenital muscular dystrophy and laminin-alpha 2 deficient congenital muscular dystrophy; is congenital muscular dystrophy a primary fibrotic disease? Biochemical and Biophysical Research Communications. 2006; 342: 489–502. doi: 10.1016/j.bbrc.2005.12.224.

[112] Mayer U, Saher G, Fassler R, Bornemann A, Echtermeyer F, von der Mark H, Miosge N, Poschl E, von der Mark K. Absence of integrin alpha 7 causes a novel form of muscular dystrophy. Nature Genetics. 1997; 17: 318–323. doi:10.1038/ng1197-318.

[113] Tan Z, Wang TH, Yang D, Fu XD, Pan JY. Mechanisms of 17beta-estradiol on the production of ET-1 in ovariectomized rats. Life Science. 2003; 73: 2665–2674.

[114] Baker NL, Morgelin M, Pace RA, Peat RA, Adams NE, Gardner RJ, Rowland LP, Miller G, De Jonghe P, Ceulemans B, Hannibal MC, Edwards M, Thompson EM, Jacobson R, Quinlivan RC, Aftimos S, Kornberg AJ, North KN, Bateman JF, Lamande SR. Molecular consequences of dominant Bethlem myopathy collagen VI mutations. Annals of Neurology. 2007; 62: 390–405. doi:10.1002/ana.21213.

[115] Leroy-Willig A, Willig TN, Henry-Feugeas MC, Frouin V, Marinier E, Boulier A, Barzic F, Schouman-Claeys E, Syrota A. Body composition determined with MR in patients with Duchenne muscular dystrophy, spinal muscular atrophy, and normal subjects. Magnetic Resonance Imaging. 1997; 15: 737–744.

[116] Goodpaster BH, Stenger VA, Boada F, McKolanis T, Davis D, Ross R, Kelley DE. Skeletal muscle lipid concentration quantified by magnetic resonance imaging. American Journal of Clinical Nutrition. 2004; 79: 748–754.

[117] Visser M, Goodpaster BH, Kritchevsky SB, Newman AB, Nevitt M, Rubin SM, Simonsick EM, Harris TB. Muscle mass, muscle strength, and muscle fat infiltration as predictors of incident mobility limitations in well-functioning older persons. Journals of Gerontology Series A: Biological Sciences and Medical Sciences. 2005; 60: 324–333.

[118] Uezumi A, Fukada S, Yamamoto N, Takeda S, Tsuchida K. Mesenchymal progenitors distinct from satellite cells contribute to ectopic fat cell formation in skeletal muscle. Nature Cell Biology. 2010; 12: 143–152. doi:10.1038/ncb2014.

[119] Greco AV, Mingrone G, Giancaterini A, Manco M, Morroni M, Cinti S, Granzotto M, Vettor R, Camastra S, Ferrannini E. Insulin resistance in morbid obesity: reversal with intramyocellular fat depletion. Diabetes. 2002; 51: 144–151.

[120] Wu N, Jansen ED, Davidson JM. Comparison of mouse matrix metalloproteinase 13 expression in free-electron laser and scalpel incisions during wound healing. Journal of Investigative Dermatology. 2003; 121: 926–932. doi:10.1046/j.1523-1747.2003.12497.x.

[121] Choi YC, Dalakas MC. Expression of matrix metalloproteinases in the muscle of patients with inflammatory myopathies. Neurology. 2000; 54: 65–71.

[122] Murphy G, Gavrilovic J. Proteolysis and cell migration: creating a path? Current Opinion in Cell Biology. 1999; 11: 614–621.

[123] Reznick AZ, Menashe O, Bar-Shai M, Coleman R, Carmeli E. Expression of matrix metalloproteinases, inhibitor, and acid phosphatase in muscles of immobilized hindlimbs of rats. Muscle & Nerve. 2003; 27: 51–59. doi:10.1002/mus.10277.

[124] Zhang Q, Joshi SK, Lovett DH, Zhang B, Bodine S, Kim H, Liu X. Matrix metalloproteinase-2 plays a critical role in overload induced skeletal muscle hypertrophy. Muscles Ligaments and Tendons Journal. 2014; 4: 362–370.

[125] El Fahime E, Torrente Y, Caron NJ, Bresolin MD, Tremblay JP. In vivo migration of transplanted myoblasts requires matrix metalloproteinase activity. Experimental Cell Research. 2000; 258: 279–287. doi:10.1006/excr.2000.4962.

[126] Lewis MP, Tippett HL, Sinanan AC, Morgan MJ, Hunt NP. Gelatinase-B (matrix metalloproteinase-9; MMP-9) secretion is involved in the migratory phase of human and murine muscle cell cultures. Journal of Muscle Research and Cell Motility. 2000; 21: 223–233.

[127] Frisdal E, Teiger E, Lefaucheur JP, Adnot S, Planus E, Lafuma C, D'Ortho MP. Increased expression of gelatinases and alteration of basement membrane in rat soleus muscle following femoral artery ligation. Neuropathology and Applied Neurobiology. 2000; 26: 11–21.

[128] Carmeli E, Moas M, Reznick AZ, Coleman R. Matrix metalloproteinases and skeletal muscle: a brief review. Muscle & Nerve. 2004; 29: 191–197. doi:10.1002/mus.10529.

[129] Guerin CW, Holland PC. Synthesis and secretion of matrix-degrading metalloproteases by human skeletal muscle satellite cells. Developmental Dynamics. 1995; 202: 91–99. doi:10.1002/aja.1002020109.

[130] Kieseier BC, Schneider C, Clements JM, Gearing AJ, Gold R, Toyka KV, Hartung HP. Expression of specific matrix metalloproteinases in inflammatory myopathies. Brain. 2001; 124: 341–351.

http://dx.doi.org/10.5772/62645

[131] Schoser BG, Blottner D, Stuerenburg HJ. Matrix metalloproteinases in inflammatory myopathies: enhanced immunoreactivity near atrophic myofibers. Acta Neurologica Scandinavica. 2002; 105: 309–313.

Chapter 3

Remodelling of Skeletal Muscle Extracellular Matrix: Effect of Unloading and Reloading

Eva-Maria Riso, Priit Kaasik and Teet Seene

Additional information is available at the end of the chapter

http://dx.doi.org/10.5772/62295

Abstract

This chapter summarizes studies that examine remodelling of extracellular matrix (ECM) and role of regulatory factors of ECM during unloading and reloading. Hypokinesia has a catabolic effect on both the contractile apparatus and ECM of the skeletal muscle, causing the formation of muscle atrophy, the decrease of the synthesis of contractile proteins and disturbance of the collagen metabolism. The metabolism of fibrillar and non-fibrillar collagens in ECM plays a crucial role in exercise and sport, influencing the strength development through transmission of contractile force in skeletal muscle. The impairment of motor activity and muscle strength is accompanied by the muscle atrophy. The muscle atrophy caused by inactivity and recovery from atrophy demonstrates the plasticity of muscle. Muscle mass and volume increase in a relatively short time, but the recovery of strength takes much longer and is related with the regeneration of the muscle structures. The recovery period of the contractile apparatus and ECM structures is different in slow-and fast-twitch skeletal muscle.

Although the muscle tissue's response to inactivity is more pronounced than the response of ECM, important changes occur in the connective tissue structures during unloading, causing the impairment of the functional characteristics of the skeletal muscle.

Keywords: ECM, regulatory factors of ECM, unloading, reloading, functional characteristics of skeletal muscle

1. Introduction

The intramuscular connective tissue accounts for 1–10% of skeletal muscle and has multiple functions [1,2]. It provides a basic mechanical support for vessels and nerves. The connective tissue ensures a passive elastic response of the muscle [1,3,4].

It is important to accept that both the tendon and the intramuscular connective tissue interact closely with the contractile elements of the skeletal muscle to transmit force [5–7]. The force transmission from the muscle fibres is not only transformed to the tendon and the subsequent bone via the myotendinous junctions, but also via the lateral transmission between neighbouring fibres and fascicles within the muscle [1,6]. The tension developed in one part of the muscle can be transmitted via shear links to other parts of the muscle. The perimysium is especially capable of transmitting tensile force [3,7].

The extracellular matrix (ECM) is formed by complex molecular networks, which determine the architecture of a tissue and regulate various biological processes [6,8]. The skeletal muscle ECM is organized in three levels: the epimysium surrounds the entire skeletal muscle, the perimysium surrounds muscle bundles consisting of a variable number of muscle cells, and the endomysium outlines the individual muscle fibres [2,7]. The ECM consists of various substances, of which collagen fibrils and proteoglycans are the most widespread [3,9]. The most abundant protein of the extracellular matrix is collagen, accounting for 20–25% of all protein in the whole body and forming more than 90% of the organic mass of bone [2,6]. At present, 26 different collagen types have been identified [2,10]. Although the impact is not well established, various isoforms of collagen exist, and they have varying strength and functional characteristics [6]. Muscular flexibility is partly provided by collagen. For this purpose the organization of fibrils and fibres is critical, because individual collagen molecules, fibrils, and fibres are intrinsically inextensible [2,11]. The extensibility of collagen results from the straightening of curved fibrils and fibres [2]. In addition to the proteoglycans, the hydrophilic ECM includes several other proteins such as noncollagen glycoproteins [9].

2. Functions and composition of ECM in skeletal muscle

2.1. Fibril forming collagen types in skeletal muscle

Type I and III collagen are the most abundant fibril forming collagens in the skeletal muscle. Type I collagen dominates in the intramuscular collagen content – reported from 30% up to 90% of total collagen [10,12]. The epimysium consists mainly of type I collagen with minor amounts of type III collagen [9–11]. Equal amounts of both collagen types are found in the perimysium. In the endomysium, type III collagen is the predominant form and only small amounts of type I collagen are found [2,11]. The smaller average diameter of oxidative muscle fibres should result in a higher endomysial connective tissue content in slow-twitch muscles, as the surface area/volume ratio of each fibre is greater than the average glycolytic fibre in fast-twitch muscle [3]. Slow muscles contain more type I collagen than type III collagen, the proportion of type III collagen is greater in fast muscles [13].

Type I collagen is the major stress-tolerant fibrillar collagen in the muscle. It has a high tensile strength and limited elasticity and is thus well-suited for force transmission [1,2]. Type III collagen, the other main fibrillar collagen, has a structure and arrangement similar to that of type I collagen, but it forms thinner and more elastic fibres. The fibres of type III collagen can also form copolymers with those of type I collagen [7]. Collagens I and III are fibril forming

http://dx.doi.org/10.5772/62295

and serve as a supportive structure in the muscle tissue. They attach myocytes and muscle bundles to each other [1,2,11]. Also nerves and capillaries are surrounded and attached to muscle by collagen [11]. Type V collagen is also fibril forming and can be found in the endo- and perimysium in smaller amounts than the collagen types I and III [2,10]. Collagens III and V are known to copolymerize with type I collagen and they may have a role in collagen fibre diameter regulation [6,10]. Type V collagen is considered to form the core of the fibrils, and collagens I and III copolymerize around this core [2,3]. Type II and XI collagens are also fibril forming and have been detected in the skeletal muscle only at mRNA level [3,10]. Type V and XI collagens form heterotypic molecules and can be considered as a single kind of collagen [2, 10]. Fibril associated collagens with interrupted helix (FACIT) types XII and XIV are located only in the perimysium [2,14]. These FACIT collagens associate with the surface of interstitial collagen fibrils and possibly act as molecular bridges among or between fibrils and other components of the ECM [2,9]. Although mRNAs of the other members of FACIT subfamily (IX, XVI, XIX, XXI) are detected in the skeletal muscle, the respective proteins have not been found [14,15]. The formation process of fibrillar collagen is depicted on **Figure 1**.

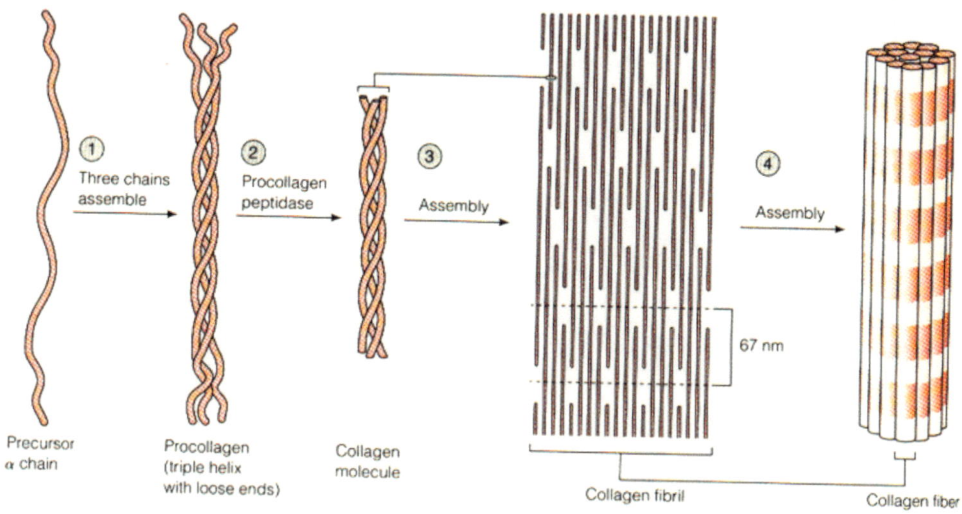

Figure 1. The formation process of fibrillar collagen.

2.2. Nonfibrillar collagen types of skeletal muscle.

Nonfibrillar collagens of the skeletal muscle are mainly located in the basement membranes. The basement membrane (BM) is a highly specialized sheet of the connective tissue surrounding individual muscle fibres, blood vessels, Schwann's cells and the spindle capsule cells. The components of the BM are the regulators of many biological activities such as cell growth, differentiation and migration which influence tissue development and repair [2,6,16]. Integrins attach muscle cells to ECM and serve as the force-transmitters between ECM and the contracting components inside the muscle cells. They connect laminin to the cell membrane to form the inner layer of basement membrane [3,17].

Type IV collagen is a major component in the basement membrane and therefore plays a critical role in the cellular arrangement in the muscle tissue. It is an integral component of basement membrane and forms a covalently stabilized polymer network around the muscle fibres [2,10]. Type IV collagen molecules form a mesh-like structure outside the laminin layer and give stability to the BM [18]. Laminin and type IV collagen are connected to each other by nidogen-1 in the muscular basement membranes [2,19]. As a part of the flexible basement membrane, type IV collagen network is interconnected with other extracellular matrix compounds and sarcolemmal proteins, being consequently exposed to stretching effects during muscle contraction [20,21]. The formation process of type IV collagen is demonstrated in **Figure 2**.

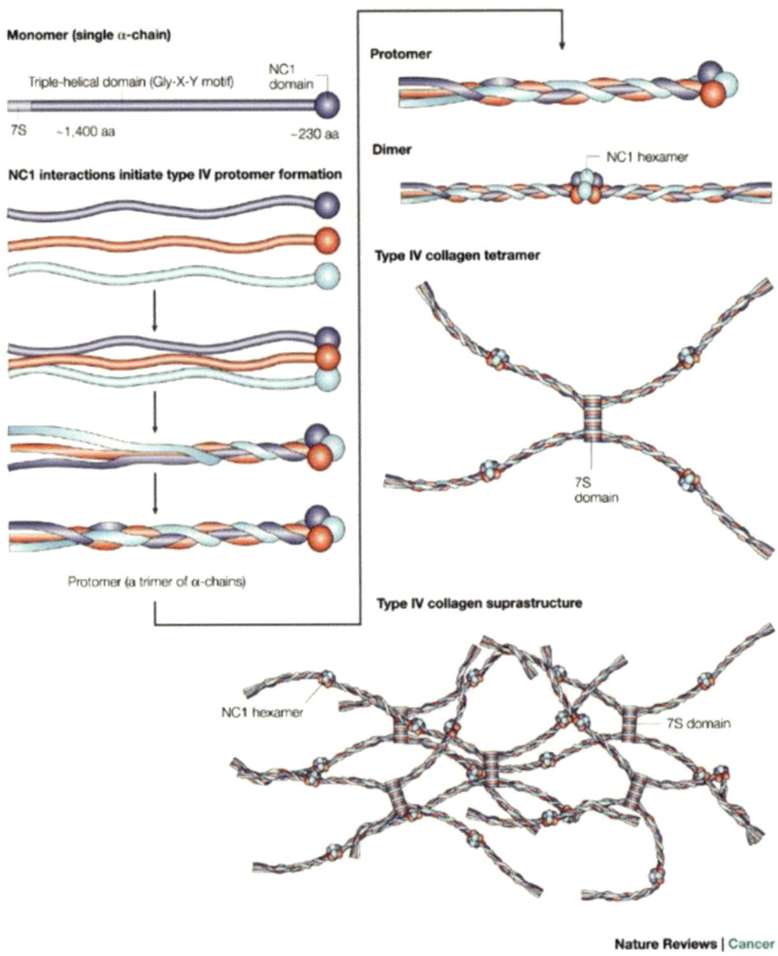

Figure 2. The formation process of type IV collagen.

http://dx.doi.org/10.5772/62295

Type VI collagen interacts with type IV and type I collagens [2,10], providing a link between the basement membranes and the surrounding matrix. Collagens XV and XVII belong to the multiplexin subfamily of nonfibrillar collagens [16] and are located in the basement membrane zone [2,10,16]. Collagens XV and XVIII may have a role in stabilizing the muscle cells [2,10]. Type XIII collagen is the transmembrane protein which is capable of binding certain basement membrane proteins [2,22]. It probably provides a link between the muscle cell and its basement membrane [10]. Type XIII collagen is concentrated in the myotendinous junctions [10,22].

3. Collagen synthesis in skeletal muscle

The synthesis of collagen is similar to other proteins, consisting of genetic transcription with messenger ribonucleic acid (mRNA) and ribosomal translation of the mRNA to prepro α-chains. In the skeletal muscle, collagens are expressed principally by fibroblasts, and their biosynthesis is characterized by the presence of an extensive number of co- and posttranslational modifications of the polypeptide chains [10,23]. Gross fractional synthesis rate for collagen is about 5% a day in the skeletal muscles of young adult rats (), whereas the fractional synthesis rate for total protein is about 11–15% /day [24].

Collagen is a protein with three polypeptide chains where each chain contains at least one stretch of the repeating amino acid sequence (Gly-X-Y)$_n$ and X and Y can be any amino acid (often proline and hydroxyproline, respectively). Both fibrillar and non-fibrillar collagens consist of three long polypeptide chains, which may or may not be identical and combine together via their (Gly-X-Y)$_n$ sequences to form a collagen triple helix. The molecular organization of different collagen types differs so that type I collagen is a heterotrimer of two identical α1(I)chains and one α2(I) chain, whereas type III collagen is a homotrimer with α1(III) chains [2,6]. The repeating unique amino acid sequence Gly-X-Y, where the glycine is in every third position, has no interruptions in the fibril-forming collagen types, whereas a considerable number of interruptions occurs in the nonfibrillar collagens [2,10]. The Gly-X-Y repeat unit gives requirements for coiling the three α-chains tightly around one another. Proline and 4-hydroxyproline residues appear frequently at the X- and Y positions, respectively, and promote the stability of the triple-helix and the structure of collagen as a whole. The structure of type IV collagen genes is distinctly different from those of fibril forming collagens. The most common form of type IV collagen consists of two α1(IV) chains and one α2(IV) chain, although the combinations of α3(IV) and α4(IV) as well as α5(IV) and α6(IV) are found in some basement membranes [2,10].

3.1. Modifications of the polypeptide chains

An exception to the synthesis of other proteins is that collagen synthesis is characterized by an extensive number of co- and posttranslational modifications of the polypeptide chains. The intracellular modifications of polypeptide chains involve hydroxylation and glycosylation reactions to form the procollagen. Hydroxylation of proline, the reaction catalyzed by prolyl 4-hydroxylase (P-4-H), influences the stability of the triple-helical structure of collagen [20]. The triple-helix formation of the pro-α-chains prevents any further hydroxylation. Intracellular events of collagen synthesis include also 3-hydroxylation of proline residues, hydroxylation of lysine residues and glycosylation of certain hydroxylysine residues of propeptides. The assay of prolyl-4-hydroxylase activity has been commonly used to estimate the changes in the rate of collagen synthesis [20]. Fibrillar collagens are secreted as soluble procollagens, which are converted to collagen by the cleavage of C- and N terminal propeptides by procollagen N- and C-proteinases. Extracellular modifications in the collagen synthesis involve an ordered self-assembly for the formation of collagen fibrils and the crosslink formation to make the fibrils stable. The stabilization of the fibrils is provided by covalent cross-links generated by the conversion of some of the lysine and hydroxylysine residues to aldehyde derivates by lysyl oxidase [2,25]. Lysyl oxidase (LO) is a key enzyme in the extracellular modification of collagen [25]. LO, an amine oxidase expressed and secreted by fibrogenic cells, plays a critical role in the formation and repair of the ECM by oxidizing lysine residues in elastin and collagen, thereby imitating the formation of covalent crosslinkages which stabilize these fibrous proteins [25]. Type IV collagen molecules form their network with different processes. A tight meshwork is formed by irregularly branching lateral associations of the triple helical regions [2,10]. The formation steps of collagen is demonstrated in **Table 1**.

NUCLEUS	Collagen gene → RNA processing → mRNA
CYTOPLASM Rough endoplasmic reticulum	Pre-procollagen chain → cleavage of signal peptide
Golgi apparatus	Procollagen α chain → hydroxylation – glycosylation – association of the C-terminal propeptides - disulfide bond formation → procollagen molecule
EXTRACELLULAR MATRIX	Cleavage of propeptides → collagen fibril self assembly → crosslinking

Table 1. The formation steps of collagen.

4. Degradation of collagens in skeletal muscle

Degradation of collagen represents the obligatory step of a turnover and the remodelling of the connective tissue and during the mechanical loading of fibroblasts and extracellular matrix structures. Both intracellular and extracellular degrading pathways are present, using either lysosomal phagocytosis or ECM proteinases, respectively [26,27]. Collagens can be degraded prior to or after their secretion from the cell. Secreted collagen is degraded mainly by two different routes: proteolytic and phagocytotic. Proteolytic degradation occurs mainly through matrix metalloproteinase (MMP) activity. Macrophages remove ECM components, although also fibroblasts are able to the phagocytosis and degradation of collagen fibrils [27]. Degradation is continued by specific proteinases and the collagen fragments are phagocytosed by cells and processed by lysosomal enzymes [28]. About 26% of newly synthesized collagen is degraded per day in young adult rats [24]. The most recently synthesized collagen seems to be more susceptible to degradation than mature collagen [10,24]. The main steps of collagen degradation are depicted in **Figure 3** and **4**.

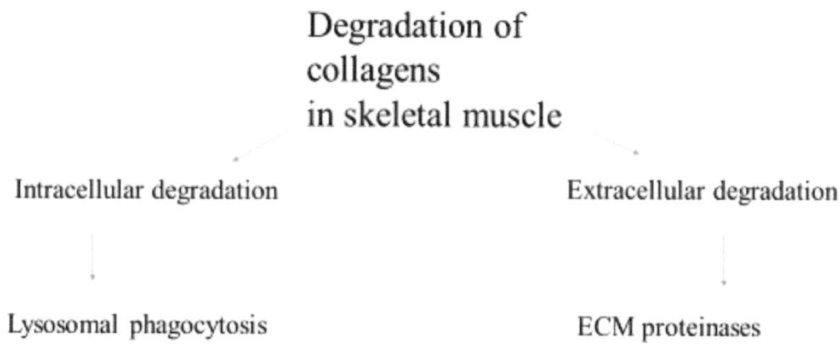

Figure 3. Degradation of collagens in skeletal muscle.

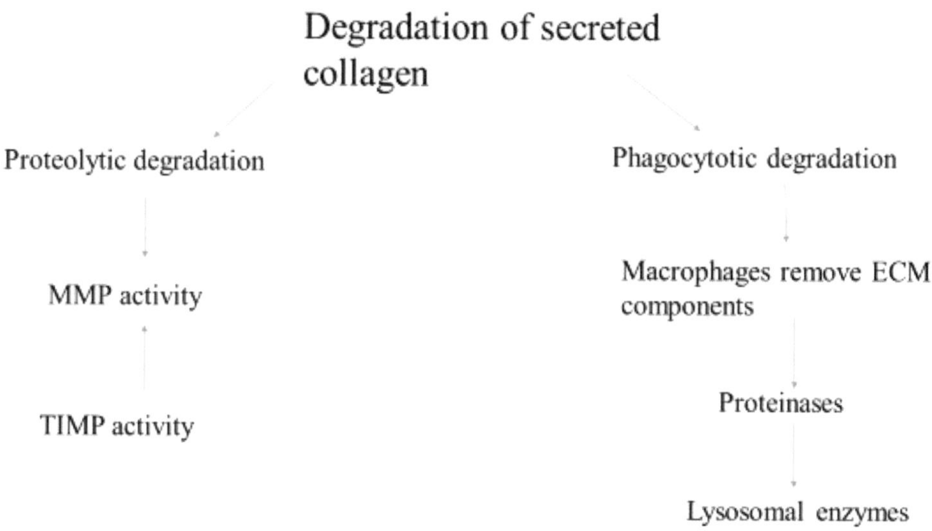

Figure 4. Degradation of secreted collagens.

4.1. Role of matrix metalloproteinases

Collagen degradation is initiated extracellularly by MMPs or matrix metalloproteinases, which are presented in tissues mostly as latent proMMPs [29,30]. MMPs are a family of zinc-dependent proteolytic enzymes that function mainly in the ECM [30,31]. The activation of specific matrix metalloproteinases has been implicated in degradative and atrophic changes in the ECM after muscle injury or in various myopathic conditions. These matrix metalloproteinases may cause structural and physiological alterations to the basal lamina and sarcolemma of myofibres, leading to uncontrolled influx and efflux of ions and subsequent myopathy [31, 32]. Because of their ability to degrade ECM components, MMPs are considered to be important components in many biological and pathological processes [30–32]. They have regulatory roles in muscle growth and development and are also important in repair processes after traumatic injury or disuse myopathy [30,31]. MMPs are mainly produced from endotendon fibroblasts and intramuscular matrix fibroblasts [30,31], although some level of expression has been found to occur also in satellite cells [33]. MMPs are secreted or released in latent form and become activated in pericellular environments [23,34]. The activities of MMPs are also under the control of enzyme tissue inhibitors of matrix metalloproteinases (TIMPs). Disturbances in the ratio of specific MMPs and their inhibitors may be manifested by physiological dysfunction, resulting in clinical disorders [31,35].

Up until now, 24 different vertebrate MMPs have been identified, of which 23 have been found in humans. MMPs are usually divided according to their main substrate into collagenases, gelatinases, stromelysins, matrilysins, membrane-type MMPs and others, although many of them have wide and overlapping substrate specificity [36].

MMP-1, MMP-8, MMP-13 and MMP-18 are collagenases, which have the ability to cleave the native helical structure of collagens I, II and III. Cleavage products are then susceptible to the action of other MMPs [36,37].

Gelatinases MMP-2 and MMP-9 degrade denatured collagen, gelatin, native type IV, V and VII collagens as well as other ECM components [36]. One of the most important MMPs associated with the function and dysfunction of the skeletal muscle appears to be MMP-2, also known as gelatinase A, or 72-kDa type IV collagenase. MMP-2, by regulating the integrity and composition of the ECM in skeletal muscle, plays essential role in myofibre proliferation and differentiation, the fibre healing after injury, and maintenance of the surrounding connective tissue [38]. MMP-2 also digests fibrillar type I and II collagens. MMP-2 and –9 are known to be overexpressed and present in higher amounts in patients with inflammatory myopathies, which may increase ECM degradation and thus facilitate lymphocyte adhesion [32,38,39].

MMP-3 and MMP-10, or stromelysin-1 and –2, both digest ECM components and activate proMMP-1. The third stromelysin, MMP-11, differs from other stromelysins by its sequence and substrate specificity [36].

Matrilysins- MMP-7 and MMP-26 are the smallest MMPs. MMP-7 can also process cell surface molecules [36].

Six membrane-type MMPs (MT-MMPs) have been characterized. Except the MT4-MMP, they all are all capable to activate proMMP-2 [36,40]. For their pericellular fibrinolytic activity, MT-MMPs have an important role in angiogenesis [36,40].

Six MMPs – MMP-12, MMP-19, MMP-20, MMP-22, MMP-23, MMP-28 are currently classified into the group of "other MMPs" [34,36].

4.2. Inhibition of matrix metalloproteinases by tissue inhibitors of metalloproteinases

MMPs and tissue inhibitors of metalloproteinases (TIMPs) have an important role in the adaptive changes in the muscle in response to local developmental, physiological, surgical, and pathological conditions [31,39]. TIMPs are the major cellular inhibitors of the MMP subfamily, exhibiting varying efficacy against different members, as well as different tissue expression patterns and modes of regulation [36,41]. Four mammalian TIMPs have been characterized and considered to regulate MMP activity during tissue remodelling [41,42]. All four TIMPs (TIMP-1, -2, -3 and –4) can inhibit all MMPs, except TIMP-1, which is a poor inhibitor of MMP-19 and most of the MT-MMPs [41,42]. Although TIMP-2 inhibits MMP-2 in high concentrations, it has an important role in activating proMMP-2 in a complex with MT1-MMP, which demonstrates an integrated response of MMPs and TIMPs [43]. In skeletal muscle, TIMP-1, TIMP-2 and TIMP-3 are expressed [33,39]. TIMP-4 appears to be cardiac-specific and has not been detected in the skeletal muscle [37].

The increased MMP activity and thus the enhanced degradation of collagen often parallels the stimulated activation of collagen synthesis. TIMPs are often activated together with MMPs in response to physical activity, indicating a simultaneous stimulation and the inhibition of degradation [44,45]. MMP activity precedes TIMP activity and thus TIMP serves as the regulator of degradation termination to ensure a limited amount of degradation [39,43].

In addition to MMP-binding activities, TIMPs have many important biological functions. TIMPs can promote or inhibit cell growth, depending on the type of the cell and the inductor [36,41].

5. Effect of unloading on the skeletal muscle

The inactivity of the skeletal muscle leads to the loss of muscle contractile proteins and strength [46,47]. The weakening of the muscle is accompanied by the loss of the muscle mass and the reduction of the size of the muscle cell [46,48]. The decrease in the protein synthesis and the increase in protein degradation appear both in the contractile apparatus and in the ECM [8,46]. The skeletal muscle atrophy attributable to the muscular inactivity has significant adverse functional consequences, nevertheless the tight connections between the contractile machinery and the ECM are still unknown [27]. Changes in the intramuscular collagen protein fraction have been shown to significantly impact mechanical properties of skeletal muscle in non-loading conditions [27,49]. Events in skeletal muscle during unloading are shown in **Figure 5**.

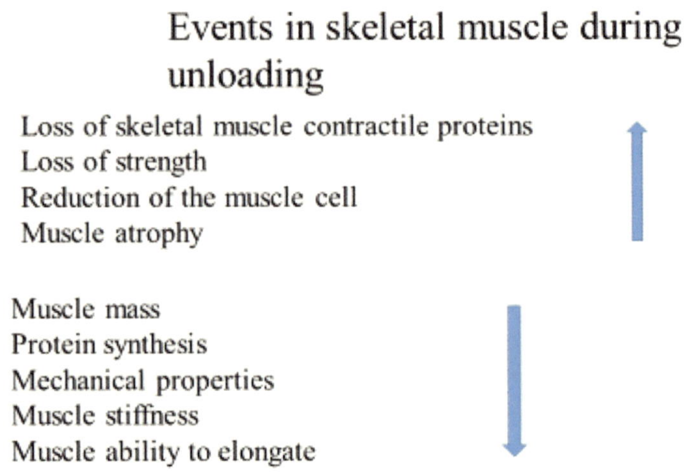

Figure 5. Events in skeletal muscle during unloading.

5.1. Effect of unloading on the extracellular matrix

The ECM of connective tissues enables links to other tissues and plays a key role in force transmission and tissue structure maintenance in tendons, ligaments, bone and muscle [18]. ECM turnover is influenced by physical activity [50,51]. Immobilization causes a marked relative increase in the endo- and perimysial connective tissue, which results in changes of the mechanical properties of skeletal muscle [11].

Fibrillar type I and III collagens are most abundant in the skeletal muscle epi- and perimysium. Non-fibrillar type IV collagen is present only in basement membranes and has a critical role in the cellular arrangement of muscle tissue [2,11]. There are differences in the collagen metabolism and the content between muscles. Slow-twitch muscles contain 40–50% more collagen than fast-twitch muscles [13].

A reduced muscular activity decreases the collagen synthesis rate in the skeletal muscle, the immobilization down-regulates the collagen synthesis at the pretranslational level, mainly among I and III collagens [18,20]. Unloading also induces a shift in the relative proportion of collagen isoform type I to III [13]. Decrease of collagen I mRNA level in slow-twitch Soleus (Sol) and fast-twitch gastrocnemius (GM) muscle during the three-week hindlimb suspension shows that the fibrillar type I collagen is more sensitive to unloading and the effect is much more long-lasting than that of fibrillar type III collagen. This finding shows that hindlimb unloading induces reduction of collagen type I [52].

Lysyl oxidase which plays an important role in the formation and regeneration of ECM by oxidizing lysine residues in elastin and collagen, initiates the formation of covalent cross-linkages which stabilize fibrous proteins [25]. From this standpoint it is understandable that a significant decrease in LO mRNA level was registered only in Sol muscle [25].

Matrix metalloproteinases are providing degradation of ECM compounds [31,36]. MMP-2 level did not change significantly during three weeks of hindlimb suspension. TIMPs are proteins which inhibit ECM degradation [35,53,54]. The mRNA level of TIMP-1 decreased in slow-twitch muscle after one-week hindlimb suspension. As both intracellular (lysosomal phagocytosis) and extracellular degrading pathways (ECM proteinases) are present in the degradation of the skeletal muscle during the unloading, it is complicated to put all the role to the MMPs in this process [6].

The biggest changes in the specific mRNA level of type I, III and IV collagen were registered in Sol and GM muscle during the three weeks of unloading mRNA level of LO decreased also in Sol muscle [52]. Changes in TIMP-1 mRNA level during first week of hindlimb suspension were contradictory in Sol and GM muscle [52].

The metabolism of fibrillar and non-fibrillar collagens in ECM plays a crucial role both in decreased locomotory activity and in exercise and sport, influencing the strength development through transmission of contractile force in skeletal muscle. Events in ECM during unloading are shown in **Figure 6**.

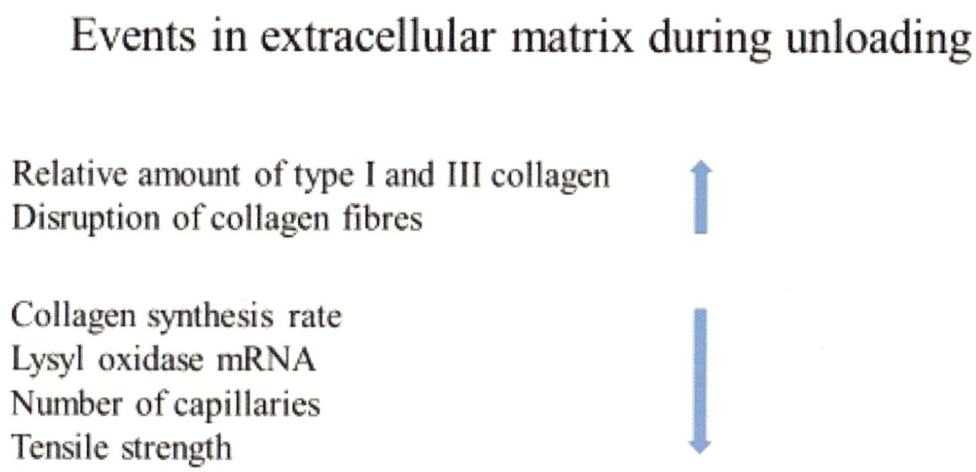

Figure 6. Events in the ECM during unloading.

5.2. Effect of unloading on the synthesis of collagen

Several quantitative and qualitative changes in the intramuscular connective tissue contribute to the deteriorated function and biomechanical properties of the immobilized skeletal muscle [6,11,51]. Muscle and tendon collagen and the connective tissue network are known to respond to altered levels of physical activity [51,55]. In contrast to physical loading, immobilization of rat limb leads to a decrease in the activities of collagen synthesizing enzymes both in skeletal muscle and tendon [20,56].

Along with the increased amount of intramuscular connective tissue, the number of capillaries decreases dramatically [57,58]. Each capillary is surrounded by a dense layer of the connective tissue fibres, isolating the capillary from the adjacent muscle fibre, which disturbs the blood supply of the muscle fibres and further increases the muscle fibre atrophy [57,58].

During immobilization, the normal three-dimensional orientation of the collagen fibres is disrupted. The normal orientation constitutes of the fibres running parallel to the muscle fibres on their surface, preventing muscle cells from over-elongation and –contraction. In addition, thin perpendicular fibres connect adjacent muscle fibres to each other [7,10,11]. As a result of decreased loading, the number of longitudinal fibres increases, the crimp angel of the collagen decreases and this diminishes the ability of the muscle to elongate [2,11] and because of that the skeletal muscle shows significantly decreased tensile strength [1,3]. In addition to changes in collagen abundance, alterations in the degree of collagen cross-linking would have a profound effect on the mechanical properties of skeletal muscle, causing a decrease in muscle stiffness [10,17].

In contrast to physical loading, immobilization leads to the decrease in the enzyme activities of collagen biosynthesis, which suggests that the biosynthesis of the collagen network decreases as a result of reduced muscular and tendinous activity [8,56]. The rate of the total collagen synthesis depends mostly on the overall protein balance of the tissue, but it seems to be positively affected by stretch in both muscle and tendon [55].

Collagen expression during immobilization has been shown to be at least partially down-regulated at the pretranslational level [55]. Although the relative amount of the connective tissue increases during immobilization, the gene expression of type I and III collagens decreases during the first three days of immobilization [20]. The content of type IV collagen was also reduced as a result of immobilization [23]. The activities of prolyl 4-hydroxylase (P 4-H) and galactosylhydroxylysyl glucosyltransferase (GGT) decrease from the first three days of immobilization up to at least three weeks, suggesting decreased collagen biosynthesis during that time [20,59]. The degradation of collagens has been found to be enhanced during immobilization, as the expression of both MMP-2 and MMP-9 increased after 30 days of immobilization [30]. The quantity of TIMP-1 was also increased after 30 days of immobilization [30].

The collagen concentration increases when expressed both as a function of muscle dry weight or muscle cross-sectional area, but this increase in muscle collagen is primarily due to the muscle atrophy induced by immobilization [13].

5.3. Effect of unloading on the contractile apparatus of skeletal muscle

As the skeletal muscle is a highly plastic tissue, the conditions associated with the disuse are accompanied by adaptation. A period of time without weight bearing cause modifications of structure and the function of skeletal muscles, of which atrophy and a slow-to-fast transition are the most prominent [60,61]. Many animal models such as the kind limb suspension, immobilization in shortened and lengthened position, spaceflight and denervation show that the removal of a mechanical load produces atrophy and contractile alterations, more evident in the slow muscle soleus than in fast muscles as extensor digitorum longus [60,62,63]. Besides space flight and bed rest experiments, long periods of muscle disuse in relation to a disease or traumatic injuries of the joints or of the bones are relatively common experience for human beings [62,63]. The inactivity causes only small increases in contractile speed and myofibrillar adenosine triphosphatase (ATPase) activity and slight elevations in the percentage of the fast type myosin heavy chain (MyHC) isoforms in fast-twitch muscles, as compared to slow-twitch muscle [64,65].

5.4. Slow-to-fast transition in skeletal muscle during unloading

The different response can be explained, considering the fact that skeletal muscles in different parts of the body are subjected to different patterns of recruitment and activity [66,67]. The anti-gravitational Sol muscle is recruited for prolonged periods at a moderate level of intensity, whereas the extensor digitorum longus muscle is less frequently recruited, performing short, high-force contractions [68,69]. It is commonly known that inactivity affects the functional and biochemical properties of antigravity muscles, causing a significant decrease in both contraction and relaxation times [66,70] and a significant increase in the maximal shortening velocity and myofibrillar ATPase activity [71]. The above-mentioned changes are considered to be a result of the gene expression, especially the genes involved in the fibre type transformation [71–73]. The co-ordinated changes in the gene expression are particularly apparent for myosin and consequently the disuse induces a slow-to-fast transition, as reflected by an increase in fast MyHC isoforms at the expense of slow MyHC in the Sol and a fast-to-faster MyHC shift in the GM muscle [74,75]. An increase in fast MyLC isoforms, an increased proportion of fast troponin subunits and hybrid fibres co-expressing fast and slow MyHC and MyLC appears during slow-to-fast transition in Sol muscle [64,71,76]. The fibre type transition results in a change in muscle metabolism, fuel use, and more fatigable muscle [77]. Several histochemical analyses have also suggested that the functional changes in immobilized muscles are due to an increase in fast-twitch IIa fibres [64]. In addition to above-mentioned facts, the increased sarcoplasmic reticulum calcium-ATPase activity and the preferential loss of thin filaments all contribute to faster contractile properties of the Sol muscle [78]. The increased shortening velocity may be an attempt to compensate for the loss of power generating capacity during unloading caused by weakening [79].

Muscle disuse is often accompanied by increased fatigability, which is caused by the reduced oxidative capacity of disused muscles [65,80,81]. Capillary loss and reduction in blood flow might contribute to the increased fatigability by an impaired supply of energy substrates and oxygen to the muscle [82].

5.5. Formation of muscle atrophy during unloading

While immobilization at shortened length induces atrophy, immobilization in lengthened position induces hypertrophy, which is largely attributable to addition of sarcomeres in the longitudinal direction [26,83]. Immobilization in the shortened position, e.g. hindlimb suspension, induces preferential transcription of fast MyHC isoforms, reminiscent of the slow-to-fast transition observed in other models of disuse [26,84]. The disuse atrophy is characterized by the loss of muscle mass and decrease of muscle diameter. In the case of muscle atrophy, there some noticeable changes in the muscle cell at the cellular level including sarcomere dissolution and endothelial degradation, a marked reduction in the number of mitochondria, the accumulation of the connective tissue, the elimination of apoptotic myonuclei and a decrease in capillary density [85–87]. Selective susceptibility of fibre types to immobilization seems to exist, while the red muscle fibres show the greatest atrophy. The decreased synthesis of protein and increased protein degradation are characteristic features to muscle atrophy. At least half of the total muscle protein is myofibrillar protein, and this fraction is lost at a faster rate than other muscle proteins during atrophy [84]. Three major proteolytic systems to skeletal mass protein loss are the cytosolic calcium-dependent calpain system, the lysosomal proteases and the ATP-dependent ubiquitin-proteasome system, which work as partners during muscle proteolysis rather than one system being used exclusively during atrophy [88]. Recent advances in cellular biology show the oxidative stress to be an important regulator of pathways leading to muscle atrophy during periods of disuse, increasing the expression of the key components of the proteasome proteolytic system. This proteolytic system is a prominent contributor to protein breakdown in skeletal muscle during periods of inactivity [37,84,86].

5.6. Effect of unloading on the skeletal muscle MyHC composition

Prolonged periods of time spent with a diminished or no-weight bearing have a deleterious effect on skeletal muscle with the decreased protein synthesis, the loss of muscle mass and alterations of biochemical parameters [74,77]. The main findings confirmed that the proportion of slow MyHC isoforms decreased and the proportion of MyHC fast isoforms increased in consequence of altered functional conditions [46,64,89].

Clinical observations show that the atrophy of the skeletal muscle occurs as a result of immobilization and is caused by the changed functional conditions in the muscular system. Skeletal muscle function depends on the intact proprioceptive activity, motor innervation, mechanical load, and joint mobility. If one of these factors is altered, the muscle will undergo adaptation. As an increased muscular activity leads to the enhancement of the structures involved in contraction, inactivity or disuse is followed by the reduction of the muscle mass [8,77].

Alterations of biochemical parameters and changes at the ultrastructural level of the contractile apparatus are considered to be characteristic of atrophied muscles. The effect of disuse on the skeletal muscle depends on the fibre type composition of the muscle. The degenerative changes in disused muscles at the ultrastructural level have been shown to be most severe in slow oxidative muscle fibres. It is suggested that the most vulnerable muscles were antigravity muscles crossing a single joint [90].

The properties of muscle contraction which depend on the MyHC isoform composition decrease in atrophied skeletal muscle [46,89,91]. Contractile activity can induce differential expression of myosin protein isoforms in skeletal muscle. MyHC composition has an important regulatory role in myosin ATPase activity and muscle fibre shortening velocity [92]. A prolonged activity causes alterations in the MyHC composition. A decrease in the mechanical load stimulates the conversion of slow myosin in muscles of mixed fibre type composition, whereas a decrease in the weight-bearing load results in a decrease in slow myosin content [46,92].

http://dx.doi.org/10.5772/62295

Comparing the changes of proportion of MyHC isoforms in the contractile apparatus during unloading among human subjects and experimental animals, we can see that they are similar in their direction and amplitude. In conclusion, the adaptation of the mammalian skeletal muscles to the unloading depends on the contractile and metabolic characteristics of skeletal muscle and is not dependent on the species of the mammal. As the certain connections exist between the contractile and metabolic characteristics of skeletal muscle it is understandable why the specific atrophy causes the decrease of the main function of the skeletal muscle.

6. Effect of reloading on the skeletal muscle

The plasticity of muscle is apparent in the phenomena such as muscle atrophy caused by inactivity, and recovery from atrophy. When atrophic muscles once again become active, the muscle mass and the volume reportedly increase in a relatively short period of time, but the recovery of the muscle strength takes much longer [93]. The recovery of motor activity after the hindlimb suspension is as fast as the recovery of the muscle strength. It is probably related with the regeneration of the muscle structures from disuse atrophy [89]. The fact that the increases in the muscular strength lag behind those in the muscular mass suggests the presence of functionally immature muscle fibres during the recovery process following disuse atrophy [89]. Several studies have shown that the increases in the muscle mass soon after reloading are attributable to oedema and do not actually represent recovery [89]. The recovery of the muscle mechanical properties depends also on the metabolism of the skeletal muscle. The two-week reloading period has shown that the Sol muscle metabolism can be restored [94]. Full recovery of slow-twitch muscle function via cross-sectional area and myonuclear domain size has been shown to need more time for restoration of neural and mechanical properties of muscle [81].

Disuse muscle atrophy can be experimentally induced by suspending animals by their tails [89,95], immobilizing joints, severing tendons or conducting muscle denervation [96]. Muscle atrophy in tail suspension is caused by hindlimb unloading, conserving the functions of nerves and joints. Reloading is thus possible after tail suspension and is suitable for investigating the recovery process following disuse muscle atrophy caused by sports injuries [89].

The reloading after hindlimb suspension shows that collagen III mRNA level at the end of the second week is higher than in control group. It has been found that in response to reload, the skeletal muscle expression of collagen I and III was markedly induced from the second day of reloading [4].

It has been shown that non-fibrillar type IV collagen mRNA level is decreasing in both, slow-twitch and fast-twitch muscles during the three weeks of hindlimb suspension, but two weeks of reloading period is obviously not enough to restore the metabolic states of this collagen in the basal lamina of the muscle fibre [26]. It was demonstrated that the reorganization of the basement membrane compounds needs certain time [97]. As type IV collagen plays a role in the regenerative process on ECM, including the matrix-associated receptors that underline muscle fibre-matrix interactions, it shows how complicated is the evaluation of the functional significance of type IV collagen metabolism [5].

The recovery of the collagen degradation markers during reloading period is different in slow- and fast-twitch skeletal muscles. MMP-2 level increased in slow-twitch soleus muscle during two weeks of reloading and in fast-twitch gastrocnemius muscle after one week reloading. The mRNA level of TIMP-1 increased in fast-twitch GM muscle after two weeks of reloading. A significant increase in mRNA level for MMP-2 was registered in Sol muscle during the reloading, showing that the reaction of MMP-2 on the pretranslational level is not fast in all muscles [52].

Concomitant to atrophy, numerous molecular events testify of a slow-to-fast transition of muscle properties [47,74]. The recovery of muscle properties effectively occurs on return to normal load [98]. It is also known that muscle fibre damage occurs during reloading, likely due to the inability of the muscle fibres to bear eccentric contractions and the consequent inflammation process [94,99]. Natural recovery seems to be most effective after reloading while several investigations show the delayed recovery of rats during running exercise [100].

The muscle tissue response to unloading seems to more pronounced than the connective tissue response. The connective structures are protected from rapid changes in tissue mass, while muscle, which is known to act as a protein store for the organism, is subject to substantial and fast changes in tissue mass. However, it should be considered that important changes occur in the connective tissue structures during unloading despite the small changes in tissue mass.

Author details

Eva-Maria Riso*, Priit Kaasik and Teet Seene

*Address all correspondence to: eva-maria.riso@ut.ee

Institute of Sports Sciences and Physiotherapy, University of Tartu, Tartu, Estonia

References

[1] Trotter JA. Structure-function considerations of muscle-tendon junctions. Comp Biochem Physiol 2002; Part A: 1127-33. doi: 10.1016/S1095-6433(02)00213-1

[2] Gelse K, Pöschl E, Aigner T. Collagens – structure, function, and biosynthesis. Advanced Drug Delivery Reviews 2003; 55: 1531-46. doi: 10.1016/j.addr.2003.08.002

[3] Purslow PP. The structure and functional significance of variations in the connective tissue within muscle. Comp Biochem Physiol 2002; Part A: 947-66. doi: 10.1016/S1095-6433(02)00141-1

[4] Heinemeier KM, Olesen JL, Haddad F, Schjerling P, Baldwin KM, Kjær M. Effect of unloading followed by reloading on expression of collagen and related growth factors

in rat tendon and muscle. J Appl Physiol 2009; 106: 178-86. DOI: 10.1152/japplphysiol.91092.2008

[5] Sanes JR. The Basement Membrane/Basal Lamina of Skeletal Muscle. J Biol Chem 2003; 278: 12601-604. DOI: 10.1074/jbc.R200027200

[6] Kjær M. Role of extracellular matrix in adaptation of tendon and skeletal muscle to mechanical loading. Physiol Rev 2004; 84: 649-98. DOI: 10.1152/physrev.00031.2003

[7] Grounds MD, Sorokin L, White J. Strength at the extracellular matrix-muscle interface. Scand J Med Sci Sports 2005; 15: 381-91. DOI: 10.1111/j.1600-0838.2005.00467.x

[8] Haus JM, Carrithers JA, Carroll CC, Tesch PA, Trappe TA. Contractile and connective tissue protein content of human skeletal muscle: effects of 35 and 90 days of simulated microgravity and exercise countermeasures. Am J Physiol Regul Integr Comp Physiol 2007a; 293: R1722-27. DOI: 10.1152/ajpregu.00292.2007

[9] Heinegård D. Proteoglucans and more – from molecules to biology. Int J Exp Path 2009; 90: 575-86. doi: 10.1111/j.1365-2613.2009.00695.x

[10] Ricard-Blum S, Ruggiero F. The collagen superfamily: from the extracellular matrix to the cell membrane. Path Biol 2005; 53: 430-42. doi: 10.1016/j.patbio.2004.12.024

[11] Järvinen TA, Józsa L, Kannus P, Järvinen TLN, Järvinen M. Organization and distribution of intramuscular connective tissue in normal and immobilized skeletal muscles. J Muscle Res Cell Motil 2002; 23: 245-54.

[12] Eyre DR. Collagen of articular cartilage. Arthritis Res 2002; 4: 30-35. doi: 10.1186/ar.380

[13] Miller TA, Lesniewski LA, Muller-Delp JM, Majors AK, Scalise D, Delp MD. Hindlimb unloading induces a collagen isoform shift in the soleus muscle of the rat. Am J Physiol 2001; 281: R1710-17.

[14] Listrat A, Lethias C, Hocquette JF, Renard G, Messinier P, Geay Y, Picard B. Age-related changes and location of types I, III, XII and XIV collagen during development of skeletal muscles from genetically different animals. Histochem J 2000; 32: 349-56.

[15] Chou YC, Li HC. Genomic organization and characterization of the human type XXI collagen (COL21A1) gene. Genomics 2002; 79: 395-401. doi: 10.1006/geno.2002.6712

[16] Erickson AC, Couchman JR. Still more complexity in mammalian basement membranes. J Histochem Cytochem 2000; 48: 1291-306. doi:10.1177/002215540004801001

[17] Wiberg C, Heinegård D, Wenglen C, Timpe R, Morgelin M. Biglycans organize collagen VI into hexagonlike networks resembling tissue structures. J Biol Chem 2002; 277: 49120-126. doi:10.1074/jbc.M206891200

[18] Takala TES, Virtanen P. Biochemical composition of muscular extracellular matrix: the effect of loading. Scand J Med Sci Sports 2000; 10: 321-25. DOI: 10.1034/j.1600-0838.2000.010006321.x

[19] Salmivirta K, Talts JF, Olsson M, Sasaki T, Timpl R, Ekblom P. Binding of mouse nidogen-2 to basement membrane components and cells and its expression in embryonic and adult tissues suggest complementary functions of the two nidogens. Exp Cell Res 2002; 279: 188-201. doi: 10.1006/excr.2002.5611

[20] Han XY, Wang W, MyllyläR, Virtanen P, Karpakka J, Takala TES. MRNA levels for α-subunit of prolyl 4-hydroxylase and fibrillar collagens in immobilized rat skeletal muscle. J Appl Physiol 1999b; 87: 90-96.

[21] Ahtikoski AM, Riso EM, Koskinen SOA, Risteli J, Takala TES. Regulation of type IV collagen gene expression and degradation in fast and slow muscles during dexamethasone treatment and exercise. Pflügers Arch Eur J Physiol 2004; 448: 123-30. DOI: 10.1007/s00424-003-1226-5

[22] Tu H, Sasaki T, Snellman A, Göhring W, PiriläP, Timpl R, Pihlajaniemi T. The type XIII collagen ectodomain is a 150-nm rod and capable of binding to fibronectin, niddogen-2, perlecan, and heparin. J Biol Chem 2002; 277: 23092-99. DOI: 10.1074/jbc.M107583200

[23] Trackman PC. Diverse biological functions of extracellular collagen processing enzymes. J Cell Biochem 2005; 96: 927-37. doi:10.1002/jcb.20605

[24] Mays PK, McAnulty RJ, Campa JS, Laurent GJ. Age related changes in collagen synthesis and degradation in rat tissues. Biochem J 1991; 276: 307-13.

[25] Kagan HM, Li W. Lysyl oxidase: Properties, specificity, and biological roles inside and outside of the cell. J Cell Biochem 2002; 88: 660-72. DOI: 10.1002/icb.10413

[26] Ahtikoski AM, Koskinen SOA, Virtanen P, Kovanen V, Risteli J, Takala TES. Synthesis and degradation of type IV collagen in rat skeletal muscle during immobilization in shortened and lengthened position. Acta Physiol Scand 2003; 177: 473-81. DOI: 10.1046/j.1365-201X.2003.01061.x

[27] Gianelli G, DeMarzo A, Marinosci F, Antonaci S. Matrix metalloproteinase imbalance in muscle disuse atrophy. Histol Histopathol 2005; 20: 99-106.

[28] Holm L, van Hall G, Rose AJ, Miller BF, Doessing S, Richter EA, Kjær M. Contraction intensity and feeding affect collagen and myofibrillar protein synthesis rates differently in human skeletal muscle. Am J Physiol Endocrinol Metab 2010; 298: 257-69. DOI: 10.1152/ajpendo.00609.2009

[29] Balbin M, Fueyo A, Knäuper V, Lopez JM, Alvarez J, Sanchez LM, Quesada V, Bordallo J, Murphy G, Lopez-Otin C. Identification and enzymatic characterization of two diverging murine counterparts of human interstitial collagenase (MMP-1) expressed at sites of embryo implantation. J Biol Chem 2001; 276: 10253-262. doi:10.1074/jbc.M00767400

[30] Reznick AZ, Menashe O, Bar-Shai M, Coleman R, Carmeli E. Expression of matrix metalloproteinases, inhibitor, and acid phosphatase in muscles of immobilized hindlimbs of rats. Muscle Nerve 2003; 27: 51-59.

[31] Carmeli E, Moas M, Reznick AZ, Coleman R. Matrix metalloproteinases and skeletal muscle: A brief review. Muscle Nerve 2004; 29: 191-97.

[32] Choi YC, Dalakas MC. Expression of matrix metalloproteinases in the muscles of patients with inflammatory myopathies. Neurology 2000, 54: 65-71.

[33] Balcerzak D, Querengesser L, Dixon WT, Baracos VE. Coordinate expression of matrix-degrading proteinases and their activators and inhibitors in bovine skeletal muscle. J Anim Sci 2001; 79: 94-107. doi: /2001.79194x

[34] Chen X, Li Y. Role of matrix metalloproteinases in skeletal muscle: migration, differentiation, regeneration and fibrosis. Cell Adh Migr 2009; 3: 337-41.

[35] Carmeli E, Kodesh E, Nemcovsky C. Tetracycline therapy for muscle atrophy due to immobilization. J Musculoskelet Neuronal Interact 2009; 9: 81-88.

[36] Visse R, Nagase H. Matrix metalloproteinase and tissue inhibitors of metalloproteinases: Structure, function and biochemistry. Circ Res 2003; 92: 827-39. doi: 10.1161/01.RES.0000070112.80711.3D

[37] Li YY, McTiernan CF, Feldman AM. Interplay of matrix metalloproteinases, tissue inhibitors of metalloproteinases and their regulators in cardiac remodelling. Cardiovasc Res 2000; 46: 214-24. DOI: http://dx.doi.org/10.1016/S0008-6363(00)00003

[38] Liu X, Lee DJ, Skittone LK, Natsuhara K, Kim HT. Role of gelatinases in disuse-induced skeletal muscle atrophy. Muscle Nerve 2010; 41: 174-78. doi:10.1002/mus.21463

[39] Koskinen SOA, Heinemeier K, Olesen JL, Langberg H, Kjær M. Physical exercise can influence local level of matrix metalloproteinases and their inhibitors in tendon-related connective tissue. J Appl Physiol 2004; 96: 861-64. DOI: 10.1152/japplphysiol.00489.2003

[40] Hernandez-Barrantes S, Bernardo M, Toth M, Fridman R. Regulation of membrane type-matrix metalloproteinases. Semin Cancer Biol 2002; 12: 131-38. doi: 10.1006/scbi.2001.0421

[41] Baker AH, Edwards DR, Murphy G. Metalloproteinase inhibitors: biological actions and therapeutic opportunities. J Cell Sci 2002; 115: 3719-27. doi: 10.1242/jcs.00063

[42] Stetler-Stevenson WG, Krutzsch HC, Liotta LA. Tissue inhibitor of metalloproteinase (TIMP-2). J Biol Chem 1989; 264: 17374-8.

[43] Wang Z, Juttermann R, Soloway PD. TIMP-2 is required for efficient activation of proMMP-2 *in vivo*. J Biol Chem 2000; 275: 26411-415. doi:10.1074/jbc.M001270200

[44] Koskinen SOA, Kjær M, Mohr T, Biering Sørensen F, Suuronen T, Takala TES. Type IV collagen and its degradation in paralysed human muscle: effect of functional electrical stimulation. Muscle Nerve 2000; 23: 580-89. DOI: 10.1002/(SICI)1097-4598(200004)23:4<580::AID-MUS18>3.0.CO;2-4

[45] Koskinen SOA, Ahtikoski AM, Komulainen J, Hesselink MKC, Drost MR, Takala TES. Short-term effects of forced eccentric contractions on collagen synthesis and degrada-

tion in rat skeletal muscle. Pflügers Arch 2002; 444: 59-72. DOI: 10.1007/s00424-002-0792-2

[46] Riley DA, Bain JL, Thompson JL, Fitts RH, Widrick JJ, Trappe SW. Thin filament diversity and physiological properties of fast and slow fiber types in astronaut leg muscles. J Appl Physiol 2002; 92: 817-25. DOI: 10.1152/japplphysiol.00717.2001

[47] Baldwin KM, Haddad F. Skeletal muscle plasticity. Cellular and molecular responses to altered physical activity paradigms. Am J Physiol Med Rehabil 2002; S1: 40-51.

[48] Akima H, Ushiyama J, Kubo J, Fukuoka H, Kanehisa H, Fukunaga T. Effect of unloading on muscle volume with and without resistance training. Acta Astronaut 2007; 60: 728-36.

[49] Lemoine JK, Haus JM, Trappe SW, Trappe TA. Muscle proteins during 60-day bedrest in women: impact of exercise or nutrition. Muscle Nerve 2009; 39: 463-71. doi: 10.1002/mus.21189

[50] Kjær M, Langberg H, Miller BF, Boushel R, Crameri R, Koskinen S, Heinemeier K, Olesen JL, Døssing S, Hansen M, Pedersen SG, Rennie MJ, Magnusson P. Metabolic activity and collagen turnover in human tendon in response to physical activity. J Musculoskeletal Neuronal Interact 2005; 5: 41-52.

[51] Kjær M, Magnusson P, Krogsgaard M, Møller JB, Olesen J, Heinemeier K, Hansen M, Haraldsson B, Koskinen S, Esmarck B, Langberg H. Extracellular matrix adaptation of tendon and skeletal muscle to exercise. J Anat 2006; 208: 445-50. doi: 10.1111/j.1469-7580.2006.00549.x

[52] Riso EM. The effect of glucocorticoid myopathy, unloading and reloading on the skeletal muscle contractile apparatus and extracellular matrix. [thesis] Dissertationes Kinesiologiae Universitatis Tartuensis. Tartu University Press; 2007. p. 9-50.

[53] Jaworsky DM, Soloway P, Catesina J, Falls WA. Tissue Inhibitor of Metalloproteinase-2 (TIMP-2) Deficient Mice Display Motor Deficits. Int J Neurobiol 2006; 66: 82-94. doi: 10.1002/neu.20205

[54] Lluri G, Langlois GD, Soloway PD, Jaworsky DM. Tissue inhibitor of metalloproteinase-2 (TIMP-2) regulates myogenesis and β1 integrin expression *in vivo*. Exp Cell Res 2008; 314: 11-24. doi: 10.1016/j.yexcr.2007.06.007

[55] Ahtikoski AM, Koskinen SOA, Virtanen P, Kovanen V, Takala TES. Regulation of synthesis of fibrillar collagens in skeletal muscle during immobilization in shortened and lengthened positions. Acta Physiol Scand 2001; 171: 131-40. DOI:10.1046/j.1365-201X.2001.00848.x

[56] Martinez DA, Vailas AC, VanderbyRJr, Grindeland RE. Temporal extracellular matrix adaptations in ligament during wound healing and hindlimb unloading. Am J Physiol Regul Integr Comp Physiol 2007; 293: R1552-60. DOI: 10.1152/ajpregu.00423.2007

[57] Tyml K, Mathieu-Costello O. Structural and functional changes in the microvasculature of disused skeletal muscle. Front Biosci 2001; 6: D45-52. dx.doi.org/10.2741/A592

[58] Desaki J, Nishida W. A further observation of the structural changes of microvessels in the extensor digitorum longus muscle of the aged rat. J Electron Micros (Tokyo) 2007; 56: 249-55. doi:10.1093/jmicro/dfm032

[59] Karpakka J, Virtanen P, Väänänen K, Orava S, Takala TES. Collagen synthesis in rat skeletal muscle during immobilization and remobilisation. J Appl Physiol 1991; 70: 1775-80.

[60] D'Antona G, Pellegrino MA, Adami R, Rossi R, Carlizzi CN, Canepari M. The effect of ageing and immobilization on structure and function of human skeletal muscle fibres. J Physiol 2003; 552: 499-511. doi:10.1113/jphysiol.2003.046276

[61] Narici MV, Maganaris CN. Plasticity of the muscle-tendon complex with disuse and ageing. Exerc Sport Sci Rev 2007; 35: 126-34. doi:10.1097/jes0b013e3180a030ec

[62] D'Antona G, Pellegrino MA, Rossi R, Carlizzi CN, Reggiani C, Bottinelli R. Disuse induced atrophy and contractile impairment of human skeletal muscle fibres. Basic Appl Myol 2000; 10: 27-32.

[63] Belavý DL, Miokovic T, Armbrecht G, Richardson CA, Rittweger J, Felsenberg D. Differential atrophy of the lower-limb musculature during prolonged bed-rest. Eur J Appl Physiol 2009; 107: 489-99. doi: 10.1007/s00421-009-1136-0

[64] Oishi Y. Relationship between myosin heavy chain IId isoform and fibre types in soleus muscle of the rat after hindlimb suspension. Eur J Appl Physiol 1993; 66: 451-54.

[65] Degens H, Always SE. control of muscle size during disuse, disease, and ageing. Int J Sports Med 2006; 27: 94-99. DOI: 10.1055/s-2005-837571

[66] Alkner BA, Tesch PA. Efficacy of a gravity –independent resistance exercise device as a countermeasure to muscle atrophy during 29-day bed-rest. Acta Physiol Scand 2004a; 181: 345-57. DOI:10.1111/j.1365-201X.2004.01293.x

[67] Trappe TA, Burd NA, Louis ES, Lee GA, Trappe SW. Influence of concurrent exercise or nutrition countermeasures on thigh and calf muscle size and function during 60 days of bed rest in women. Acta Physiol (OXF) 2007b; 191: 147-59. DOI:10.1111/j.1748-1716.2007.01728.x

[68] Mendis MD, Hides JA, Wilson SJ, Grimaldi DL, Belavý DL, Stanton W, Felsenberg D, Rittweger J, Richardson C. Effect of prolonged bed rest on the anterior hip muscles. Gait and Posture 2009; 30: 533-37. doi:10.1016/j.gaitpost.2009.08.002

[69] Schuenke MD, Reed DW, Kraemer WJ, Staron RS, Volck JS, Hymer WC, Gordon S, Koziris LP. Effects of 14 days of microgravity on fast hindlimb and diaphragm muscles of the rat. Eur J Appl Physiol 2009; 106: 885-92. doi:10.1007/s00421-009-1091-9

[70] Adams GR, Caiozzo VJ, Baldwin KM. Skeletal Muscle unweighting: spaceflight and ground-based models. J Appl Physiol 2003; 95: 2185-2201. DOI: 10.1152/japplphysiol. 00346.2003

[71] Stevens L, Bastide B, Bozzo C, Mounier Y. Hybrid fibres under slow-to-fast transformations: expression of myosin heavy and light chains in rat soleus muscle. Pflügers Arch 2004; 448: 507-14. DOI:10.1007/s00424-004-1287-0

[72] Isfort RJ, Wang F, Greis KD, Sun Y, Keough TW, Farrar RP, Bodine SC, Anderson NL. Proteomic analysis of rat soleus muscle undergoing hindlimb suspension-induced atrophy and reweighting hypertrophy. Proteomics 2002; 2: 543-50.

[73] Wittwer M, Flück M, Hoppeler H, Muller S, Desplanches D, Billeter R. Prolonged unloading of rat soleus muscle causes distinct adaptations of the gene profile. FASEB J 2002; 16: 884-86. doi:10.1096/fj.01-0792fjc

[74] Stevens L, Firinga C, Gohlsch B, Bastide B, Mounier Y, Pette D. Effects of unweighting and clenbuterol on myosin light and heavy chains in fast and slow muscles of rat. Am J Physiol 2000; 279: C1558-63.

[75] Talmadge RJ. Myosin heavy chain isoform expression following reduced neuromuscular activity: potential regulatory mechanisms. Muscle Nerve 2000; 10: 321-25. DOI: 10.1002/(SICI)1097-4598(200005)23:5<661::AID-MUS>3.0.CO.2-J

[76] Stevens L, Bastide B, KischelPPette D, Mounier Y. Time-dependent changes in expression of troponin subunit isoforms in unloaded rat soleus muscle. Am J Physiol 2002; 282: C1025-30. DOI: 10.1152/ajpcell.00252.2001

[77] Trappe T. Influence of aging and long-term unloading on the structure and function of human skeletal muscle. J Appl Physiol Nutr Metab 2009; 34: 459-64. doi: 10.1139/ H09-041

[78] Fisher JS, Hasser EM, Brown M. Effects of ovaryectomy and hindlimb unloading on skeletal muscle. J Appl Physiol 1998; 1316-21.

[79] Fitts RH, Riley DR, Widrick JJ. Functional and structural adaptations of skeletal muscle to microgravity. J Exp Biol 2001; 204: 3201-08.

[80] Ohira M, Hanada H, Kawano F, Ishihara A, Nonaka I, Ohira Y. Regulation of the properties of rat hindlimb muscles following gravitational unloading. Jpn J Physiol 2002; 52: 235-45. doi.org/10.2170/jphysiol.52.235

[81] Ohira Y, Yoshinaga T, Ohira M, Kawano F, Wang XD, Higo Y, Terada M, Matsuko Y, Roy RR, Edgerton VR. The role of neural and mechanical influences in maintaining normal fast and slow muscle properties. Cells Tissues Organs 2006; 182: 129-42. DOI: 10.1159/000093963

[82] Kano Y, Shimegi S, Takahashi H, Masuda K, Katsuta S. Changes in capillary luminal diameter in rat soleus muscle after hindlimb suspension. Acta Physiol Scand 2000; 169: 271-76. DOI:10.1046/j.1365-201x.2000.00743.x

[83] Pontén E, Fridén J. Immobilization of the rabbit tibialis anterior muscle in a lengthened position causes addition of sarcomeres in series and extracellular matrix proliferation. J Biomech 2008; 41: 1801-4. doi:10.1016/j.jbiomech.2008.03.006

[84] Zhang P, Chen X, Fan M. Signaling mechanisms involved in disused muscle atrophy. Med Hypothesis 2007; 69: 310-21. doi:10.1016/j.mehy.2006.11.043

[85] Mujika I, Padilla S. Muscular characteristics of detraining in humans. Med Sci Sports Exerc 2001; 33: 1297-1303.

[86] Adhihetty PJ, Irrcher I, Joseph AM, Ljubicic V, Hood DA. Plasticity of skeletal muscle mitochondria in response to contractile activity. Exp Physiol 2003; 88: 99-107.

[87] Smith HK, Maxwell L, Martyn JA, Bass JJ. Nuclear DNA fragmentation and morphological alterations in adult rabbit skeletal muscle after short-term immobilization. Cell Tissue Res 2000; 302: 235-41.

[88] Jackman RW, Kandarian SC. The molecular basis of skeletal muscle atrophy. Am J Physiol 2003; 287: C834-C843. DOI:10.1152/ajpcell.00579.2003

[89] Itai Y, Kariya Y, Hoshino Y. Morphological changes in rat hindlimb muscle fibres during recovery from disuse atrophy. Acta Physiol Scand 2004; 181: 217-24. DOI: 10.1111/j.1365-201X.2004.01271.x

[90] Lieber RL, Friden JO, Hargens AR, Danzig LA, Gershuni DA. Differential response of the dog quadriceps muscle to external skeletal fixation of the knee. Muscle Nerve 1988; 11: 193-201.

[91] Carroll CC, Carrithers JA, Trappe TA. Contractile protein concentrations in human single muscle fibres. J Muscle Res Cell Motil 2004; 25: 55-59.

[92] Pette D, Staron RS. Transition of muscle fiber phenotypic profiles. Histochem Cell Biol 2001; 115: 359-72.

[93] Pottle D, Gosselin LE. Impact of mechanical load on functional recovery after muscle reloading. Med SciSports Exerc 2000; 32: 2012-17.

[94] Desaphy JF, Pierno A, Liantonio A, DeLuca A, Didonna MP, Frigeri A, Nicchia GP, Svelto M, Camerino C, Zallone A, Camerino DC. Recovery of the soleus muscle after short- and longterm disuse induced by hindlimb unloading: effects on the electrical properties and myosin heavy chain profile. Neurobiol Dis 2005; 18: 356-65. doi: 10.1016/j.nbd.2004.09.016

[95] Morey-Holton ER, Globus RK. Hindlimb unloading rodent model: technical aspects. J Appl Physiol 2002; 92: 1367-77. DOI:10.1152/japplphysiol.00969.2001

[96] Oishi Y, Ishihara A, Talmadge RJ. Expression of heat-shock protein 72 in atrophied rat skeletal muscle. Acta Physiol Scand 2001; 172: 123-30. DOI:10.1046/j.1365-201X.2001.00847.x

[97] Koskinen SOA, Wang W, Ahtikoski AM, Kjær M, Han XY, Komulainen J, Kovanen V, Takala TES. Acute exercise induced changes in rat skeletal muscle mRNAs and proteins regulating type IV collagen content. Am J Physiol 2001; 280: R1292-300.

[98] Thomason DB, Herrick RE, Surdyka D, Baldwin KM. Time course of soleus muscle myosin expression during hindlimb suspension and recovery. J Appl Physiol 1987; 62: 2180-86.

[99] Kasper CE, Talbot LA, Gaines JM. Skeletal muscle damage and recovery. AACN Clin Issues 2002; 13: 237-47.

[100] Lee K, Lee YS, Lee M, Yamashita M, Choi I. Mechanics and fatigability of the rat soleus muscle during early reloading. Yonsei Med J 2004; 45: 690-702. doi:org/10.3349/ymj.2004.45.4.690

Chapter 4

The Extracellular Matrix Complexome from Skeletal Muscle

Sandra Murphy and Kay Ohlendieck

Additional information is available at the end of the chapter

http://dx.doi.org/10.5772/62253

Abstract

The various layers of the extracellular matrix, forming the endomysium, perimysium and epimysium of skeletal muscles, provide essential structural and mechanical support to contractile fibres. Crucial aspects of muscle elasticity and fibre contractility are dependent on proper cell–matrix interactions. A complex network of collagen fibres, non-fibrillar collagens, proteoglycans, matricellular proteins, matrix metalloproteinases, adhesion receptors and signalling molecules maintain the physical structure for force transmission within motor units, embed critical cellular structures such as capillaries and motor neurons, and enable essential sarcolemma-matrix adhesion processes and signalling cascades. The systems biological concept of protein complexomes, which assumes the existence of interconnectivities between large protein assemblies, can be readily applied to the proteins within the extracellular space of muscles. Recent proteomic studies confirm that the extracellular matrix complexome has considerable influence on the integrity and cellular functions of skeletal muscle fibres. Adaptations or changes in the organization of the extracellular matrix play a crucial role during fibre regeneration following injury, extensive neuromuscular activity or pathophysiological insults. This chapter outlines the molecular components of the matrisome from skeletal muscles and discusses the extracellular matrix in relation to myogenesis, maturation of motor units, adaptation to changed functional demands and myofibrosis in muscular disorders.

Keywords: collagen, fibrosis, matricellular, matrisome, proteoglycan

1. Introduction

In addition to the universal physiological, metabolic and regulatory challenges of almost all cellular entities in the body, muscle tissues have to constantly adapt to a variety of biological issues related to high energy demand, elevated levels of cellular stress and enormous physical strains during excitation–contraction–relaxation cycles [1]. The survival of skeletal muscle fibres therefore depends heavily on (i) a high degree of physiological adaptability, (ii) a unique level of tissue plasticity, (iii) efficient molecular chaperoning to prevent proteotoxic insults and (iv) a sophisticated repair machinery that can counter-act frequent cellular injuries [2]. A crucial stabilizing element that is intrinsically involved in this continuous maintenance of contractile tissues is the extracellular matrix (ECM) [3]. On the one hand, the complex layers of the muscle ECM provide the physical structure for force transmission between contracting fibres and their surrounding tissue environment [4], and on the other hand the ECM functions as an embedding medium for essential supportive components of muscles such as capillaries and motor neurons [5].

The composition and organization of the ECM adapts considerably in response to changed functional or structural demands during myogenesis, fibre maturation and exercise-induced changes [6]. During the natural aging process and in association with a variety of muscular disorders, a hyperactive connective tissue may trigger myofibrosis with a detrimental impact on muscle elasticity and fibre contractility [7]. Physiological or pathological changes in the muscle ECM frequently mirror the different phases of altered muscle structure and function [8]. The main component of the ECM is represented by collagen, which exits in a large number of isoforms that connect with proteoglycans, matricellular proteins and adhesion receptors to form an elaborate extracellular network and tight cell–matrix interactions [9].

This chapter provides an overview of the molecular components of the ECM from skeletal muscle and describes the proteomic concept of the ECM complexome. The formation, maturation and flexibility within the various layers of the ECM in developing, maturing and adapting skeletal muscles is outlined, as well as the crucial role of myofibrosis in neuromuscular pathology.

2. Molecular and cellular structure of the extracellular matrix

In skeletal muscles, the ECM is involved in a variety of processes during development, contractile maturation, fibre regeneration following injury, physiological adaptations to changed functional demands and the natural aging process [8]. As outlined in **Figure 1**, the ECM is a highly dynamic non-cellular system that undergoes frequent cycles of modifications, degradation and reassembly. The muscle ECM functions on the one hand as an embedding and stabilizing structural support and on the other hand as a cellular interaction and signalling medium. The molecular lattice of collagens and proteoglycans with its associated matricellular proteins, enzyme systems and adhesion receptors mediates various physiological and biochemical mechanisms, including (i) the overall maintenance of muscle tissue stability and elasticity, (ii) the mechanical transduction of force from the contractile fibres to their anchoring tissues, (iii) cytoskeletal coupling to enable the efficient execution of frequent excitation–contraction–relaxation cycles, (iv) the provision of signalling pathways at the fibre periphery, (v) the preservation of neuromuscular homeostasis, and (vi) the physical scaffold and embedding medium of cellular constituents, such as contractile fibres, capillaries, motor neurons and satellite cell pools. Cell–matrix interactions are of central importance during cell adhesion and cell migration and thus essential during both embryonic and adult myogenesis.

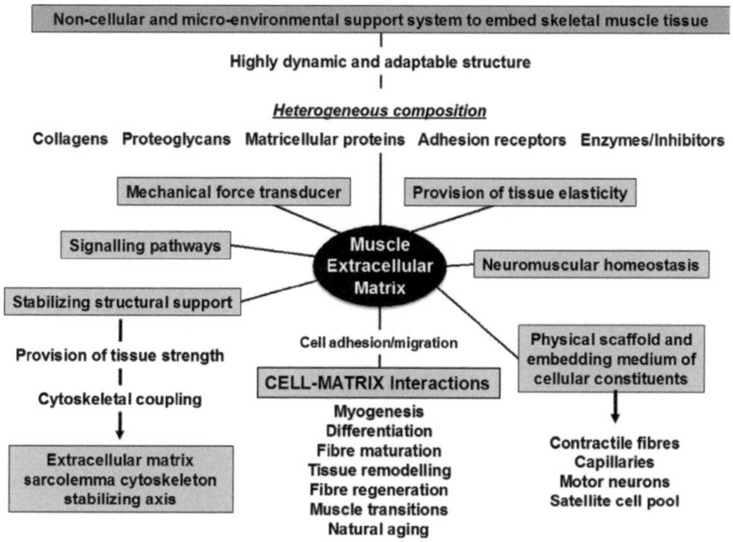

Figure 1. Overview of the complex functions and interactions of the extracellular matrix from skeletal muscles.

The distinct layers of the ECM that surround muscle fibres, muscle fascicles and the entire skeletal muscle are formed by the endomysium, perimysium and epimysium, respectively [3]. **Figure 2** shows diagrammatically the arrangements of the ECM from skeletal muscles and lists the main molecular constituents, including various isoforms of collagen, proteoglycans, matricellular proteins, crosslinking proteins and matrix metalloproteinases. Crucial adhesion systems that maintain sarcolemma–matrix interactions are marked. They include the collagen-laminin-α/β-dystroglycan-dystrophin/utrophin axis and the collagen–fibronectin–integrin axis that link the basal lamina via the plasmalemma to the underlying membrane cytoskeleton [10].

The stabilizing linkage between the outside and inside of muscle cells, provided by the ECM–sarcolemma–cytoskeleton axis, is of critical importance for maintaining normal contractile functions [8]. Primary or secondary abnormalities in individual binding partners of these surface complexes may result in severe neuromuscular disorders, as discussed in below section on the role of the ECM in skeletal muscle pathology and myofibrosis.

Collagens form tight helical structures and function as the main structural protein species in the extracellular space. Muscle-associated collagens are highly abundant in the interstitial matrix, ECM microfibrils and the basal lamina [11]. Collagen isoform COL I, the most abundant protein in the mammalian body [12], is the primary collagen in the perimysium and tendon. The interstitial matrix contains mostly collagen COL I, COL III and COL V. Minor types of

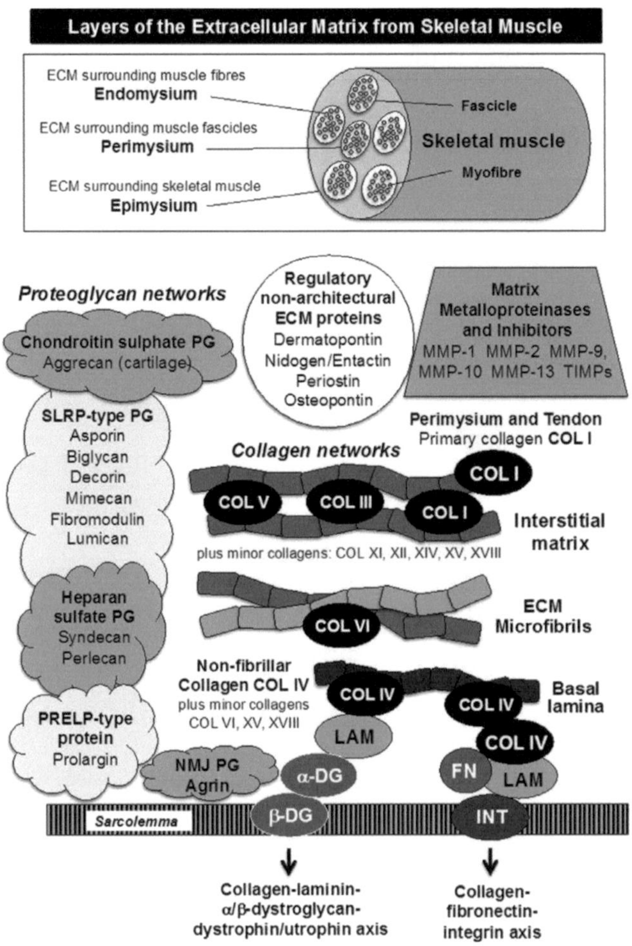

Figure 2 Diagrammatic presentation of the various layers of the extracellular matrix (ECM) from skeletal muscle and an overview of the molecular components that form the interstitial matrix, microfibrils and the basal lamina. Abbreviations: COL, collagen; DG, dystroglycan; FN, fibronectin; LAM, laminin; INT, integrin; MMP, matrix metalloproteinase; NMJ, neuromuscular junction; PG, proteoglycan; PRELP, proline-arginine-rich end leucine-rich repeat protein; SLRP, small leucine-rich repeat proteoglycans; TIMP, tissue inhibitors of matrix metalloproteinases.

collagens in this extracellular region are COL XI, XII, XIV, XV and XVIII and are mostly expressed during muscle development [3]. The main structural constituent in ECM microfibrils is collagen isoform COL VI. Microfibrils provide structural support during the physical strains of continuing contraction–relaxation cycles. Directly overlaying the sarcolemma membrane is the basal lamina consisting of the main non-fibrillar collagen isoform COL IV and a few minor constituents, including COL VI, XV and XVIII. The collagen network of the basement membrane interacts with two crucial plasmalemma adhesion complexes, the integrin complex and the dystrophin–glycoprotein complex [10]. These sarcolemma-bridging protein assemblies form the laminin–dystroglycan axis [13] and the fibronectin-integrin ($\alpha7\beta1$) axis [14], whereby the ECM glycoprotein fibronectin mediates the connection between laminin-211 (α2β1γ1 merosin) and collagen COL IV [15].

A variety of regulatory ECM proteins are involved in matrix assembly and the modulation of cell–matrix interactions, such as dermatopontin, nidogen/entactin, periostin (osteoblast-specific factor OSF-2) and osteopontin [16–18]. Matricellular proteins represent non-architectural ECM components and are crucial factors during muscle development and fibre repair. Matrix metalloproteinases are an important class of ECM-associated enzymes that regulate the degradation of ECM proteins and support tissue integrity during phases of collagen deposition and muscle regeneration [19]. The main isoforms present in skeletal muscles are matrix metalloproteinases MMP-1, MMP-2, MMP-9, MMP-10 and MMP-13 [20,21]. The tissue inhibitors of matrix metalloproteinases, named TIMP, are endogenous regulatory factors involved in the formation, adaptation and controlled degradation of the ECM [22]. TIMP molecules play a crucial role in the migration and differentiation of muscle stem cells during regeneration following cellular injury [19].

A large variety of proteoglycans fill the gaps between collagen molecules and thereby form an integral part of the highly complex ECM structure. Proteoglycan molecules are highly glycosylated with glycosaminoglycans at multiple sites along the peptide backbone [23]. Skeletal muscles contain small leucine-rich repeat proteoglycans (SLRP), heparan sulfate proteoglycans and chondroitin sulphate proteoglycans. Muscle-associated proteoglycans of the type SLRP are asporin, biglycan, decorin, mimecan (osteoglycin), fibromodulin and lumican. Asporin is mostly found in the cartilage matrix. Biglycan is a small SLRP-type proteoglycan that interacts with α-sarcoglycan and γ-sarcoglycan of the dystrophin-glycoprotein complex [24]. Decorin is the primary proteoglycan molecule of the perimysium and tendon structures [25]. Fibromodulin is involved in collagen fibril formation, which is illustrated by the biomedical fact that fibromodulin-deficient tendons exhibit abnormal collagen fibrils [26]. The heparan sulfate proteoglycan syndecan is transiently up-regulated during tissue differentiation and is involved in stem cell maintenance and muscle regeneration [27]. Perlecan is located to the basement membrane and its expression is also transiently increased during muscle differentiation [28]. The chondroitin sulphate proteoglycan named aggrecan forms large aggregates in cartilage [29]. A proline-arginine-rich end leucine-rich repeat protein (PRELP) is presented by prolargin of the basal lamina. At the highly differentiated neuromuscular junction region, the large proteoglycan molecule agrin is present and forms via α-dystroglycan a tight linkage to the utrophin-glycoprotein complex. Agrin is essential for the normal development of the neuromuscular junction and agrin-induced clustering processes are crucial for the anchoring of the acetylcholine receptor complex in the junctional folds [30].

3. Skeletal muscle development and the extracellular matrix

Skeletal muscle fibres derive from the mesoderm [31] and represent one of the most abundant cell types in the body. Myofibres play a key physiological role in the provision and regulation of locomotion, breathing, postural control, heat homeostasis and metabolic integration [32–

34]. The development of contractile fibres is a highly complex process and changes in the ECM play an essential role during myogenesis. Adaptations occur at the level of the basal lamina, the interstitial ECM and the collagen-rich tendon and encompass a variety of fundamental processes of development, such as the determination of cell fate, as well as proliferation, cell division, patterning and tissue transitions [35]. The most critical developmental processes occur during the initial activation of precursor cells, various differentiation steps and the final maturation of innervated myofibres. These major developmental mechanisms are regulated by a large number of genetic and signalling factors [36]. Many specialized extracellular components are involved in muscle development, which is reflected by the transient expression patterns of certain ECM molecules, such as the minor collagen isoforms COL XI, XII, XIV, XV and XVIII [3].

The molecular and cellular events that occur during the establishment of the myogenic cell lineage and resulting formation of multi-nucleated contractile fibres entail initially a fibrillar and fibronectin-rich matrix. In developmental terms, almost all myofibres of the skeletal musculature derive from mesodermal structures named somites that develop during early embryonic segmentation on both sides of the neural tube [37–39]. The developing ECM of somites is represented by a fibronectin core, a basement membrane and an outer fibronectin-containing matrix [35]. **Figure 3** outlines the initiation and control of embryonic and adult myogenesis by myogenic factors and through a complex series of spatio-temporal dependent signalling cascades. Fibronectin and its interactions with the integrin complex play a central role in polarizing and guiding somitic cells [40–42] and the ECM is crucial for somite formation and as a guiding cue during morphogenesis [43–45]. The small SLRP-type proteoglycan decorin was shown to be majorly involved in skeletal muscle development by promoting proliferation and differentiation of muscle cells through suppressing myostatin activity [46–48].

The key regulator of early myogenesis that initiates the developmental commitment into the myogenic cell lineage is the paired-type homeobox gene *PAX3* [49–51]. Interestingly, another member of the PAX transcription factor family, the *PAX3* orthologue *PAX7*, acts as a regulator of mature skeletal muscle regeneration and postnatal growth mechanisms [52]. Following the induction of mesodermal precursors, the subsequent segmentation into somites and the formation of the primary myotome involves a variety of signalling molecules and transcription factors, such as the secreted and lipid-modified family of Wnt-glycoproteins and the large group of myogenic basic helix-loop-helix muscle regulatory factors, such as MyoD, Myf5, MRF4 and myogenin [53–55]. Myogenic factors act at multiple regulatory points during embryonic myogenesis, whereby the overall genetic pathway that is responsible for the transcriptional activation of skeletal muscle-specific genes is highly complex and partially redundant [56]. A key step during muscle development is the fusion of myogenic cells that result in the formation of innervated and multi-nucleated myofibres (**Figure 3**). Although specific regulatory processes differ between embryonic, foetal, postnatal and mature regenerative myogenesis [57], the basic biological mechanisms of skeletal muscle development and injury-related adult muscle regeneration are extraordinarily similar [58,59]. A large pool of satellite cells provides a high level of regenerative capacity of the matured post-mitotic fibres and is located between the sarcolemma and the basal lamina [60]. These mono-nucleated myogenic stem cells are activated during cycles of fibre regeneration and cellular maintenance. To provide maximum skeletal muscle performance, activated stem cells undergo a complex pattern of proliferation, differentiation and cellular fusion to form multi-nucleated and functional contractile myofibres [61]. Following maturation, the tissue mass of skeletal muscles is then regulated by catabolic and anabolic mechanisms that include signalling factors such as NF-kB and FoxO, as well as the mTOR pathway [62].

The systematic profiling of myogenesis using mass spectrometry-based proteomics has covered postnatal growth and development [63,64], but focused mostly on studying cell culture models [65–67] and specifically the skeletal muscle secretome during myoblast differentiation and myotube formation [68–72]. The concept that skeletal muscle cells act as secretory tissues has recently been reviewed by Pedersen [73]. Secreted myokines probably exert autocrine, paracrine or endocrine effects within the neuromuscular system and also in relation to other organ systems [74]. The skeletal muscle secretome is estimated to consist of several hundred muscle-derived peptides and proteins [75]. During muscle development,

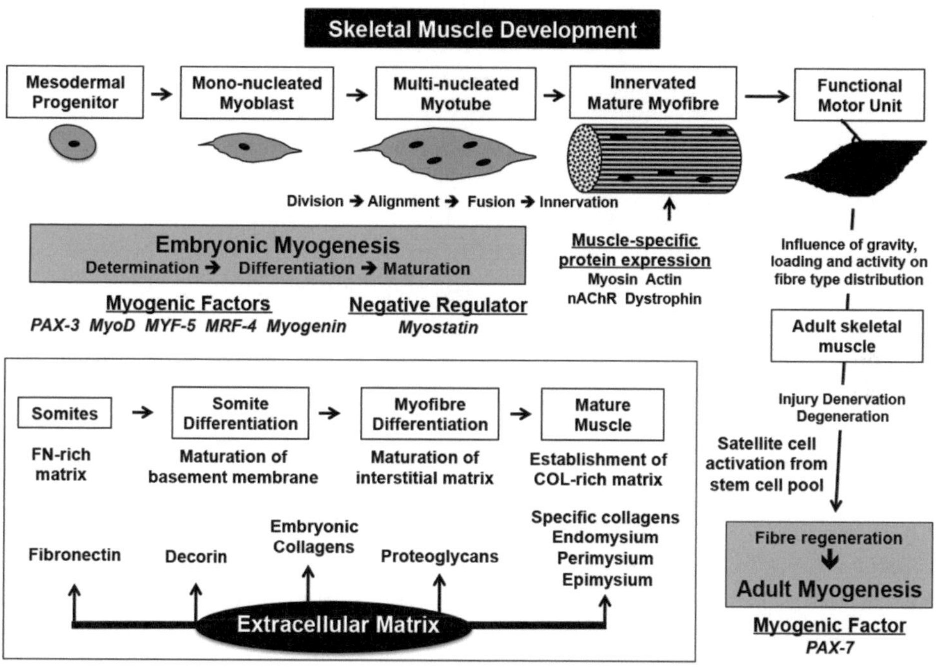

Figure 3. Outline of the initiation and control of embryonic and adult myogenesis. Shown is the transition from mesodermal progenitor cells to myoblasts, followed by cellular fusion to produce multi-nucleated myotubes and finally innervated and functional myofibres. The critical role of stem cells during both embryonic development and regeneration of injured adult muscle fibres is highlighted, as well as changes in the ECM. Fibronectin (FN) and its interactions with the integrin complex are of central importance for polarizing and guiding somitic cells and the proteoglycan decorin is majorly involved in myogenesis by promoting proliferation and differentiation through supressing myostatin activity. Changes in the ECM and the formation of a collagen (COL) matrix during muscle development are marked.

changes in ECM proteins have recently been studied in relation to human myogenesis and established that altered protein expression underlies the dramatic phenotypic conversion of primary mono-nucleated muscle cells during differentiation to form multi-nucleated myotubes [76]. The temporal profiling of the human myoblast proteome during *in vitro* differentiation highlighted the importance of ECM rearrangement during early myogenesis and showed a drastic increase in key ECM components, including several α-isoforms of collagen COL VI and COL XVIII, as well as the heparan sulfate proteoglycan HSPG2 of the basement membrane, the elastin–microfibril interface–located ECM glycoprotein EMILIN2 and nidogen isoform NID2 [76]. These findings confirm the developmental concept that enhanced synthesis of ECM proteins occurs during the transition from myoblasts to syncytial myotubes [72] and that complex interactions at the cell–ECM interface facilitate the fusion of myoblasts [35].

4. Mature fibres, skeletal muscle plasticity and the extracellular matrix

Both, proteomic cataloguing studies of various skeletal muscle specimens [77–82] and the comparative expression profiling of crude skeletal muscle preparations or subcellular fractions [83–86] routinely identify ECM proteins that form the core complexes of the basal lamina, microfibrils and interstitial matrix [87]. **Figure 4** shows a bioinformatics STRING [88] map of core ECM components from skeletal muscles. In mature skeletal muscles, frequently identified ECM molecules include collagens of the interstitial matrix (COL I, COL III and COL V), the microfibrillar collagen isoform COL VI, the non-fibrillar collagen isoform COL IV of the basement membrane, components of cell–ECM adhesion complexes (laminin, fibronectin, integrins, dystrophin, utrophin), regulatory ECM proteins (dermatopontin, nidogen, periostin and osteopontin), SLRP-type proteoglycans (asporin, biglycan, decorin, mimecan, fibromodulin and lumican), heparan sulfate proteoglycans (syndecan and perlecan), the chondroitin sulphate proteoglycan aggrecan, the PRELP-type proteoglycan prolargin and the neuromuscular junction-specific proteoglycan agrin, as well as matrix metalloproteinases and their inhibitors (MMP-1, MMP-2, MMP-9, MMP-10, MMP-13 and TIMPs) [77–86]. **Figure 4** includes the protein products of the following genes: *BGN, SDC1, HSPG2, AGRN, FN1, DAG1, ACAN, DPT, POSTN, MMP1, MMP2, MMP9, MMP10, MMP13, TIMP1, COL1A1, COL1A2, COL6A1, COL6A2, COL3A1, COL5A1, COL4A4, LAMA2, LAMB1, LAMB2, LAMC1, NID1, ASPN, PRELP, PGS2, LUM, OGN, FMOD, ITGA7, ITGB1*. This encompasses essential members of the various collagen networks found in the basement membrane, the microfibrillar structures and the interstitial matrix. In addition, major proteoglycans, matricellular proteins and matrix metalloproteinases are shown, as well as the interaction sites between sarcolemmal adhesion receptor complexes and the ECM.

Individual skeletal muscles are characterized by their fibre type distribution pattern whereby the proportion of fast-twitching fibres, slow-twitching fibres and hybrid fibres is highly adaptable and changes according to specific physiological, biochemical and/or metabolic demands [89]. Alterations in physical activity affect the molecular and cellular composition of the neuromuscular system, including hypertrophy, i.e. the increase in fibre size and hyperplasia, i.e. the increase in fibre number [90]. The plasticity of the neuromuscular system is a well-established physiological concept. Endurance training is associated with an increased aerobic capacity and elevated utilization of fatty acid oxidation [91]. In contrast, sprint training triggers higher activities of the glycolytic and phosphocreatine pathways and enhances carbohydrate metabolism [92]. Prior to systematic proteomic studies, a large number of biochemical, cell biological and physiological studies have established ECM changes in response to exercise [93–95], including collagens, adhesion receptors, growth factors, matricellular proteins and matrix metalloproteinases [96–100]. Neuromuscular unloading clearly depresses collagen COL I and COL III production and reloading enhances collagen expression in fast muscles [101]. The mass spectrometric analysis of exercise indicates proteome-wide changes in the graded response of skeletal muscles to physical exercise using different training regimes [102,103]. While moderate-intensity exercise causes a shift to a more fatigue-resistant and a slower-contracting skeletal muscle phenotype, interval-exercise training is associated with changes in post-translational modifications of metabolic enzymes [104–106]. The proteomic analysis of skeletal muscle plasticity in relation to acute versus chronic exercise was recently determined using human *vastus lateralis* muscle biopsy specimens and label-free LC-MS/MS analysis [107]. While structural and mitochondrial proteins were shown to be increased after long-term exercise, components related to energy metabolism were decreased following short-term exercise. Moderate ECM changes were described for several α-chains of collagen VI, fibronectin and decorin [107].

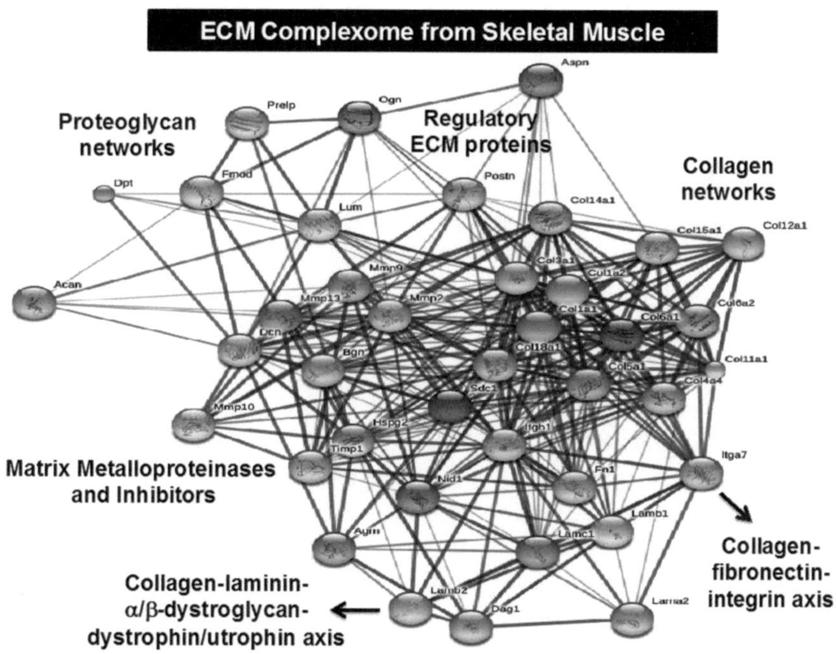

Figure 4. Bioinformatics STRING map of major components that form the core of the extracellular matrix (ECM) complexome. Shown are key proteins belonging to the collagen network of the basal lamina, microfibrils and the interstitial matrix, proteoglycans, matricellular proteins and matrix metalloproteinases. The interaction sites of the ECM with sarcolemmal adhesion receptor complexes are shown.

An interesting non-physiological system is presented by external chronic low-frequency stimulation of fast muscles. This electro-stimulation method causes the complete activation of all affected motor units to a maximum extent [108]. During fast-to-slow transitions, skeletal muscles show a remarkable adaptation and transform physiologically and biochemically into motor units with an improved resistance to fatigue [109]. Chronic low-frequency stimulated fast muscles are characterized by decreased fibre calibres, an increase in the time-to-peak twitch tension, an increase in half-relaxation time and a significant elevation of aerobic-oxidative capacity [110]. The proteomic analysis of continuous electro-stimulation at 10 Hz has demonstrated complex biochemical changes with a significant shift from glycolytic to more aerobic-oxidative metabolism [111,112]. The ECM of transforming skeletal muscle undergoes distinct changes and exhibits increased collagen levels [113,114]. In the stimulated *latissimus dorsi* model for testing the suitability of dynamic cardiomyoplasty to treat heart failure, the collagen content was shown to be significantly elevated in the paced muscle. Although the chronically electro-stimulated muscle increased the level of fatigue resistance, distal regions of the paced *latissimus dorsi* muscle were characterized by muscular atrophy and myofibrosis [114].

5. Neuromuscular disorders and the extracellular matrix

A general myopathological parameter of a variety of acquired and inherited muscle diseases [115], as well as the gradual loss of contractile strength during the natural aging process [116], is the progressive accumulation of ECM components, especially collagens [7]. Inflammatory processes and tissue infiltration often accompany the loss of skeletal muscle fibres. Increased levels of non-contractile entities, such as fibrous connective and fatty tissue, within the neuromuscular system are a key pathological factor in the dysregulation of skeletal muscle function. Myofibrosis often correlates with poor motor outcome in neuromuscular disorders, such as the progressive loss of muscle strength and concomitant endomysial changes in the X-linked inherited disorder Duchenne muscular dystrophy [117]. In muscle pathology, changes in ECM components can be differentiated as being a consequence of a primary defect in the matrisome of muscles, such as the ECM diseases Collagen IV myopathy [118,119] and LAMA2-related congenital myopathy [120,121], or a secondary response in the form of reactive myofibrosis, as is seen in dystrophinopathies [122,123].

The systematic profiling of changes in ECM components in Collagen IV myopathy and X-linked muscular dystrophy has resulted in interesting new findings in relation to primary ECM defects versus reactive fibrosis. Mutations in the genes encoding collagen isoform COL VI are the underlying cause of the severe UCMD type of Ullrich congenital muscular dystrophy and the milder BM type of Bethlem myopathy. Both disorders are characterized by skeletal muscle wasting, cycles of cellular degeneration and regeneration, and the substitution of contractile fibres with fat and connective tissue [119]. The cell biological and proteomic analysis of mouse models and biopsy material from patients afflicted with Collagen IV myopathy revealed metabolic dysregulation, enhanced cellular stress, autophagic impairment and alterations in mechano-transduction signalling pathways [124,125]. In the case of X-linked muscular dystrophy, a large number of proteomic studies have surveyed secondary changes down-

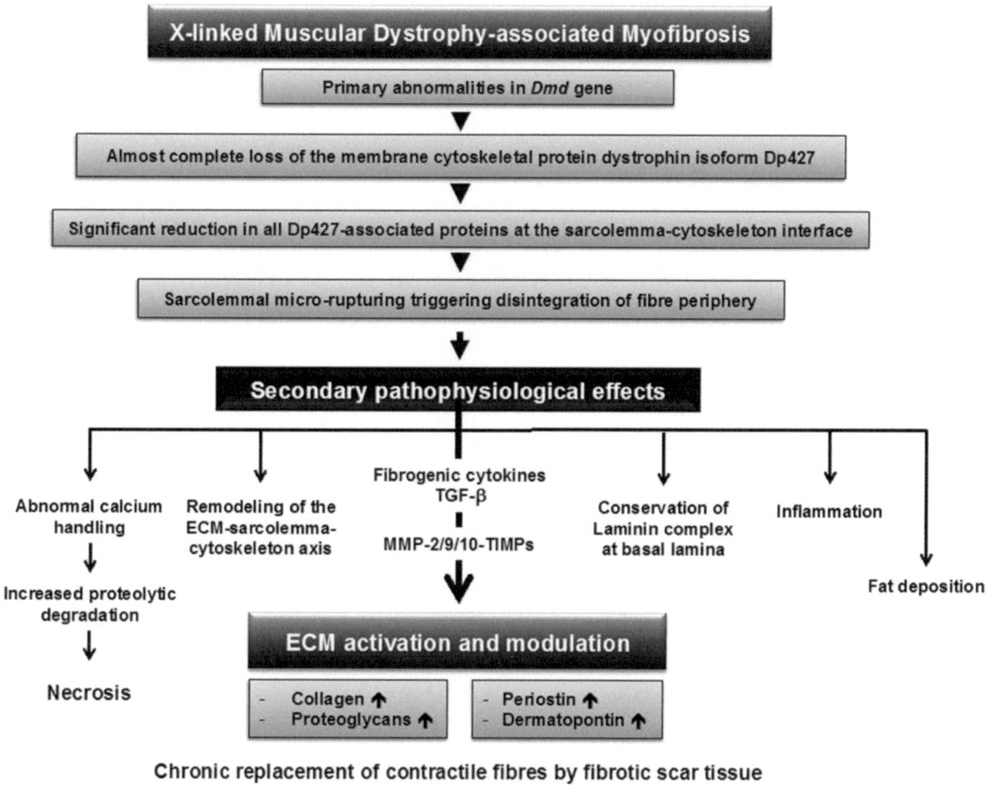

Figure 5. Overview of reactive myofibrosis in X-linked muscular dystrophy. The transforming growth factor TGF-β and the activity of matrix metalloproteinases (MMP) and tissue inhibitors (TIMPs) play a crucial role in ECM activation and modulation, which is characterized by increased levels of various collagens, proteoglycans, dermatopontin and periostin. Progressive ECM accumulation triggers a chronic replacement of muscle fibres by fibrotic scar tissue leading to a loss of muscle elasticity and contractile strength.

stream of the primary abnormality in the membrane cytoskeletal protein dystrophin [126]. Muscular dystrophy-related changes affect energy metabolism, cellular signalling, the excitation–contraction–relaxation cycle, the stress response, the cytoskeletal network and the ECM [127]. The recent proteomic profiling of established genetic animal models of dystrophinopathy has revealed a drastic increase in various collagens, proteoglycans and matricellular proteins [83–86,128,129]. **Figure 5** outlines the histo- and pathobiochemical consequences of muscular changes and a hyperactive connective tissue in dystrophin-deficient skeletal muscles.

The simultaneous mass spectrometric analysis of dystrophin isoform Dp427 and collagen in moderately dystrophic *mdx-4cv* leg muscles revealed significant increases in collagens and associated ECM proteins, such as fibronectin, biglycan, asporin, decorin, prolargin, mimecan and lumican [85]. The pathoproteomic signature of the severely dystrophic *mdx-4cv* diaphragm included a significant increase in collagens and the related ECM proteins asporin, decorin, dermatopontin, prolargin and periostin [83]. Especially interesting was the proteomic identification of dermatopontin and periostin [83,128,129]. Dermatopontin, also named tyrosine-rich acidic matrix protein TRAMP [130], is involved in matrix assembly and cell–matrix interactions [17] via interactions with decorin, TGF-β and fibronectin [131]. High levels of dermatopontin in the dystrophic *mdx* diaphragm most likely reflect an increased demand for collagen matrix organization within Dp427-deficient fibres [128]. Periostin is a crucial matricellular protein of 93 kDa [18] that is involved in the regulation of the biomechanical properties of connective tissues and collagen fibrillogenesis [132]. Normally periostin is only temporally expressed in the muscle ECM during cellular differentiation and regeneration processes [133], making its drastic up-regulation a characteristic feature of dystrophic muscles [83,86]. Of diagnostic and therapeutic importance is the fact that muscle biopsies from Duchenne patients exhibit an elevated concentration of periostin and that the deletion of periostin clearly reduces dystrophic symptoms and myofibrosis in mice by modulating the TGF-β pathway [134]. Interestingly, laminin-deficient muscular dystrophy also shows dysregulation of matricellular proteins as an early pathophysiological feature [120,135]. Therefore altered levels of periostin and related matricellular proteins are good biomarker candidates for the characterization of myofibrosis in relation to inherited muscular dystrophies [136]. Although the natural aging process is also associated with increased collagen levels [116], which were also shown by proteomics [137], the collagen accumulation is much less pronounced in senescent muscles as compared to muscular dystrophy.

6. Conclusions

In conclusion, a dynamic balance exists within the ECM system from skeletal muscles. Highly regulated cycles of protein deposition, accumulation, remodelling and degradation occur in response to development, fibre transformation, neuromuscular loading, mechanical unloading, disease processes or aging. The main components of the ECM are represented by collagen fibres, non-fibrillar collagens, matricellular proteins, proteoglycans, matrix metalloproteinases, signalling molecules and adhesion complexes. The muscle ECM forms a structural scaffold that plays a central role in the maintenance of the physical structure of motor units and provides the framework for force transmission. The ECM is also involved in signalling cascades and adhesion processes at the sarcolemma–matrix interface. The systematic mass spectrometry-based proteomic analysis of the muscle ECM has established an enormous complexity and interconnectivity of matrix proteins and confirmed the dynamic nature of collagen networks and its associated proteoglycans in health and disease. Swift adaptations or alterations in the arrangement of the ECM have been established to occur during myogenesis, fibre regeneration, increased neuromuscular activity or pathological muscle wasting. Hence, concentration changes in ECM proteins are useful indicators for studying basic cell biological or pathophysiological processes in skeletal muscles. In the future, distinct ECM molecules may be useful for designing improved diagnostic, prognostic or therapy-monitoring approaches to study neuromuscular alterations.

http://dx.doi.org/10.5772/62253

Acknowledgements

Research in the author's laboratory was supported by project grants from the Deutsche Duchenne Stiftung *aktion benni & co e.V.*, Muscular Dystrophy Ireland and the Hume Scholarship programme of Maynooth University, as well as the Irish Higher Education Authority (HEA). The Programme for Research in Third Level Institutions PRTLI Cycle 5 is co-funded by the Irish Government and the European Union under Ireland's EU Structural Funds Programme 2007-2013. We thank Prof. Dieter Swandulla (University of Bonn), Prof. Heinrich Brinkmeier (University of Greifswald) and Dr. Paula Meleady (Dublin City University) for their continued support of our studies into the mechanisms of muscle fibrosis.

Author details

Sandra Murphy and Kay Ohlendieck*

*Address all correspondence to: kay.ohlendieck@nuim.ie

Department of Biology, Maynooth University, National University of Ireland, Maynooth, County Kildare, Ireland

References

[1] Pette D. The adaptive potential of skeletal muscle fibers. *Canadian Journal of Applied Physiology*. 2002 Aug;27(4):423-48.

[2] Blaauw B, Schiaffino S, Reggiani C. Mechanisms modulating skeletal muscle phenotype. *Comprehensive Physiology*. 2013 Oct;3(4):1645-87.

[3] Gillies AR, Lieber RL. Structure and function of the skeletal muscle extracellular matrix. *Muscle & Nerve*. 2011 Sep;44(3):318-31.

[4] Kjaer M. Role of extracellular matrix in adaptation of tendon and skeletal muscle to mechanical loading. *Physiological Reviews*. 2004 Apr;84(2):649-98.

[5] Purslow PP. The structure and functional significance of variations in the connective tissue within muscle. *Comparative Biochemistry and Physiology Part A: Molecular & Integrative Physiology*. 2002 Dec;133(4):947-66.

[6] Lund DK, Cornelison DD. Enter the matrix: shape, signal and superhighway. *FEBS Journal*. 2013 Sep;280(17):4089-99.

[7] Lieber RL, Ward SR. Cellular mechanisms of tissue fibrosis. 4. Structural and functional consequences of skeletal muscle fibrosis. *American Journal of Physiology: Cell Physiology*. 2013 Aug;305(3):C241-52

[8] Goody MF, Sher RB, Henry CA. Hanging on for the ride: adhesion to the extracellular matrix mediates cellular responses in skeletal muscle morphogenesis and disease. *Developmental Biology*. 2015 May;401(1):75-91.

[9] Naba A, Hoersch S, Hynes RO. Towards definition of an ECM parts list: an advance on GO categories. *Matrix Biology*. 2012 Sep;31(7):371-2.

[10] Grounds MD, Sorokin L, White J. Strength at the extracellular matrix-muscle interface. *Scandinavian Journal of Medical Science in Sports*. 2005 Dec;15(6):381-91.

[11] Sweeney SM, Orgel JP, Fertala A, Mc Auliffe JD, TurnerKR, Di Lullo GA, Chen S, Antipova O, Perumal S, Ala-Kokko L, Forlino A, Cabral WA, Barnes AM, Marini JC, San Antonio JD. Candidate cell and matrix interaction domains on the collagen fibril, the predominant protein of vertebrates. *Journal of Biological Chemistry*. 2008 Jul;283(30): 21187-97.

[12] Stefanovic B. RNA protein interactions governing expression of the most abundant protein in human body, type I collagen. *Wiley Interdisciplinary Reviews: RNA*. 2013 Sep; 4(5):535-45.

[13] Ohlendieck K. Towards an understanding of the dystrophin–glycoprotein complex: linkage between the extracellular matrix and the membrane cytoskeleton in muscle fibers. *European Journal of Cell Biology*. 1996 Jan;69(1):1-10.

[14] Burkin DJ, Kaufman SJ. The alpha7beta1 integrin in muscle development and disease. *Cell and Tissue Research*. 1999 Apr;296(1):183-90.

[15] Holmberg J, Durbeej M. Laminin-211 in skeletal muscle function. *Cell Adhesion & Migration*. 2013 Jan;7(1):111-21.

[16] Funanage VL, Smith SM, Minnich MA. Entactin promotes adhesion and long-term maintenance of cultured regenerated skeletal myotubes. *Journal of Cellular Physiology*. 1992 Feb;150(2):251-7.

[17] Kato A, Okamoto O, Ishikawa K, Sumiyoshi H, Matsuo N, Yoshioka H, Nomizu M, Shimada T, Fujiwara S. Dermatopontin interacts with fibronectin, promotes fibronectin fibril formation, and enhances cell adhesion. *Journal of Biological Chemistry*. 2011 Apr; 286(17):14861-9.

[18] Norris RA, Damon B, Mironov V, Kasyanov V, Ramamurthi A, Moreno-Rodriguez R, Trusk T, Potts JD, Goodwin RL, Davis J, Hoffman S, Wen X, Sugi Y, Kern CB, Mjaatvedt CH, Turner DK, Oka T, Conway SJ, Molkentin JD, Forgacs G, Markwald RR. Periostin regulates collagen fibrillogenesis and the biomechanical properties of connective tissues. *Journal of Cellular Biochemistry*. 2007 Jun;101(3):695-711.

[19] Alameddine HS. Matrix metalloproteinases in skeletal muscles: friends or foes? *Neurobiology of Disease*. 2012 Dec;48(3):508-18.

[20] Chen X, Li Y. Role of matrix metalloproteinases in skeletal muscle: Migration, differentiation, regeneration and fibrosis. *Cell Adhesion & Migration*. 2009 Oct; 3(4): 337–341.

[21] Lei H, Leong D, Smith LR, Barton ER. Matrix metalloproteinase 13 is a new contributor to skeletal muscle regeneration and critical for myoblast migration. *American Journal of Physiology: Cell Physiology*. 2013 Sep;305(5):C529-38.

[22] Bellayr I, Holden K, Mu X, Pan H, Li Y. Matrix metalloproteinase inhibition negatively affects muscle stem cell behavior. *International Journal of Clinical and. Experimental Pathology*. 2013;6(2):124-41.

[23] Heinegård D. Proteoglycans and more - from molecules to biology. *International Journal of Experimental Pathology*. 2009 Dec;90(6):575-86.

[24] Nastase MV, Young MF, Schaefer L. Biglycan: a multivalent proteoglycan providing structure and signals. *Journal of Histochemistry & Cytochemistry*. 2012 Dec;60(12):963-75.

[25] Goetsch KP, Kallmeyer K, Niesler CU. Decorin modulates collagen I-stimulated, but not fibronectin-stimulated, migration of C2C12 myoblasts. *Matrix Biology*. 2011 Mar; 30(2):109-17.

[26] Chakravarti S. Functions of lumican and fibromodulin: lessons from knockout mice. *Glycoconjugate Journal*. 2002 May;19(4-5):287-93.

[27] Pisconti A, Bernet JD, Olwin BB. Syndecans in skeletal muscle development, regeneration and homeostasis. *Muscles Ligaments and Tendons Journal*. 2012 Jun 17;2(1):1-9.

[28] Zoeller JJ, McQuillan A, Whitelock J, Ho SY, Iozzo RV. A central function for perlecan in skeletal muscle and cardiovascular development. *Journal of Cell Biology*. 2008 Apr; 181(2):381-94.

[29] Matuszewski PE, Chen YL, Szczesny SE, Lake SP, Elliott DM, Soslowsky LJ, Dodge GR. Regional variation in human supraspinatus tendon proteoglycans: decorin, biglycan, and aggrecan. *Connective Tissue Research*. 2012;53(5):343-8.

[30] Campanelli JT, Roberds SL, Campbell KP, Scheller RH. A role for dystrophin-associated glycoproteins and utrophin in agrin-induced AChR clustering. *Cell*. 1994 Jun;77(5): 663-74.

[31] Buckingham M, Bajard L, Chang T, Daubas P, Hadchouel J, Meilhac S, Montarras D, Rocancourt D, Relaix F. The formation of skeletal muscle: from somite to limb. *Journal of Anatomy*. 2003 Jan;202(1):59-68.

[32] Frontera WR, Ochala J. Skeletal muscle: a brief review of structure and function. *Calcified Tissue International*. 2015 Mar;96(3):183-95.

[33] Mayeuf-Louchart A, Staels B, Duez H. Skeletal muscle functions around the clock. *Diabetes, Obesity and Metabolism*. 2015 Sep;17 Suppl 1:39-46.

[34] Baskin K, Winders BR, Olson EN. Muscle as a "mediator" of systemic metabolism. *Cell Metabolism*. 2015 Feb 3;21(2):237-48.

[35] Thorsteinsdóttir S, Deries M, Cachaço AS, Bajanca F. The extracellular matrix dimension of skeletal muscle development. *Developmental Biology*. 2011 Jun;354(2):191-207.

[36] Comai G, Tajbakhsh S. Molecular and cellular regulation of skeletal myogenesis. *Current Topics in Developmental Biology*. 2014;110:1-73.

[37] Christ B, Ordahl CP. Early stages of chick somite development. *Anatomy and Embryology (Berlin)*. 1995 May;191(5):381-96.

[38] PourquiéO. Vertebrate somitogenesis. *Annual Review of Cell and Developmental Biology*. 2001;17:311-50.

[39] Hollway G, Currie P. Vertebrate myotome development. *Birth Defects Research Part C Embryo Today*. 2005 Sep;75(3):172-9.

[40] Ostrovsky D, Cheney CM, Seitz AW, Lash JW. Fibronectin distribution during somitogenesis in the chick embryo. *Cell Differentiation*. 1983 Nov;13(3):217-23.

[41] Ostrovsky D, Sanger JW, Lash JW. Somitogenesis in the mouse embryo. *Cell Differentiation*. 1988 Mar;23(1):17-25.

[42] Lash JW, Ostrovsky D, Mittal B, Sanger JW. Alpha actinin distribution and extracellular matrix products during somitogenesis and neurulation in the chick embryo. *Cell Motility*. 1985;5(6):491-506.

[43] Martins GG, Rifes P, Amândio R, Rodrigues G, Palmeirim I, Thorsteinsdóttir S. Dynamic 3D cell rearrangements guided by a fibronectin matrix underlie somitogenesis. *PLoS One*. 2009 Oct;4(10):e7429.

[44] Girós A, Grgur K, Gossler A, Costell M. $\alpha5\beta1$ integrin-mediated adhesion to fibronectin is required for axis elongation and somitogenesis in mice. *PLoS One*. 2011;6(7):e22002.

[45] Lackner S, Schwendinger-Schreck J, Jülich D, Holley SA. Segmental assembly of fibronectin matrix requires rap1b and integrin $\alpha5$. *Developmental Dynamics*. 2013 Feb; 242(2):122-31.

[46] Kishioka Y, Thomas M, Wakamatsu J, Hattori A, Sharma M, Kambadur R, Nishimura T. Decorin enhances the proliferation and differentiation of myogenic cells through suppressing myostatin activity. *Journal of Cellular Physiology*. 2008 Jun;215(3):856-67. doi: 10.1002/jcp.21371.

[47] Velleman SG, Shin J, Li X, Song Y. Review: The skeletal muscle extracellular matrix: Possible roles in the regulation of muscle development and growth. *Canadian Journal of Animal Science*. 2012;92:1-10.

[48] Zeng QJ, Wang LN, Shu G, Wang SB, Zhu XT, Gao P, Xi QY, Zhang YL, Zhang ZQ, Jiang QY. Decorin-induced proliferation of avian myoblasts involves the myostatin/ Smad signaling pathway. *Poultry Science*. 2014 Jan;93(1):138-46.

[49] Relaix F, Rocancourt D, Mansouri A, Buckingham M. A Pax3/Pax7-dependent population of skeletal muscle progenitor cells. *Nature*. 2005 Jun;435(7044):948-53.

[50] Buckingham M, Relaix F. The role of Pax genes in the development of tissues and organs: Pax3 and Pax7 regulate muscle progenitor cell functions. *Annual Review of Cell and Developmental Biology*. 2007;23:645-73.

[51] Buckingham M, Relaix F. PAX3 and PAX7 as upstream regulators of myogenesis. *Seminars in Cell & Developmental Biology*. 2015 Aug;44:115-25.

[52] Oustanina S, Hause G, Braun T. Pax7 directs postnatal renewal and propagation of myogenic satellite cells but not their specification. *EMBO Journal*. 2004 Aug;23(16): 3430-9.

[53] Rawls A, Olson EN. Myo D meets its maker. *Cell*. 1997 Apr;89(1):5-8.

[54] Molkentin JD, Olson EN. Defining the regulatory networks for muscle development. *Current Opinion in Genetics & Development*. 1996 Aug;6(4):445-53.

[55] Cisternas P, Henriquez JP, Brandan E, Inestrosa NC. Wnt signaling in skeletal muscle dynamics: myogenesis, neuromuscular synapse and fibrosis. *Molecular Neurobiology*. 2014;49(1):574-89.

[56] Tajbakhsh S, Rocancourt D, Cossu G, Buckingham M. Redefining the genetic hierarchies controlling skeletal myogenesis: Pax-3 and Myf-5 act upstream of MyoD. *Cell*. 1997 Apr;89(1):127-38.

[57] Kang JS, Krauss RS. Muscle stem cells in developmental and regenerative myogenesis. *Current Opinion in Clinical Nutrition & Metabolic Care*. 2010 May;13(3):243-8.

[58] Wagers AJ, Conboy IM. Cellular and molecular signatures of muscle regeneration: current concepts and controversies in adult myogenesis. *Cell*. 2005 Sep;122(5):659-67.

[59] Karalaki M, Fili S, Philippou A, Koutsilieris M. Muscle regeneration: cellular and molecular events. *In Vivo*. 2009 Sep;23(5):779-96.

[60] ChargéSB, Rudnicki MA. Cellular and molecular regulation of muscle regeneration. *Physiological Reviews*. 2004 Jan;84(1):209-38.

[61] Yablonka-Reuveni Z, Day K, Vine A, Shefer G. Defining the transcriptional signature of skeletal muscle stem cells. *Journal of Animal Science*. 2008 Apr;86(14 Suppl):E207-16.

[62] McCarthy JJ, Esser KA. Anabolic and catabolic pathways regulating skeletal muscle mass. *Current Opinion in Clinical Nutrition & Metabolic Care*. 2010 May;13(3):230-5.

[63] Sun H, Zhu T, Ding F, Hu N, Gu X. Proteomic studies of rat tibialis anterior muscle during postnatal growth and development. *Molecular and Cellular Biochemistry*. 2009 Dec;332(1):161-71.

[64] Xu Y, Qian H, Feng X, Xiong Y, Lei M, Ren Z, Zuo B, Xu D, Ma Y, Yuan H. Differential proteome and transcriptome analysis of porcine skeletal muscle during development. *Journal of Proteomics*. 2012 Apr;75(7):2093-108.

[65] Tannu NS, Rao VK, Chaudhary RM, Giorgianni F, Saeed AE, Gao Y, Raghow R,. Comparative proteomes of the proliferating C(2)C(12) myoblasts and fully differentiated myotubes reveal the complexity of the skeletal muscle differentiation program. *Molecular and Cellular Proteomics*. 2004 Nov;3(11):1065-82.

[66] Casadei L, Vallorani L, Gioacchini AM, Guescini M, Burattini S, D'Emilio A, Biagiotti L, Falcieri E, Stocchi V. Proteomics-based investigation in C2C12 myoblast differentiation. *European Journal of Histochemistry*. 2009 Dec;53(4):261-8.

[67] Kislinger T, Gramolini AO, Pan Y, Rahman K, Mac Lennan DH, Emili A. Proteome dynamics during C2C12 myoblast differentiation. *Molecular and Cellular Proteomics*. 2005 Jul;4(7):887-901.

[68] Cui Z, Chen X, Lu B, Park SK, Xu T, Xie Z, Xue P, Hou J, Hang H, Yates JR, 3rd, Yang F. Preliminary quantitative profile of differential protein expression between rat L6 myoblasts and myotubes by stable isotope labeling with amino acids in cell culture. *Proteomics*. 2009 Mar;9(5):1274-92.

[69] Chan XC, McDermott JC, Siu KW. Identification of secreted proteins during skeletal muscle development. *Journal of Proteome Research*. 2007 Feb;6(2):698-710.

[70] Henningsen J, Rigbolt KT, Blagoev B, Pedersen BK, Kratchmarova I. Dynamics of the skeletal muscle secretome during myoblast differentiation. *Molecular and Cellular Proteomics*. 2010 Nov;9(11):2482-96.

[71] Chan CY, Masui O, Krakovska O, Belozerov VE, Voisin S, Ghanny S, Chen J, Moyez D, Zhu P, Evans KR, McDermott JC, Siu KW. Identification of differentially regulated secretome components during skeletal myogenesis. *Molecular and Cellular Proteomics*. 2011 May;10(5):M110.004804.

[72] Le Bihan MC, Barrio-Hernandez I, Mortensen TP, Henningsen J, Jensen SS, Bigot A, Blagoev B, Butler-Browne G, Kratchmarova I. Cellular Proteome Dynamics during Differentiation of Human Primary Myoblasts. *Journal of Proteome Research*. 2015 Aug; 14(8):3348-61.

[73] Pedersen BK. Muscle as a secretory organ. *Comprehensive Physiology*. 2013 Jul;3(3): 1337-62.

[74] Iizuka K, Machida T, Hirafuji M. Skeletal muscle is an endocrine organ. *Journal of Pharmacological Science*. 2014;125(2):125-31.

[75] Hartwig S, Raschke S, Knebel B, Scheler M, Irmler M, Passlack W, Muller S, Hanisch FG, Franz T, Li X, Dicken HD, Eckardt K, Beckers J, de Angelis MH, Weigert C, Häring HU, Al-Hasani H, Ouwens DM, Eckel J, Kotzka J, Lehr S. Secretome profiling of primary human skeletal muscle cells. *Biochimica et Biophysica Acta.* 2014 May;1844(5):1011-7.

[76] Le Bihan MC, Bigot A, Jensen SS, Dennis JL, Rogowska-Wrzesinska A, Lainé J, Gache V, Furling D, Jensen ON, Voit T, Mouly V, Coulton GR, Butler-Browne G. In-depth analysis of the secretome identifies three major independent secretory pathways in differentiating human myoblasts. *Journal of Proteomics.* 2012 Dec;77:344-56.

[77] Højlund K, Yi Z, Hwang H, Bowen B, Lefort N, Flynn CR, Langlais P, Weintraub ST, Mandarino LJ. Characterization of the human skeletal muscle proteome by one-dimensional gel electrophoresis and HPLC-ESI-MS/MS. *Molecular and Cellular Proteomics.* 2008;7:257-67.

[78] Parker KC, Walsh RJ, Salajegheh M, Amato AA, Krastins B, Sarracino DA, Greenberg SA. Characterization of human skeletal muscle biopsy samples using shotgun proteomics. *Journal of Proteome Research.* 2009;8:3265-77.

[79] Drexler HC, Ruhs A, Konzer A, Mendler L, Bruckskotten M, Looso M, Günther S, Boettger T, Krüger M, Braun T. On marathons and Sprints: an integrated quantitative proteomics and transcriptomics analysis of differences between slow and fast muscle fibers. *Molecular and Cellular Proteomics.* 2012;11:M111.010801.

[80] Burniston JG, Connolly J, Kainulainen H, Britton SL, Koch LG. Label-free profiling of skeletal muscle using high-definition mass spectrometry. *Proteomics.* 2014;14:2339-44.

[81] Murgia M, Nagaraj N, Deshmukh AS, Zeiler M, Cancellara P, Moretti I, Reggiani C, Schiaffino S, Mann M. Single muscle fiber proteomics reveals unexpected mitochondrial specialization. *EMBO Reports.* 2015;16:387-95.

[82] Deshmukh AS, Murgia M, Nagaraja N, Treebak JT, Cox J, Mann M. Deep proteomics of mouse skeletal muscle enables quantitation of protein isoforms, metabolic pathways and transcription factors. *Molecular and Cellular Proteomics.* 2015:14:841-53.

[83] Holland A, Dowling P, Meleady P, Henry M, Zweyer M, Mundegar RR, Swandulla D, Ohlendieck K. Label-free mass spectrometric analysis of the mdx-4cv diaphragm identifies the matricellular protein periostin as a potential factor involved in dystrophinopathy-related fibrosis. *Proteomics.* 2015;15:2318-31.

[84] Murphy S, Henry M, Meleady P, Zweyer M, Mundegar RR, Swandulla D, Ohlendieck K. Simultaneous Pathoproteomic Evaluation of the Dystrophin-Glycoprotein Complex and Secondary Changes in the mdx-4cv Mouse Model of Duchenne Muscular Dystrophy. *Biology (Basel)* 2015;4:397-423.

[85] Murphy S, Zweyer M, Mundegar RR, Henry M, Meleady P, Swandulla D, Ohlendieck K. Concurrent Label-Free Mass Spectrometric Analysis of Dystrophin Isoform Dp427

and the Myofibrosis Marker Collagen in Crude Extracts from mdx-4cv Skeletal Muscles. Proteomes 2015;3:298-327.

[86] Holland A, Murphy S, Dowling P, Ohlendieck K. Pathoproteomic profiling of the skeletal muscle matrisome in dystrophinopathy associated myofibrosis. *Proteomics.* 2016 Jan;16(2):345-66.

[87] Smith LR, Meyer G, Lieber RL. Systems analysis of biological networks in skeletal muscle function. *Wiley Interdisciplinary Reviews: Systems Biology and Medicine.* 2013 Jan-Feb;5(1):55-71.

[88] Franceschini A, Szklarczyk D, Frankild S, Kuhn M, Simonovic M. STRING v9.1: Protein–protein interaction networks, with increased coverage and integration. *Nucleic Acids Research.* 2013;41:D808-15.

[89] Booth FW, Thomason DB. Molecular and cellular adaptation of muscle in response to exercise: perspectives of various models. *Physiological Reviews.* 1991;71(2):541-85.

[90] Wisdom KM, Delp SL, Kuhl E. Use it or lose it: multiscale skeletal muscle adaptation to mechanical stimuli. *Biomechanics and Modeling in Mechanobiology.* 2015 Apr;14(2): 195-215.

[91] Hawley JA. Adaptations of skeletal muscle to prolonged, intense endurance training. *Clinical and Experimental Pharmacology and Physiology.* 2002 Mar;29(3):218-22.

[92] Ross A, Leveritt M. Long-term metabolic and skeletal muscle adaptations to short-sprint training: implications for sprint training and tapering. *Sports Medicine.* 2001;31(15):1063-82.

[93] Kjaer M, Magnusson P, Krogsgaard M, Boysen Møller J, Olesen J, Heinemeier K, Hansen M, Haraldsson B, Koskinen S, Esmarck B, Langberg H. Extracellular matrix adaptation of tendon and skeletal muscle to exercise. *Journal of Anatomy.* 2006 Apr; 208(4):445-50.

[94] Coutinho EL, DeLuca C, Salvini TF, Vidal BC. Bouts of passive stretching after immobilization of the rat soleus muscle increase collagen macromolecular organization and muscle fiber area. *Connect Tissue Research.* 2006;47(5):278-86.

[95] Hyldahl RD, Nelson B, Xin L, Welling T, Groscost L, Hubal MJ, Chipkin S, Clarkson PM, Parcell AC. Extracellular matrix remodeling and its contribution to protective adaptation following lengthening contractions in human muscle. *FASEB Journal.* 2015 Jul;29(7):2894-904.

[96] Heinemeier KM, Olesen JL, Haddad F, Langberg H, Kjaer M, Baldwin KM, Schjerling P. Expression of collagen and related growth factors in rat tendon and skeletal muscle in response to specific contraction types. *Journal of Physiology.* 2007 Aug;582(3):1303-16.

[97] Rullman E, Rundqvist H, Wågsäter D, Fischer H, Eriksson P, Sundberg CJ, Jansson E, Gustafsson T. A single bout of exercise activates matrix metalloproteinase in human skeletal muscle. *Journal of Applied Physiology.* 2007 Jun;102(6):2346-51.

[98] Heinemeier KM, Bjerrum SS, Schjerling P, Kjaer M. Expression of extracellular matrix components and related growth factors in human tendon and muscle after acute exercise. *Scandinavian Journal of Medicine & Science in Sports.* 2013 Jun;23(3):e150-61.

[99] Ogasawara R, Nakazato K, Sato K, Boppart MD, Fujita S. Resistance exercise increases active MMP and β1-integrin protein expression in skeletal muscle. *Physiological Reports.* 2014 Nov;2(11):e12212.

[100] Pincu Y, Linden MA, Zou K, Baynard T, Boppart MD. The effects of high fat diet and moderate exercise on TGFβ1 and collagen deposition in mouse skeletal muscle. *Cytokine.* 2015 May;73(1):23-9.

[101] Kaasik P, Riso EM, Seene T. Extracellular matrix and myofibrils during unloading and reloading of skeletal muscle. *International Journal of Sports Medicine.* 2011 Apr;32(4): 247-53.

[102] Burniston JG, Hoffman EP. Proteomic responses of skeletal and cardiac muscle to exercise. *Expert Review of Proteomics.* 2011 Jun;8(3):361-77.

[103] Ohlendieck K. Proteomics of exercise-induced skeletal muscle adaptations. *OA Sports Medicine.* 2013 Mar;1(1):3.

[104] Burniston JG. Changes in the rat skeletal muscle proteome induced by moderate-intensity endurance exercise. *Biochimica et Biophysica Acta.* 2008 Jul;1784(7):1077-86.

[105] Egan B, Dowling P, O'Connor PL, Henry M, Meleady P, Zierath JR, O'Gorman DJ. 2-D DIGE analysis of the mitochondrial proteome from human skeletal muscle reveals time course-dependent remodelling in response to 14 consecutive days of endurance exercise training. *Proteomics.* 2011 Apr;11(8):1413-28.

[106] Holloway KV, O'Gorman M, Woods P, Morton JP, Evans L, Cable NT, Goldspink DF, Burniston JG. Proteomic investigation of changes in human vastus lateralis muscle in response to interval-exercise training. *Proteomics.* 2009 Nov;9(22):5155-74.

[107] Schild M, Ruhs A, Beiter T, Zügel M, Hudemann J, Reimer A, Krumholz-Wagner I, Wagner C, Keller J, Eder K, Krüger K, Krüger M, Braun T, NießA, Steinacker J, Mooren FC. Basal and exercise induced label-free quantitative protein profiling of m. vastus lateralis in trained and untrained individuals. *Journal of Proteomics.* 2015 Jun;122:119-32.

[108] Ljubicic V, Adhihetty PJ, Hood DA. Application of animal models: chronic electrical stimulation-induced contractile activity. *Canadian Journal of Applied Physiology.* 2005 Oct;30(5):625-43.

[109] Pette D, Staron RS. Transitions of muscle fiber phenotypic profiles. *Histochemistry & Cell Biology.* 2001 May;115(5):359-72

[110] Hicks A, Ohlendieck K, Göpel SO, Pette D. Early functional and biochemical adaptations to low-frequency stimulation of rabbit fast-twitch muscle. *American Journal of Physiology.* 1997 Jul;273(1):C297-305.

[111] Donoghue P, Doran P, Dowling P, Ohlendieck K. Differential expression of the fast skeletal muscle proteome following chronic low-frequency stimulation. *Biochimica et Biophysica Acta.* 2005 Sep;1752(2):166-76.

[112] Donoghue P, Doran P, Wynne K, Pedersen K, Dunn MJ, Ohlendieck K. Proteomic profiling of chronic low-frequency stimulated fast muscle. *Proteomics.* 2007 Sep;7(18): 3417-30.

[113] Henriksson J, Chi MM, Hintz CS, Young DA, Kaiser KK, Salmons S, Lowry OH. Chronic stimulation of mammalian muscle: changes in enzymes of six metabolic pathways. *American Journal of Physiology.* 1986 Oct;251(4):C614-32.

[114] Kratz JM, Johnson WS, Mukherjee R, Hu J, Crawford FA, Spinale FG. The relation between latissimus dorsi skeletal muscle structure and contractile function after cardiomyoplasty. *Journal of Thoracic and Cardiovascular Surgery.* 1994 Mar;107(3):868-78.

[115] Carmignac V, Durbeej M. Cell–matrix interactions in muscle disease. *Journal of Pathology.* 2012 Jan;226(2):200-18.

[116] Kragstrup TW, Kjaer M, Mackey AL. Structural, biochemical, cellular, and functional changes in skeletal muscle extracellular matrix with aging. *Scandinavian Journal of Medicine & Science in Sports.* 2011 Dec;21(6):749-57.

[117] Desguerre I, Mayer M, Leturcq F, Barbet JP, Gherardi RK, Christov C. Endomysial fibrosis in Duchenne muscular dystrophy: a marker of poor outcome associated with macrophage alternative activation. *Journal of Neuropathology & Experimental Neurology.* 2009 Jul;68(7):762-73.

[118] Merlini L, Bernardi P. Therapy of collagen VI-related myopathies (Bethlem and Ullrich). *Neurotherapeutics.* 2008 Oct;5(4):613-8.

[119] Bönnemann CG. The collagen VI-related myopathies: muscle meets its matrix. Nat Rev Neurol. 2011 Jun 21;7(7):379-90.

[120] Mehuron T, Kumar A, Duarte L, Yamauchi J, Accorsi A, Girgenrath M. Dysregulation of matricellular proteins is an early signature of pathology in laminin-deficient muscular dystrophy. *Skeletal Muscle.* 2014 Jul;4:14.

[121] Løkken N, Born AP, Duno M, Vissing J. LAMA2-related myopathy: Frequency among congenital and limb-girdle muscular dystrophies. *Muscle & Nerve.* 2015 Oct;52(4): 547-53.

[122] Klingler W, Jurkat-Rott K, Lehmann-Horn F, Schleip R. The role of fibrosis in Duchenne muscular dystrophy. *Acta Myologica.* 2012 Dec;31(3):184-95.

[123] Kharraz Y, Guerra J, Pessina P, Serrano AL, Muñoz-Cánoves P. Understanding the process of fibrosis in Duchenne muscular dystrophy. *Biomedical Research International.* 2014;2014:965631.

[124] De Palma S, Leone R, Grumati P, Vasso M, Polishchuk R, Capitanio D, Braghetta P, Bernardi P, Bonaldo P, Gelfi C. Changes in muscle cell metabolism and mechanotransduction are associated with myopathic phenotype in a mouse model of collagen VI deficiency. *PLoS One.* 2013;8(2):e56716.

[125] De Palma S, Capitanio D, Vasso M, Braghetta P, Scotton C, Bonaldo P, Lochmüller H, Muntoni F, Ferlini A, Gelfi C. Muscle proteomics reveals novel insights into the pathophysiological mechanisms of collagen VI myopathies. *Journal of Proteome Research.* 2014 Nov;13(11):5022-30.

[126] Holland A, Carberry S, Ohlendieck K. Proteomics of the dystrophin-glycoprotein complex and dystrophinopathy. *Current Protein and Peptide Science.* 2013 Dec;14(8): 680-97.

[127] Dowling P., Holland A., Ohlendieck K., Mass spectrometry-based identification of muscle-associated and muscle-derived proteomic biomarkers of dystrophinopathies. *Journal of Neuromuscular Disorders.* 2014;1:15-40.

[128] Carberry S, Zweyer M, Swandulla D, Ohlendieck K. Proteomics reveals drastic increase of extracellular matrix proteins collagen and dermatopontin in the aged mdx diaphragm model of Duchenne muscular dystrophy. *International Journal of Molecular Medicine.* 2012 Aug;30(2):229-34.

[129] Carberry S, Zweyer M, Swandulla D, Ohlendieck K. Application of fluorescence two-dimensional difference in-gel electrophoresis as a proteomic biomarker discovery tool in muscular dystrophy research. *Biology (Basel).* 2013 Dec;2(4):1438-64.

[130] Forbes EG, Cronshaw AD, MacBeath JR and Hulmes DJ: Tyrosine-rich acidic matrix protein (TRAMP) is a tyrosine-sulphated and widely distributed protein of the extracellular matrix. *FEBS Letters.* 1994 Sep;351(3):433-6.

[131] Okamoto O and Fujiwara S: Dermatopontin, a novel player in the biology of the extracellular matrix. *Connective Tissue Research.* 2006;47(4):177-89.

[132] Conway S.J., Izuhara K, Kudo Y, Litvin J, Markwald R, Ouyang G, Arron JR, Holweg CT, Kudo A. The role of periostin in tissue remodeling across health and disease. *Cellular and Molecular Life Science.* 2014 Apr;71(7):1279-88.

[133] Ozdemir C., Akpulat U., Sharafi P., Yıldız Y. Onbaşılar I, Kocaefe C. Periostin is temporally expressed as an extracellular matrix component in skeletal muscle regeneration and differentiation. *Gene.* 2014 Dec;553(2):130-9.

[134] Lorts A, Schwanekamp JA, Baudino TA, McNally EM, Molkentin JD. Deletion of periostin reduces muscular dystrophy and fibrosis in mice by modulating the trans-

forming growth factor-β pathway. *Proceedings of the National Acadamy of Science USA.* 2012 Jul;109(27):10978-83.

[135] Menezes de Oliveira, B., Matsumura C.Y., Fontes-Oliveira C.C., Gawlik K.I. Acosta H, Wernhoff P, Durbeej M. Quantitative proteomic analysis reveals metabolic alterations, calcium dysregulation and increased expression of extracellular matrix proteins in laminin $\alpha 2$ chain-deficient muscle. *Molecular and Cellular Proteomics.* 2014 Nov;13(11): 3001-13.

[136] Pessina P., Cabrera D., Morales M.G., Riquelme C.A., Gutiérrez J, Serrano AL, Brandan E, Muñoz-Cánoves P. Novel and optimized strategies for inducing fibrosis in vivo: focus on Duchenne Muscular Dystrophy. *Skeletal Muscle.* 2014 Aug;4:7.

[137] Carberry S, Ohlendieck K. Gel electrophoresis-based proteomics of senescent tissues. *Methods in Molecular Biology.* 2013;1048:229-46.

Section 2

Role of ECM in Nervous System

Chapter 5

The Extracellular Matrix in the Nervous System: The Good and the Bad Aspects

Elena Vecino and Jessica C. F. Kwok

Additional information is available at the end of the chapter

http://dx.doi.org/10.5772/62527

Abstract

The study of extracellular matrix (ECM) in the nervous system has longed been focused on the molecules promoting growth and migration. This is well supported by the work in the developing nervous system. However, the discovery of Nogo and chondroitin sulphate proteoglycans (CSPGs) in the injured nervous system in late 1980s has shifted some of the focus to inhibitory molecules. One of the biggest hurdles in neural regeneration is the formation of glial scar and the highly up-regulated inhibitory molecules present in the area. Apart from Nogo and CSPGs, other myelin-associated inhibitors, tenascins and semaphorins have been found associated with neuronal inhibition. Together with the identification of their receptors, we now have a better understanding on the mechanism of how these molecules control and limit regeneration in the central nervous system (CNS). Recent focus has been put on designing strategies in neutralizing these inhibitions for promoting regeneration after injury, and some are showing promising results. Moreover, latest studies also show that rehabilitation in injured animal models demonstrated drastic remodeling of ECM favoring regeneration. This review shall discuss all these different aspects and the importance of matrix remodeling in the CNS and the implication of ECM in some retinal pathologies.

Keywords: retina, regeneration, perineuronal nets, chondroitin sulfate, integrins

1. Introduction

The extracellular matrix (ECM) constitutes a three-dimensional network that surrounds the cells and conform the structure and characteristics to tissues. It has become increasingly evident that once being considered as a bystander between cells, the ECM indeed performs significant functions and involves in controlling various physiological responses in the CNS. The importance of ECM is at three levels: it acts as biological scaffold for the structure of the CNS and controls the diffusion and availability of molecules for biochemical signaling and communication and, finally, the various polymers and molecular interactions in the ECM control the biomechanical properties of the central nervous system (CNS) [1, 2]. In addition, regenerative capacity of tissues is also directly related to the ECM. Disorders in mechano-transduction or alterations in the composition of ECM will drive to a loss of the regenerative ability of the tissue and cells [3, 4]. Moreover, a proper immune and toxic response to infections is in accordance with the correct equilibrium in the ECM components [5].

In the nervous system, the ECM are synthesized and secreted by both neurons and glia. In the present chapter, we shall introduce the main key components of the ECM present in the brain and the main implications of these molecules associated to the normal and pathological CNS, including the spinal cord injury and in retina [6, 7]. While axon–glia interaction helps to determine the extent and direction of axon outgrowth, the growth of axons are also directed by factors present in the ECM. The growth enhancing cues such as laminin and fibronectin will encourage the growth and extension of neurites, while the inhibitory cues such as chondroitin sulfate proteoglycans (CSPGs) and semaphorins (Sema) serve as barriers in precise locations to prevent the growth of certain axon pathways into inappropriate areas.

2. ECM components

The ECM in the nervous system is mainly provided by macroglia and is an important source of supporting and signaling factors [8]. ECM components are key mediators of glial activation and have the capacity to evoke both regenerative and degenerative effects on glia and neurons [9]. The production of ECM components changes drastically during reactive gliosis [10]. The activated astrocytes and microglia increases the synthesis of various ECM molecules, including the re-expression of some extracellular glycoproteins, which are down-regulated after development [8, 10]. These include a complex mixture of proteins, proteoglycans (PGs), and glycoproteins (GPs) that confer the structural properties of cells and tissues.

2.1. Proteoglycans

Proteoglycans are macromolecules composed of a core protein on which multiple glycosaminoglycan (GAG) chains are attached (**Figure 1**). The GAG chains are linear unbranched polymers of repeating disaccharides composed of hexosamine and an uronic acid [11]. These molecules have remarkable physical properties attributable to the abundance of carboxyl, hydroxyl, and sulfate groups [11]. Their electrostatic properties make them "osmotically active". Their net negative charge attracts Na^+ and thus, draws water in causing the interstitial spaces where GAGs reside to swell. This swelling opens up pathways that promote the invasion and migration of cells as has been suggested for the non-sulfated GAG hyaluronan (HA), which is correlated with an initiation of cell migration during development [12]. There are five groups of GAGs based on the composition of the repeating disaccharides [11]. They include hyaluronan, chondroitin sulfate (CS), dermatan sulphate, heparan sulfate and keratan

http://dx.doi.org/10.5772/62527

sulfate. Except HA, all GAGs are covalently attached to a core protein and form proteoglycan (PG) (**Figure 1**).

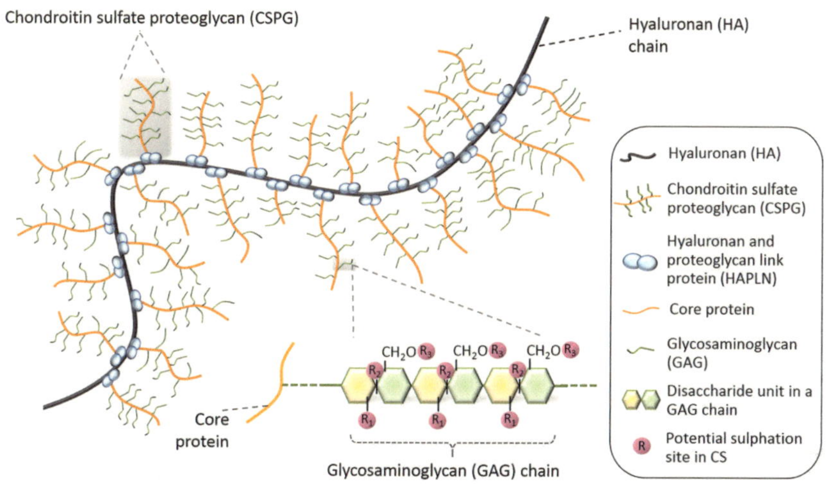

Figure 1. A schematic diagram of CSPGs and their aggregation into macromolecular matrix in the CNS-ECM. Each CSPG is composed of a core protein (orange line) decorated with different number of CS-GAG chains (green lines). The CSPGs then interact with a hyaluronan chain (grey thick line) in the ECM, forming a large molecular aggregate. Each CS GAG chain is composed of repeating disaccharide units which sulfations (pink circles) can be added on. The type of core protein, number and length of GAG chains, and different patterns of sulfations contribute to the big heterogeneity of CSPGs.

Chondroitin sulfate (CS) and its proteoglycan CSPG constitute the major population of proteoglycans in the CNS [13, 14]. There are at least sixteen CS core proteins expressed in the CNS and many of them are produced by astrocytes [15]. The disaccharides in the CS chains can be sulfated at various positions resulting in different isoforms of CSs, including chondroitin 4-sulfate and chondroitin 6-sulfate, the most CS sulfation in an adult CNS (**Figure 1**) [16]. Together with the diversity of core proteins, the CS chain length, the number of chains attached to the core proteins, these factors give a huge diversity with a wide functional heterogeneity of CSPGs (**Figure 1**) [15].

CSPGs are known for their inhibitory influences on neurite extension [17, 18]. It was first demonstrated in dorsal root ganglion (DRG) neurons and subsequently being recognized in the CNS [18, 19]. CSPGs, such as NG2 and neurocan, are strongly up-regulated in the glial scar after injury [10]. Their up-regulation induces growth cone collapse and form a strong barrier for nerve regeneration [20]. CSPGs are primarily produced locally by the reactive astrocytes which are attracted to the peri-lesioned area after injury [9]. Chondroitinase ABC (ChABC), an enzyme which digests the CS chains into disaccharides, effectively removes this inhibition both in the developing CNS and after injury in adult [21–23]. ChABC removes the CSs in the developing hindbrain and promotes the neurite extension of commissural vestibular neurons in developing embryos [21]. Similarly, ChABC is very effective in removing the inhibition in the glial scar, encouraging sprouting and regrowth, and conferring functional recovery after injury [22, 24]. Studies in recent years have been focusing on finding new strategies for sustained CS down-regulation, based on the success of ChABC, to provide a sufficient time window for regeneration to occur. The expression of ChABC using lentiviral vector show a down-regulation of CSs for up to 8 weeks, and the animal demonstrates an enhanced axonal sprouting and a superior functional recovery in the forelimb after a cervical contusion injury [25, 26].

One of the mechanisms of CSPGs mediating their functions is through binding to receptors or growth factors. Contactin-1, Nogo receptors (NgRs), and RPTPσ are the identified CSPG receptors [27–29]. The transmembrane receptor RPTPσ binds to CSPGs neurocan and aggrecan in the CNS [28]. RPTPσ induces growth cone collapse *in vitro* [30]. Mice with RPTPσ knockout show a robust regeneration, with cortical spinal tract fibers extend for a long distance after a spinal hemisection [31] Blocking the binding of CSPGs to RPTPσ using membrane-permeable peptide mimetics promotes serotonergic innervation, and the animal demonstrates enhanced functional recovery after spinal cord injury [30, 31]. This membrane-permeable peptide mimetics hold a strong promise to modulate CSPG-mediated inhibition in replacing ChABC. Apart from RPTPσ, NgRs also bind to CSs. Double mutant of Ngr1/Ngr3 showed an enhanced axon regeneration in optic nerve crush injury [29]. Contactin-1 binds specifically to highly-sulfated CS, CS-E, on the contrary to other CSPG receptor, the binding of CS-E to contactin-1 promotes neurite extension [27]. Neuro2a cells with recombinant expression of contactin-1 demonstrate extensive neurite sprouting when cultured on CS-E [27]. CSPGs also interact with other growth factors, chemokines, and guidance molecules in the developing brain. By doing so, they control the availability of these factors to cells. While fibroblast growth factor (FGF)-2, FGF-10, FGF-16, and FGF-18 bind to highly sulfated CS chains, guidance proteins such as slit2, netrin1, ephrin A1 and A5, semaphorin (Sema) 3A, 5A, and 5B bind to CS chains in a sulfation-dependent manner [32].

CSPGs in the CNS also aggregate into a macromolecular structure on the surface of neurons called perineuronal nets (PNNs) [33, 34]. PNNs are dense matrix structures formed by four families of brain ECM molecules, including CSPGs, hyaluronan, hyaluronan, and proteoglycan link proteins (HAPLNs) and tenascins [33]. PNNs wrap the neuronal surface and are crucial in controlling synaptic and neuronal plasticity in the developing and injury CNS [4, 35]. PNNs form toward the end of the critical period for plasticity, and the formation is activity dependent [36, 37]. Dark rearing from birth delays the formation of PNNs in the visual cortex [36]. Similar observation is also reported with whisker trimming from birth in the barrel cortex of mice [38]. ChABC treatment removes this layer of CSPG-enriched PNN matrix and reactivates plasticity in the adult CNS, this includes spinal cord injury, visual cortex plasticity, cuneate nucleus plasticity, and more recently, on memory enhancement [35, 39, 40]. PNNs mediate their functions, in part, through binding to other molecules such as chemo-repulsive molecule sema3A and soluble transcription factor Otx2 [41, 42]. Both Sema3A and Otx-2 bind to the PNNs via the highly sulfated CS-E. Upon binding, Otx-2 is internalized into the cells and regulates the gene expression for the maturation of PNN-positive parvalbumin neurons in the cortex [42, 43]. Binding of Sema3A to PNN-glycans potentiates the inhibitory properties of PNN-glycans to the outgrowth of DRG neurons *in vitro* [41].

2.2. Class 3 semaphorins

Semaphorins (Sema) are a family of axon guidance molecules during CNS development [44]. The family is divided into eight classes and only five out of these eight Sema are expressed in vertebrates. Unlike other types of vertebrate Sema which are either transmembrane or membrane-anchored, class 3 Sema (Sema3) is the only secreted Sema in vertebrates [45]. To date, Sema3-A, –B, –C, –E, and -F have been identified in an injured CNS [46]. They are produced by the meningeal fibroblasts migrating into the lesion area and up-regulated the expression of different members of Sema3 [46, 47]. The binding of Sema3 to its receptors, neuropilins or plexins, induces a strong growth cone collapse in DRG neurons [48]. Apart from acting through the corresponding receptors, Sema3 also mediate their function though binding to the PNNs. We have previously shown that Sema3A binds to PNNs and that this binding is mediated by a specific CS structure in the PNNs, CS-E. Blocking the binding of Sema3A to PNN glycans reduces the inhibition imposed by PNN to DRG neurons [41].

2.3. Tenascin-C and -R

Tenascin (Tn) family has four members and two of them are expressed in the CNS, including Tn-C and Tn-R [49, 50]. They are both up-regulated after CNS injury [51, 52]. Tn-C is expressed by astrocytes, radial glia, and subsets of developing retinal and hippocampal neurons during early CNS development. In adults, Tn-C is restricted to areas of high plasticity including the olfactory bulb, the cerebellum and the retina. Tn-C interacts with other ECM molecules such as integrins, proteoglycans, and collagen [50]. Tn-C up-regulates after CNS injury and interacts with the different CSPGs in the glial scar. This interaction has been implicated to the failure of axon growth beyond the injury site [53, 54]. Expression of an appropriate integrin isoform, which binds and uses Tn-C as substrate, elicits an enhancement in regeneration [55, 56].

Tn-R is trimer which is expressed in both the developing and adult brains, primarily by neurons, including the horizontal cells from the retina [57]. In adults, Tn-R is found in the PNNs [58, 59]. The trimeric TnR interacts with the core protein in the CSPGs, consolidating the PNN structure [60]. Tn-R has negative influence on axonal growth [49]. Knockout mice of Tn-R, which forms disorganized PNNs, demonstrates enhanced motor recovery after spinal cord injury suggesting that 1) Tn-R is important for PNN structure and that PNNs dissolution enhances plasticity for functional recovery [61].

2.4. Laminins

Laminins are large heterotrimeric glycoproteins that contain an alpha chain, a beta chain and a gamma chain joined together in a coiled–coiled structure. Sixteen isoforms have been identified *in vivo*, and are differentially expressed both temporally and spatially in various tissues [62]. Genetic disruptions of laminin chains lead disruptions in various tissues and also functional properties in the CNS [63, 64]. The major receptors for laminins are classified as integrins and non-integrins [65]. We shall discuss the role of integrins in later sections of this chapter. For non-integrins receptors such as dystroglycan and GM1 gangliosides, the binding of laminin serves critical functions in the peripheral nervous system (PNS) including myelination by Schwann cells, neurite outgrowth, and the integrity of blood–brain barrier [66–70]. In the CNS, laminin is primarily present in the basement membrane and is up-regulated by astrocytes after injury [71], although reports of individual isoforms of laminin have also been reported [72]. This suggests an neuronal heterogeneity of laminin isoforms in the adult brain.

Laminin provides a positive guidance to axons during development [73], and act as a supporting substrate to adult retinal ganglion cell [74] and retinal pigment epithelial cells *in vitro* [75]. The expression of laminin decreases during maturation of the optic system [76] even though, in our recent study, we observed that laminin is still present in adult retinas and optic nerves [74].

2.5. Collagens

Collagens are the most abundant proteins in the animal kingdom, there are now 29 known collagens [77, 78]. A triple-helical organization of component pro- α-chains defines the collagens and 49 distinct collagen α-chain gene products have been described [79]. Collagens are classified into both fibrillar and non-fibrillar forms and can also be assembled into reticular networks and sheets [80]. The organization, distribution, and density of fibrils and networks vary with tissue type and the direction and magnitude of forces to which are given tissue is subjected.

Collagens expressed in the PNS provide a scaffold for the growth and attachment of Schwann cells which also guide the neurite extension [78, 81]. After injury, there is an over up-regulation of collagen which changes the mechanical properties of the lesion area, hinders, and delays regeneration to occur [78]. Collagen is implicated in the progression of glaucoma, a visual neurological disease. One of the characteristics of certain glaucoma is the increment of the intraocular pressure in the anterior chamber of the eye. The heightened pressure is transmitted to the posterior eye chamber, pressing the retinal ganglion cells and eventually driving them to death [82]. Since these cells are the neurons responsible for transmitting the visual signal from the eye to the brain, their cell death leads to inevitable blindness. One of the reasons for the increasing pressure in the anterior chamber is due to an obstruction of the filtering tissue present in the trabecular meshwork, where the aqueous humor flows. The cells of this tissue have the ability to secrete the extracellular matrix. In an attempt to adapt to a biomechanical insult, the cells in the trabecular meshwork increase the synthesis of ECM, including collagen, thus blocking the flow of the aqueous humor and leading to an elevation of intraocular pressure. This mechanism has been proposed as possible cause of the origin of glaucoma [82, 83].

Other important implication of collagens in glaucoma is found in the lamina cribosa, a structure located in the optic nerve head where axons exit from the retina to the optic nerve [84]. A dysregulation in collagen secretion at this point implicates an interruption of axonal transport from the retinal ganglion cell in the retina to the visual areas in the brain. Studies have shown that activated astrocytes are cells responsible for the collagen synthesis and alterations here [85].

The third implication of collagen in glaucoma is the stiffness of the sclera and its implication in the lack of elasticity of the eye [86]. The sclera is the structure that supports the attachment of ocular components including the retina and optic nerve head. A complex network of collagen fibers forms the sclera's major component and is a major influence on the tissue's biomechanical response to changes in the intraocular pressure. It has been proposed that the mechanical influence of the sclera may be a key on the eye injury after elevation of the intraocular pressure due to the alterations in the thickness, mechanics and matrix ultrastructure provided by the arrangement of scleral collagen and fiber orientation [87].

2.6. Integrins

Integrins are a family of cell surface receptors that are important for cell adhesion to ECM proteins. They are the principal receptors on animal cells for mediating most ECM attachment and signaling. They connect the extracellular environment to intracellular cytoskeleton and are responsible for the activation of many intracellular signaling pathway [88]. All integrins are heterodimeric molecules containing two subunits, α and β. Each $\alpha\beta$ combination has its own specificity and signaling properties [89]. Most integrins recognize several ECM proteins. Conversely, individual matrix proteins, such as fibronectin, laminins, collagens, and vitronectin bind to several integrins [90, 91].

The binding of ECM to integrins is regulated by integrin conformation which is determined by the activity inside the cell (inside-out signaling), while upon binding to the ECM molecule, integrin also changes its conformation and elicits signals that are transmitted into the cell (outside-in signaling) [92]. The best understood binding site for integrins is the Arg-Gly-Asp (RGD), which is also found in fibronectin, vitronectin, tenascin, and other ECM proteins. There are 24 types of integrins in humans, formed by the 18 different α-chains and 8 different β-chains, dimerized in different combinations [88]. Each integrin dimer has distinctive properties and functions. Moreover, because the same integrin molecule in different cell types can have different ligand-binding specificities, it is likely that additional cell-specific factors interact with integrins to modulate their binding activity [93]. One example of the variability in integrin expression is in the adult retinal ganglion cells. Cells growing on different ECM *in vitro* express several distinct combinations of integrins although they are activated and signaled through the focal adhesion kinase pathway [74].

We and the others have previously shown that integrin activation encourages glial cells or neurons to traverse inhibitory areas. Non-specific activation using manganese or specific beta-1 integrin activating antibody, we were able to promote the migration of Schwann cells over CSPG substrate *in vitro* and encourage axonal outgrowth of DRG neurons [94, 95]. Similarly, integrin activation by over-expressing an integrin mediator kindlin-1 encourages axonal extension on CSPG substrate in DRG neurons *in vitro* and the growth of DRG neurons passing the lesion site into the spinal cord *in vivo* [96].

In visual system, integrins are equally important in mediating the cellular response to ECM. Integrin activation using manganese or beta-1 integrin activating antibody enhances the attachment of retinal pigment epithelial cells (RPE) on pathological Bruch's membrane, which contains a high level of Tn-C, in aged macular degeneration [55]. In retina, we and the others have also shown that adult retinal neurons have the capacity to grow on multiple ECM substrates including collagen I, collagen IV, fibronectin and laminin with different affinity. This differential binding influences the degree of branching and elongation of neurites [75]. Moreover, we demonstrated that much of effects act through integrins activation.

3. ECM degradation

Matrix components are degraded by extracellular proteolytic enzymes (proteases) acting in the close proximity around the cells after secretion. Many of these proteases belong to two general classes—matrix metalloproteinases (e.g., MMPs and ADAMTSs) and serine proteases (e.g., trypsin, chymotrypsin, elastase) [97–99]. Matrix metalloproteinases represent the largest group with about 50 members identified in vertebrates. Their activity is depended on the binding of $Ca_2{}^+$ or $Zn_2{}^+$ ions [100, 101]. The second group of matrix degrading enzyme is the serine proteases, which have a highly reactive serine in their active site. Protease activity is generally confined to the cell surface by specific anchoring proteins, by membrane-associated activators, and by the production of specific protease inhibitors in regions where protease activity is not needed [102]. Their activity is important for the homeostasis and turnover of the ECM.

4. ECM implications in retinal pathologies

One of the most specialized forms of ECM is the basement membrane, a flexible, tough, and thin sheet of very well-organized components of the ECM. The functions of basement membranes are to act as platforms for cell adhesion, to provide structural support to a tissue, to divide tissues into compartments, and to regulate cell behavior including polarity. Although small in volume and very thin (typically 40–120 nm), it has a critical role in the architecture of the body [103, 104]. Although the precise composition of the mature basal lamina varies from tissue to tissue and even from region to region in the same lamina, it typically contains the glycoproteins, laminin, type IV collagen and nidogen (also called entactin), along with perlacan [105]. Other common basal lamina components are fibronectin and type XVIII collagen. Interactions of cells with basement membranes are mediated by trans-membrane cell surface receptors, which connect the cytoskeleton of the cell with the extracellular environment, leading to the formation of focal adhesions [88, 106].

The mature polarized retina is structurally and functionally supported by two basement membranes that act as boundaries for the neural retina (**Figure 2**). The two basement membranes are (i) the Bruch's membrane, at the interface of the RPE and the choroid and (ii) the inner limiting membrane (ILM) at the interface of the neural retina formed by the endfeet of Müller cells and the vitreous body [107]. Changes in the organization or composition of these basement membranes lead to various pathologies including diabetic retinopathy, age-related macular degeneration, proliferative vitreoretinopathy or retinal detachment [108–111].

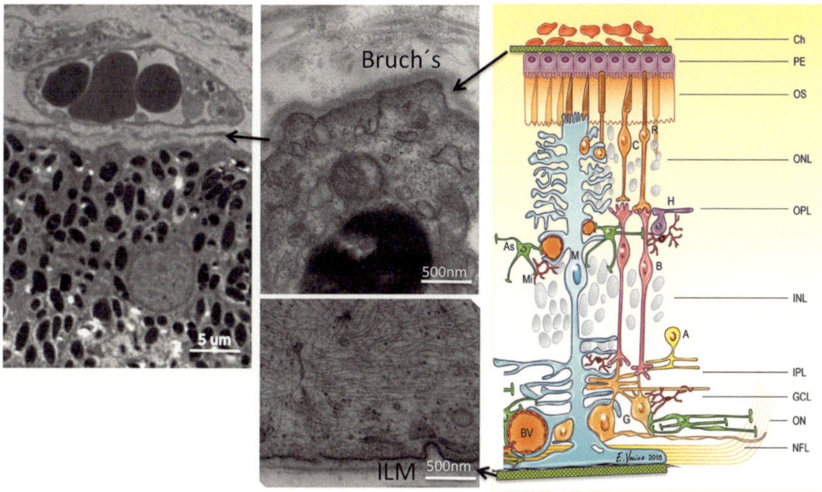

Figure 2. Schematic drawing of the cellular components of the retina, glia neurons. The different cell types are placed in the location of a standard large mammalian retina. Note the interactions between the cells and the blood vessels (BV). Amacrine cells (A), astrocytes in green (AS), bipolar cells (B), cones (C), ganglion cells (G), horizontal cells (H), Müller cells in blue (M), microglia in red (Mi), rods (R). Note the location of the different layers of the retina, from the most internal: optic nerve (ON), nerve fiber layer (NFL), ganglion cell layer (GCL), inner plexiform layer (IPL), inner nuclear layer (INL), outer plexiform layer (OPL), outer nuclear layer (ONL), outer segment layer (OS), pigment epithelium (PE), choroid (Ch). In green the two limiting membranes the Bruch's and the inner limiting membrane (ILM). In parallel to the drawing electron microscopy pictures of pig retinas showing both membranes.

The Brunch´s membrane is not only providing physical support for RPE, it also regulates RPE differentiation and acts as a barrier that prevents choroidal neovascularization, a process in which choroidal vascular cells inappropriately invade the retina [112, 113]. Alterations in the composition or organization of the Brunch´s membrane compromises the normal function of RPE cells and this disruption results in retinal pathologies including age-related macular degeneration [113].

The ILM lies on the vitreal side of the retina which is the opposite side of the retina from Bruch ´s membrane. The ILM constitutes the interface between the retina and the vitreous but is also responsible for organizing and maintaining the laminated structure of the retina and guiding astrocyte migration during vascular development [114]. Disruptions or changes in the ILM are associated with retinal dysplasia as well as retinal pathologies such as diabetic retinopathy, proliferative vitreo-retinopathy [114, 115].

5. Conclusion

This review aims to provide an overview of the major ECM partners and their functions in the CNS, PNS, and retina. As demonstrated above, ECM is not a passive by-stander present in the inter-cellular space, it actively takes part in controlling the penetration and diffusion of molecules in the extracellular space, sequestration and release of chemokines and growth factors to neurons, area where it should be permissive and inhibitory to specific population of neurons at a specific time. Dysregulation of their synthesis and production in pathological conditions such as spinal cord injury and glaucoma impede the regeneration in both systems. A better understanding of their spatial and temporal expression, molecular assembly and interaction, and production and degradation will be crucial to harness them for encouraging functional recovery in different pathological conditions.

Acknowledgements

Grupos Consolidados Gobierno Vasco GOBE (IT995-16)

Author details

Elena Vecino[1*] and Jessica C. F. Kwok[2]

*Address all correspondence to: elena.vecino@ehu.es

1 Department Cell Biology and Histology, University of the Basque Country UPV/EHU, Leioa, Vizcaya, Spain

2 School of Biomedical Sciences, Faculty of Biological Sciences, University of Leeds, Leeds, United Kingdom

References

[1] Cowman M, Schmidt T, Raghavan P, Stecco A. Viscoelastic properties of hyaluronan in physiological conditions Version 1. F1000Resarch. 2015; 4: 622.

[2] Franco SJ, Müller U. Extracellular matrix functions during neuronal migration and lamination in the mammalian central nervous system. Developmental Neurobiology. 2011;71(11):889–900.

[3] Franze K, Janmey PA, Guck J. Mechanics in neuronal development and repair. Annual Review of Biomedical Engineering. 2013;15:227–51.

[4] Carulli D, Pizzorusso T, Kwok JC, Putignano E, Poli A, Forostyak S, et al. Animals lacking link protein have attenuated perineuronal nets and persistent plasticity. Brain. 2010;133(Pt 8):2331–47.

[5] Morwood SR, Nicholson LB. Modulation of the immune response by extracellular matrix proteins. Archivum Immunologiae et Therapiae Experimentalis. 2006;54(6):367–74.

[6] Vecino E, Rodriguez FD, Ruzafa N, Pereiro X, Sharma SC. Glia–neuron interactions in the mammalian retina. Progress in Retinal and Eye Research. 2016;51:1-40.

[7] Asher RA, Morgenstern DA, Moon LDF, Fawcett JW. Chondroitin sulphate proteoglycans: Inhibitory components of the glial scar. Progress in Brain Research. Elsevier; 2001; 132:611–9.

[8] Faissner A, Pyka M, Geissler M, Sobik T, Frischknecht R, Gundelfinger ED, et al. Contributions of astrocytes to synapse formation and maturation—Potential functions of the perisynaptic extracellular matrix. Brain Research Reviews. 2010;63(1–2):26–38.

[9] Busch SA, Silver J. The role of extracellular matrix in CNS regeneration. Current Opinion in Neurobiology. 2007;17(1):120–7.

[10] Fawcett JW, Asher RA. The glial scar and central nervous system repair. Brain Research Bulletin. 1999;49(6):377–91.

[11] Rauch U. Brain matrix: Structure, turnover and necessity. Biochemical Society Transactions. 2007;35(4):656–60.

[12] Toole BP. Hyaluronan in morphogenesis. Seminars in Cell Developmental Biology. 2001;12(2):79–87.

[13] Margolis RK, Margolis RU. Nervous tissue proteoglycans. Experientia. 49(5):429–46.

[14] Ruoslahti E. Brain extracellular matrix. Glycobiology. 1996;6(5):489–92.

[15] Herndon ME, Lander AD. A diverse set of developmentally regulated proteoglycans is expressed in the rat central nervous system. Neuron. 1990;4(6):949–61.

[16] Kwok JC, Warren P, Fawcett JW. Chondroitin sulfate: A key molecule in the brain matrix. International Journal of Biochemistry & Cell Biology. 2012;44(4):582–6.

[17] Chiu AYM, Matthew WD, Patterson PH. A monoclonal antibody that blocks the activity of a neurite regeneration-promoting factor: Studies on the binding site and its localization in vivo. The Journal of Cell Biology. 1986;103(4):1383–98.

[18] Challacombe JF, Snow DM, Letourneau PC. Dynamic microtubule ends are required for growth cone turning to avoid an inhibitory guidance cue. The Journal of Neuroscience. 1997;17(9):3085–95.

[19] McKeon RJ, Höke A, Silver J. Injury-induced proteoglycans inhibit the potential for laminin-mediated axon growth on astrocytic scars. Experimental Neurology. 1995;136(1):32–43.

[20] Silver J, Miller JH. Regeneration beyond the glial scar. Nature Review Neuroscience. 2004;5(2):146–56.

[21] Kwok JC, Yuen YL, Lau WK, Zhang FX, Fawcett JW, Chan YS, et al. Chondroitin sulfates in the developing rat hindbrain confine commissural projections of vestibular nuclear neurons. Neural Development. 2012;7:6.

[22] Moon LDF, Asher RA, Rhodes KE, Fawcett JW. Regeneration of CNS axons back to their target following treatment of adult rat brain with chondroitinase ABC. Nature Neuroscience. 2001;4(5):465–6.

[23] Cafferty WB, Yang SH, Duffy PJ, Li S, Strittmatter SM. Functional axonal regeneration through astrocytic scar genetically modified to digest chondroitin sulfate proteoglycans. Journal of Neuroscience. 2007;27(9):2176–85.

[24] Garcia-Alias G, Barkhuysen S, Buckle M, Fawcett JW. Chondroitinase ABC treatment opens a window of opportunity for task-specific rehabilitation. Nature Neuroscience. 2009;12(9):1145–51.

[25] Zhao R-R, Muir EM, Alves JN, Rickman H, Allan AY, Kwok JC, et al. Lentiviral vectors express chondroitinase ABC in cortical projections and promote sprouting of injured corticospinal axons. Journal of Neuroscience Methods. 2011;201(1):228–38.

[26] James ND, Shea J, Muir EM, Verhaagen J, Schneider BL, Bradbury EJ. Chondroitinase gene therapy improves upper limb function following cervical contusion injury. Experimental Neurology. 2015;271:131–5.

[27] Mikami T, Yasunaga D, Kitagawa H. Contactin-1 is a functional receptor for neuroregulatory chondroitin sulfate-E. Journal of Biological Chemistry. 2009;284(7):4494–9.

[28] Shen Y, Tenney AP, Busch SA, Horn KP, Cuascut FX, Liu K, et al. PTP sigma is a receptor for chondroitin sulfate proteoglycan, an inhibitor of neural regeneration. Science. 2009;326(5952):592–6.

[29] Dickendesher TL, Baldwin KT, Mironova YA, Koriyama Y, Raiker SJ, Askew KL, et al. NgR1 and NgR3 are receptors for chondroitin sulfate proteoglycans. Nature Neuroscience. 2012;15(5):703–12.

[30] Lang BT, Cregg JM, DePaul MA, Tran AP, Xu K, Dyck SM, et al. Modulation of the proteoglycan receptor PTPsigma promotes recovery after spinal cord injury. Nature. 2015;518(7539):404–8.

[31] Fry EJ, Chagnon MJ, López-Vales R, Tremblay ML, David S. Corticospinal tract regeneration after spinal cord injury in receptor protein tyrosine phosphatase sigma deficient mice. Glia. 2010;58(4):423–33.

[32] Maeda N. Structural variation of chondroitin sulfate and its roles in the central nervous system. Central Nervous System Agents in Medicinal Chemistry. 2010;10(1):22–31.

[33] Kwok JC, Dick G, Wang D, Fawcett JW. Extracellular matrix and perineuronal nets in CNS repair. Developmental Neurobiology. 2011;71(11):1073–89.

[34] Kwok JCF, Tan CL, Wang D, Heller J, Fawcett JW. Chondroitin sulfates in axon regeneration and plasticity. Trends in Glycoscience and Glycotechnology. 2011;23(133): 201–11.

[35] Wang D, Ichiyama RM, Zhao R, Andrews MR, Fawcett JW. Chondroitinase combined with rehabilitation promotes recovery of forelimb function in rats with chronic spinal cord injury. Journal of Neuroscience. 2011;31(25):9332–44.

[36] Pizzorusso T, Medini P, Berardi N, Chierzi S, Fawcett JW, Maffei L. Reactivation of ocular dominance plasticity in the adult visual cortex. Science. 2002;298(5596):1248–51.

[37] Dityatev A, Bruckner G, Dityateva G, Grosche J, Kleene R, Schachner M. Activity-dependent formation and functions of chondroitin sulfate-rich extracellular matrix of perineuronal nets. Developmental Neurobiology. 2007;67(5):570–88.

[38] McRae PA, Rocco MM, Kelly G, Brumberg JC, Matthews RT. Sensory deprivation alters aggrecan and perineuronal net expression in the mouse barrel cortex. Journal of Neuroscience. 2007;27(20):5405–13.

[39] Pizzorusso T, Medini P, Landi S, Baldini S, Berardi N, Maffei L. Structural and functional recovery from early monocular deprivation in adult rats. Proceedings of the National Academy of Sciences USA. 2006;103(22):8517–22.

[40] Romberg C, Yang S, Melani R, Andrews MR, Horner AE, Spillantini MG, et al. Depletion of perineuronal nets enhances recognition memory and long-term depression in the perirhinal cortex. Journal of Neuroscience. 2013;33(16):7057–65.

[41] Dick G, Tan CL, Alves JN, Ehlert EM, Miller GM, Hsieh-Wilson LC, et al. Semaphorin 3A binds to the perineuronal nets via chondroitin sulfate type E motifs in rodent brains. Journal of Biological Chemistry. 2013;288(38):27384–95.

[42] Beurdeley M, Spatazza J, Lee HH, Sugiyama S, Bernard C, Di Nardo AA, et al. Otx2 binding to perineuronal nets persistently regulates plasticity in the mature visual cortex. Journal of Neuroscience. 2012;32(27):9429–37.

[43] Spatazza J, Lee HH, Di Nardo AA, Tibaldi L, Joliot A, Hensch TK, et al. Choroid-plexus-derived Otx2 homeoprotein constrains adult cortical plasticity. Cell Reports. 2013;3(6): 1815–23.

[44] Jongbloets BC, Pasterkamp RJ. Semaphorin signalling during development. Development. 2014;141(17):3292–7.

[45] Kolodkin AL, Matthes DJ, Goodman CS. The semaphorin genes encode a family of transmembrane and secreted growth cone guidance molecules. Cell. 1993;75(7):1389–99.

[46] De Winter F, Oudega M, Lankhorst AJ, Hamers FP, Blits B, Ruitenberg MJ, et al. Injury-induced class III semaphorin expression in the rat spinal cord. Experimental Neurology. 2002;175(1):61–75.

[47] Mecollari V, Nieuwenhuis B, Verhaagen J. A perspective on the role of class III semaphorin signaling in central nervous system trauma. Frontiers in Cellular Neuroscience. 2014;8:328.

[48] Fan J, Mansfield SG, Redmond T, Gordon-Weeks PR, Raper JA. The organization of F-actin and microtubules in growth cones exposed to a brain-derived collapsing factor. The Journal of Cell Biology. 1993;121(4):867–78.

[49] Anlar B, Gunel-Ozcan A. Tenascin-R: Role in the central nervous system. The International Journal of Biochemistry & Cell Biology. 2012;44(9):1385–9.

[50] Götz M, Bolz J, Joester A, Faissner A. Tenascin-C synthesis and influence on axonal growth during rat cortical development. European Journal of Neuroscience. 1997;9(3):496–506.

[51] Apostolova I, Irintchev A, Schachner M. Tenascin-R restricts posttraumatic remodeling of motoneuron innervation and functional recovery after spinal cord injury in adult mice. The Journal of Neuroscience. 2006;26(30):7849–59.

[52] Tang X, Davies JE, Davies SJA. Changes in distribution, cell associations, and protein expression levels of NG2, neurocan, phosphacan, brevican, versican V2, and tenascin-C during acute to chronic maturation of spinal cord scar tissue. Journal of Neuroscience Research. 2003;71(3):427–44.

[53] Laywell ED, Dörries U, Bartsch U, Faissner A, Schachner M, Steindler DA. Enhanced expression of the developmentally regulated extracellular matrix molecule tenascin following adult brain injury. Proceedings of the National Academy of Sciences of the United States of America. 1992;89(7):2634–8.

[54] Pindzola RR, Doller C, Silver J. Putative inhibitory extracellular matrix molecules at the dorsal root entry zone of the spinal cord during development and after root and sciatic nerve lesions. Developmental Biology. 1993;156(1):34–48.

[55] Afshari FT, Kwok JC, Andrews MR, Blits B, Martin KR, Faissner A, et al. Integrin activation or alpha 9 expression allows retinal pigmented epithelial cell adhesion on Bruch's membrane in wet age-related macular degeneration. Brain. 2010;133(Pt 2):448–64.

[56] Eva R, Fawcett J. Integrin signalling and traffic during axon growth and regeneration. Current Opinion in Neurobiology. 2014;27:179–85.

[57] Becker T, Anliker B, Becker CG, Taylor J, Schachner M, Meyer RL, et al. Tenascin-R inhibits regrowth of optic fibers in vitro and persists in the optic nerve of mice after injury. Glia. 2000;29(4):330–46.

[58] Galtrey CM, Kwok JC, Carulli D, Rhodes KE, Fawcett JW. Distribution and synthesis of extracellular matrix proteoglycans, hyaluronan, link proteins and tenascin-R in the rat spinal cord. European Journal of Neuroscience. 2008;27(6):1373–90.

[59] Morawski M, Dityatev A, Hartlage-Rübsamen M, Blosa M, Holzer M, Flach K, et al. Tenascin-R promotes assembly of the extracellular matrix of perineuronal nets via clustering of aggrecan. Philosophical Transactions of the Royal Society of London B: Biological Sciences. 2014;369(1654):20140046.

[60] Bruckner G, Morawski M, Arendt T. Aggrecan-based extracellular matrix is an integral part of the human basal ganglia circuit. Neuroscience. 2008;151(2):489–504.

[61] Weber P, Bartsch U, Rasband MN, Czaniera R, Lang Y, Bluethmann H, et al. Mice deficient for tenascin-R display alterations of the extracellular matrix and decreased axonal conduction velocities in the CNS. The Journal of Neuroscience. 1999;19(11): 4245–62.

[62] Yamada M, Sekiguchi K. Molecular basis of laminin–integrin interactions (Chapter 6). In: Jeffrey HM, editor. Current topics in membranes. Volume 76: Academic Press, Elsevier; 2015. pp. 197–229.

[63] Liu YB, Tewari A, Salameh J, Arystarkhova E, Hampton TG, Brashear A, et al. A dystonia-like movement disorder with brain and spinal neuronal defects is caused by mutation of the mouse laminin β1 subunit, Lamb1. eLife. 2016;4: e11102

[64] Maselli RA, Arredondo J, Ferns MJ, Wollmann RL. Synaptic basal lamina–associated congenital myasthenic syndromes. Annals of the New York Academy of Sciences. 2012;1275(1):36–48.

[65] Durbeej M. Laminins. Cell and Tissue Research. 2009;339(1):259–68.

[66] Yamada H, Chiba A, Endo T, Kobata A, Anderson LVB, Hori H, et al. Characterization of dystroglycan–laminin interaction in peripheral nerve. Journal of Neurochemistry. 1996;66(4):1518–24.

[67] Masaki T, Matsumura K. Biological role of dystroglycan in Schwann cell function and its implications in peripheral nervous system diseases. Journal of Biomedicine and Biotechnology. 2010;2010:17.

[68] Patton BL, Chiu AY, Sanes JR. Synaptic laminin prevents glial entry into the synaptic cleft. Nature. 1998;393(6686):698–701.

[69] Ichikawa N, Iwabuchi K, Kurihara H, Ishii K, Kobayashi T, Sasaki T, et al. Binding of laminin-1 to monosialoganglioside GM1 in lipid rafts is crucial for neurite outgrowth. Journal of Cell Science. 2009;122(2):289–99.

[70] Tian M, Jacobson C, Gee SH, Campbell KP, Carbonetto S, Jucker M. Dystroglycan in the cerebellum is a laminin α2-chain binding protein at the glial–vascular interface and is expressed in Purkinje cells. European Journal of Neuroscience. 1996;8(12):2739–47.

[71] Liesi P. Laminin-immunoreactive glia distinguish regenerative adult CNS systems from non-regenerative ones. The EMBO Journal. 1985;4(10):2505–11.

[72] Hagg T, Portera-Cailliau C, Jucker M, Engvall E. Laminins of the adult mammalian CNS; laminin-α2 (merosin M-) chain immunoreactivity is associated with neuronal processes. Brain Research. 1997;764(1–2):17–27.

[73] Bates CA, Meyer RL. The neurite-promoting effect of laminin is mediated by different mechanisms in embryonic and adult regenerating mouse optic axonsin vitro. Developmental Biology. 1997;181(1):91–101.

[74] Vecino E, Heller JP, Veiga-Crespo P, Martin KR, Fawcett JW. Influence of extracellular matrix components on the expression of integrins and regeneration of adult retinal ganglion cells. Plos One. 2015;10(5):e0125250.

[75] Heller JP, Kwok JC-F, Vecino E, Martin KR, Fawcett JW. A method for the isolation and culture of adult rat retinal pigment epithelial (RPE) cells to study retinal diseases. Frontiers in Cellular Neuroscience. 2015;9: 449

[76] Cohen J, Burne JF, Winter J, Bartlett P. Retinal ganglion cells lose response to laminin with maturation. Nature. 1986;322(6078):465–7.

[77] Di Lullo GA, Sweeney SM, Körkkö J, Ala-Kokko L, San Antonio JD. Mapping the ligand-binding sites and disease-associated mutations on the most abundant protein in the human, type I collagen. Journal of Biological Chemistry. 2002;277(6):4223–31.

[78] Koopmans G, Hasse B, Sinis N. The role of collagen in peripheral nerve repair (Chapter 19). International Review of Neurobiology. Volume 87: Academic Press, Elsevier; 2009. pp. 363–79.

[79] Gordon MK, Hahn RA. Collagens. Cell Tissue Research. 2010;339(1):247–57.

[80] Hermanns S, Klapka N, Müller HW. The collagenous lesion scar—an obstacle for axonal regeneration in brain and spinal cord injury. Restorative Neurology and Neuroscience. 2001;19(1):139–48.

[81] Gao X, Wang Y, Chen J, Peng J. The role of peripheral nerve ECM components in the tissue engineering nerve construction. Reviews in the Neurosciences. 2013; 24(4):443-53

[82] Campbell IC, Coudrillier B, Ross Ethier C. Biomechanics of the posterior eye: A critical role in health and disease. Journal of Biomechanical Engineering. 2014;136(2):021005.

[83] Vecino E, GaldosmM BA, Rodríguez FD, Micó C, Sharma SC. Elevated intraocular pressure induces ultrastructural changes in the trabecular meshwork. Journal of Cytology & Histology. 2015;S3:007.

[84] Sawaguchi S, Yue BYJT, Fukuchi T, Abe H, Suda K, Kaiya T, et al. Collagen fibrillar network in the optic nerve head of normal monkey eyes and monkey eyes with laser-induced glaucoma—A scanning electron microscopic study. Current Eye Research. 1999;18(2):143–9.

[85] Hernandez MR. The optic nerve head in glaucoma: Role of astrocytes in tissue remodeling. Progress in Retinal and Eye Research. 2000;19(3):297–321.

[86] Watson PG, Young RD. Scleral structure, organisation and disease: A review. Experimental Eye Research. 2004;78(3):609–23.

[87] Pijanka JK, Coudrillier B, Ziegler K, Sorensen T, Meek KM, Nguyen TD, et al. Quantitative mapping of collagen fiber orientation in non-glaucoma and glaucoma posterior human scleraefiber orientation in posterior human sclera. Investigative Ophthalmology & Visual Science. 2012;53(9):5258–70.

[88] Hynes RO. Integrins: A family of cell surface receptors. Cell. 1987;48(4):549–54.

[89] Tawil NJ, Wilson P, Carbonetto S. Expression and distribution of functional intergrins in rat CNS glia. Journal of Neuroscience Research. 1994;39(4):436–47.

[90] Roca-Cusachs P, Iskratsch T, Sheetz MP. Finding the weakest link–exploring integrin-mediated mechanical molecular pathways. Journal of Cell Science. 2012;125(13):3025–38.

[91] Wierzbicka-Patynowski I, Schwarzbauer JE. The ins and outs of fibronectin matrix assembly. Journal of Cell Science. 2003;116(16):3269–76.

[92] Giancotti FG, Ruoslahti E. Integrin signaling. Science. 1999;285(5430):1028–33.

[93] Carson AE, Barker TH. Emerging concepts in engineering extracellular matrix variants for directing cell phenotype. Regenerative Medicine. 2009;4(4):593–600.

[94] Afshari FT, Kwok JC, White L, Fawcett JW. Schwann cell migration is integrin-dependent and inhibited by astrocyte-produced aggrecan. Glia. 2010;58(7):857–69.

[95] Tan CL, Kwok JC, Patani R, Ffrench-Constant C, Chandran S, Fawcett JW. Integrin activation promotes axon growth on inhibitory chondroitin sulfate proteoglycans by enhancing integrin signaling. Journal of Neuroscience. 2011;31(17):6289–95.

[96] Tan CL, Andrews MR, Kwok JCF, Heintz TGP, Gumy LF, Fässler R, et al. Kindlin-1 enhances axon growth on inhibitory chondroitin sulfate proteoglycans and promotes sensory axon regeneration. The Journal of Neuroscience. 2012;32(21):7325–35.

[97] Stawarski M, Stefaniuk M, Wlodarczyk JM. Matrix metalloproteinase-9 involvement in the structural plasticity of dendritic spines. Frontiers in Neuroanatomy. 2014;8:68.

[98] Nita M, Strzałka-Mrozik B, Grzybowski A, Mazurek U, Romaniuk W. Age-related macular degeneration and changes in the extracellular matrix. Medical Science Monitor: International Medical Journal of Experimental and Clinical Research. 2014;20:1003–16.

[99] Wang Y, Luo W, Reiser G. Trypsin and trypsin-like proteases in the brain: Proteolysis and cellular functions. Cellular and Molecular Life Sciences. 2008;65(2):237–52.

[100] Yang Y, Rosenberg GA. Matrix metalloproteinases as therapeutic targets for stroke. Brain Research. 2015;1623:30–8.

[101] Wang W-J, Yu X-H, Wang C, Yang W, He W-S, Zhang S-J, et al. MMPs and ADAMTSs in intervertebral disc degeneration. Clinica Chimica Acta. 2015;448:238–46.

[102] Alberts B, Johnson A, Lewis J, Morgan D, Raff M, Roberts K, Walter P. Cell junctions and the extracellular matrix. Molecular Biology of the Cell. 6th edition: Garland Science; 2015.

[103] Osidak MS, Osidak EO, Akhmanova MA, Domogatsky SP, Domogatskaya AS. Fibrillar, fibril-associated and basement membrane collagens of the arterial wall: Architecture, elasticity and remodeling under stress. Current Pharmaceutical Design. 2015;21(9):1124–33.

[104] Paulsson MM. Basement membrane proteins: Structure, assembly, and cellular interactions. Critical Reviews in Biochemistry and Molecular Biology. 1992;27(1–2):93–127.

[105] Halfter W, Oertle P, Monnier CA, Camenzind L, Reyes-Lua M, Hu H, et al. New concepts in basement membrane biology. FEBS Journal. 2015;282(23):4466–79.

[106] Tan CL, Kwok JCF, Heller JPD, Zhao R, Eva R, Fawcett JW. Full length talin stimulates integrin activation and axon regeneration. Molecular and Cellular Neuroscience. 2015;68:1–8.

[107] Pichi F, Lembo A, Morara M, Veronese C, Alkabes M, Nucci P, et al. Early and late inner retinal changes after inner limiting membrane peeling. International Ophthalmology. 2013;34(2):437–46.

[108] Tang J, Mohr S, Du YD, Kern TS. Non-uniform distribution of lesions and biochemical abnormalities within the retina of diabetic humans. Current Eye Research. 2003;27(1): 7–13.

[109] Booij JC, Baas DC, Beisekeeva J, Gorgels TGMF, Bergen AAB. The dynamic nature of Bruch's membrane. Progress in Retinal and Eye Research. 2010;29(1):1–18.

[110] Francke M, Weick M, Pannicke T, Uckermann O, Grosche J, Goczalik I, et al. Upregulation of extracellular ATP-induced Müller cell responses in a dispase model of proliferative vitreoretinopathy. Investigative Ophthalmology & Visual Science. 2002;43(3):870–81.

[111] Lewis GP, Fisher SK. Müller cell outgrowth after retinal detachment: Association with cone photoreceptors. Investigative Ophthalmology & Visual Science. 2000;41(6):1542–5.

[112] Mukai R, Sato T, Kishi S. A hyporeflective space between hyperreflective materials in pigment epithelial detachment and Bruch's membrane in neovascular age-related macular degeneration. BMC Ophthalmology. 2014;14(1):1–8.

[113] Lee JH, Lee WK. Choroidal neovascularization associated with focal choroidal excavation. American Journal of Ophthalmology. 2014;157(3):710–8.e1.

[114] Edwards MM, Mammadova-Bach E, Alpy F, Klein A, Hicks WL, Roux M, et al. Mutations in Lama1 disrupt retinal vascular development and inner limiting membrane formation. The Journal of Biological Chemistry. 2010;285(10):7697–711.

[115] Romano MR, Romano V, Vallejo-Garcia JL, Vinciguerra R, Romano M, Cereda M, et al. Macular hypotrophy after internal limiting membrane removal for diabetic macular edema. Retina. 2014;34(6):1182–9.

Chapter 6

The Mosaic of Extracellular Matrix in the Central Nervous System as a Determinant of Glial Heterogeneity

Cory M. Willis and Stephen J. Crocker

Additional information is available at the end of the chapter

http://dx.doi.org/10.5772/62706

Abstract

Accumulating evidence points to a primary role for non-myelinating glia as principal mediators of homeostasis in the central nervous system (CNS). However, the origins of the basis for glial heterogeneity are not well understood. Our recent studies contribute to an emerging view that the extracellular matrix (ECM) provides clues to glia underling their specialized functions and, more importantly, the nature of how glia change in relation to neuropathology. In this review, we discuss how the dynamic mosaic of CNS ECM impacting CNS health and disease. Specifically, we focus on the roles of select extracellular matrix proteins, namely fibronectin (Fn), vitronectin (Vn), laminin (Ln) and tenascin-c (TnC), as prototypes for how ECM can modulate glial functions. We discuss the differences in expression patterns in the developing and adult CNS and relate these ECM molecules to specific changes in glial functions in neurological diseases. We also discuss how experiments have revealed the role of ECM molecules' influence on CNS development and the response of glia to injury and inflammation. We provide a new model to explain the nature of glial diversity as an adaptive response to the extracellular milieu, and provide a different approach to understand the complex nature of glia heterogeneity.

Keywords: astrocyte, fibronectin, tenascin-c, laminin, vitronectin

1. Introduction

Tissues are not made up solely of cells. A substantial part of Tissue volume is extracellular space, which is largely filled by an intricate network of macromolecules constituting the extracellular matrix (ECM). The vertebrate extracellular matrix was once thought to serve mainly as a relatively inert scaffold to stabilize the physical structure of tissues. But It is now clear that the matrix has a far more active and complex role in regulating the behavior of the cells that contact it. Throughout the body, the ECM provides structure and organization to tissues through an intricately arranged scaffold comprised of a variety of secreted proteins and complex polysaccharides that are secreted locally and assembled into an organized meshwork in close association with the surface of the cells that produced them. Variations in both the relative amounts of the different types of matrix macromolecules and the way in which they are organized in the extracellular matrix give rise to an amazing diversity of forms, each adapted to the functional requirements of the particular tissue that influences their survival, development, migration, proliferation, shape, intercellular communication, and function. The extracellular matrix has a correspondingly complex molecular composition. Although our understanding of its organization is still incomplete, there has been rapid progress in characterizing many of its major components.

In this chapter, we will focus on a select group of ECM proteins—tenascin-C, fibronectin, vitronectin, and laminin—and their patterns of expression and influence on the response and function of glia in the developing and adult central nervous system (CNS). We will then provide a detailed discussion on the differences in the patterns of expression of these factors to specific changes observed in the context of neurological diseases using studies that have pioneered this new approach to understanding the contributions of glia to injury and inflammation. About 20% of the total volume of the adult CNS is extracellular space [1, 2] that contains

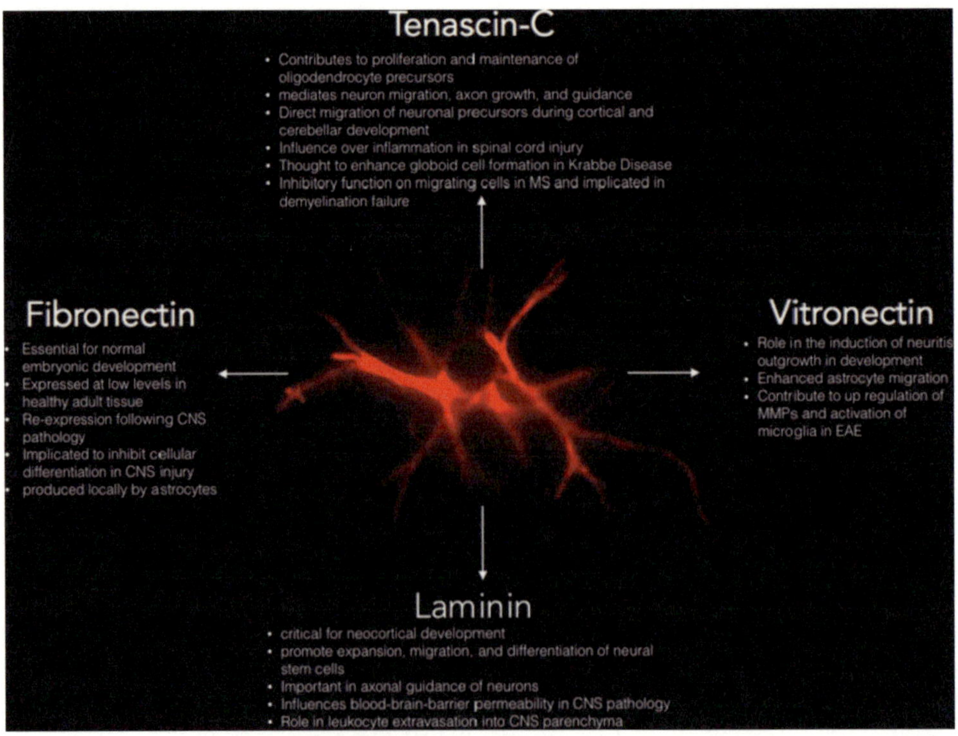

Figure 1. Schematic summary depicting the diverse impacts of select ECM proteins on astrocytes. Image shown is of a murine glial fibrillary acidic protein (GFAP+) astrocyte (red; center).

highly organized ECM [3]. As in peripheral tissues, the ECM is composed of both interstitial and basement membrane proteins of the ECM family; however, in the CNS, the ECM composition is remarkably different. Whereas the interstitial ECM of most peripheral tissues is enriched in collagen, laminin, and fibronectin, the ECM of the adult CNS is primarily a loose meshwork of hyaluronan, sulfated proteoglycans, and tenascin-R [4, 5]. The significance of these ECM proteins in the adult CNS has been extensively considered in recent reviews [6–8] and is beyond the focus of this chapter. Instead, we will focus on the Aforementioned ECM proteins and their significance for astrocyte function (**Figure 1**).

1.1. Fibronectin

Fibronectin (Fn) is a high-molecular weight, insoluble glycoprotein dimer that consists of three types of repeating amino acid modules, named type I, type II, and type III [9]. The structure of Fn varies, depending on whether it is secreted into the plasma or synthesized by resident cells. The majority of plasma Fn is produced by hepatocytes and is detectable in human blood at a concentration of 300 μg/ml [10, 11]. In contrast, cellular Fn contains the alternatively spliced extra domain A and/or extra domain B (the nomenclature for humans; for rodents: EIIA and EIIIB). In addition, Fn has been shown to be a critical component in other ECM proteins, including heparin, collagen, and fibrin, and together these protein networks contribute to the formation of the ECM [9]. One of the main functions of Fn is to serve as a scaffold for cell adhesion and migration, which influences the regulation of cell proliferation and differentiation [9]. A myriad of small proteins, such as growth factors, have been found to support these functions of Fn when they accumulate in the Fn network. As a result, local concentrations of these small proteins are seen to increase. The Fn matrix has been found to be essential for normal embryonic development by studying Fn-knockout mice [12]. In healthy adult tissue, Fn is expressed at low levels. Transient Fn re-expression either through plasma leakage and/or synthesis from resident cells is a common "default" response to tissue injury, ranging from skin wounds to joint inflammation [13] and myelin degradation [14]. In the CNS, myelin damage (demyelination) elicits the production of a temporary Fn matrix [14–18]. In this injury scenario, the Fn matrix is a result of plasma leaking into the CNS parenchyma [15, 16], and cellular Fn is secreted by resident astrocytes, microglia, and endothelial cells [14]. The generation of a temporary Fn scaffold comprised of both plasma and cellular Fn is a common response to tissue injury. We will discuss Fn re-expression during glial scar formation in multiple sclerosis (MS) and how clearance of the temporary Fn matrix is disturbed, which results in incomplete remyelination.

1.2. Tenascin-C

Tenascin-C (TnC) is a glycoprotein that is expressed in the ECM of various tissues where it has been found to regulate processes such as cell growth, migration, and adhesion during development, and represents 25% of the class of proteins that form the basic constituents of the brain ECM [19–22]. Tenascins are very large multimeric glycoproteins whose structure is well-conserved among vertebrates. TnC is built up in a modular fashion and consists of a cysteine-rich amino-terminus, EGF-like domains followed by fibronectin type III domains, and a carboxyterminal domain resembling fibrinogen-b [23]. Tenascin-C binds and interacts with a wealth of extracellular matrix and cell surface ligands [20], which is heavily mediated by Fn type III modules. Integrins, cell-surface heparin sulfate proteoglycans, and cell adhesion molecules of the immunoglobulin superfamily have been found to be the major cellular receptors of TnC [24]. TnC commonly binds to other ECM proteins, such as fibronectin, phosphacan, and, particularly, leticans. During CNS development, TnC is first expressed and accumulates around the fibrous processes of radial and Bergmann glial cells, which direct the migration of neuronal precursors during cortical and cerebellar development, respectively [25, 26]. During the later stages of development, TnC is expressed primarily by astrocytes, where it is thought to exert autocrine effects that regulate the proliferation of astrocyte progenitor cells [27]. TnC has also been found to modulate the stem cell compartment in the subventricular niche, where it is specifically enriched in the environment of mouse neural stem cell precursor cells (NSPCs) at embryonic day E14–E15 [28]. For example, TnC has been found to contribute heavily to the maturation of NSPCs [29], as well as the proliferation and maintenance of oligodendrocyte precursor cells [30–32]. *In vivo* and *in vitro* studies have demonstrated that TnC encodes for both permissive and inhibitory cues, which mediate neuron migration and axon growth and guidance by way of neuron-glia interactions [33–39]. Two to three weeks after birth, TnC expression decreases continuously, maintaining only a significant expression level in the neurogenetically active areas of the adult brain that encompass the subependymal zone and the hippocampus, as well as regions of plasticity in the hypothalamus [40–44].

1.3. Laminin

Laminins are major components of the basal lamina [45] and are also present in the ventricular zone (VZ) of the developing neocortex [46, 47]. Additionally, laminins were one of the first ECM proteins to be implicated in nervous system development as they were found to promote neurite outgrowth in an integrin-dependent manner [48–52]. During development, the extracellular matrix forms a basal lamina (BL) surrounding the brain and blood vessels throughout the CNS [53, 54]. In the neocortex, the BL at the pial surface is contacted by the end-feet of radial glial cells. A number of studies have shown how crucial the pial BL is for neocortical development. Removal of the BL leads to the detachment of radial glial cell fibers, which affects radial glial cell survival and proper cortical lamination [55–58]. Laminins have also been shown to promote the expansion, migration, and differentiation of neural stem cells (NSCs) *in vitro* [46, 59–67]. Mice lacking laminin $\alpha 1$ die during embryogenesis [68]; mice bearing a mutation that only affects the laminin $\alpha 1$ nidogen-binding domain survive until birth and display disruptions of the pial basal lamina as well as neuronal ectopias [69]. Additionally, laminin $\alpha 1$inactivation in a subset of cortical neurons has been observed to cause cortical lamination defects [70]. However, defects in the maintenance and/or differentiation of NSCs has not been reported in these mutants. *In vivo* evidence for a role of laminins in controlling NSC behavior comes from studies of their dystroglycan and integrin receptors. In human patients, mutations within enzymes that glycosylate dystroglycan have been shown to produce cortical neuronal ectopias [71, 72]. Mice lacking dystroglycan in the CNS or bearing mutations in dystroglycan glycosyltransferase display BL disruptions and neuronal migration defects [73–75]. Laminins have also been found to play a role in axonal guidance *in vivo* [76].

In mice, laminin $\alpha 1$ deficiency results in the abnormal branching of myelinated axons from the corpus callosum [70]. These mutants also show abnormal neuronal migration, impaired activation of integrin downstream effectors, such as focal adhesion kinase and paxillin, and disrupted AKT/GSK-3 signaling, which has been implicated in neurite outgrowth [77]. The exact mechanisms underlying these abnormalities remain unknown. In the CNS, oligodendrocytes derive mainly from precursors residing in the ventral VZ and ganglionic eminences. They proliferate and migrate before becoming mature, myelinating cells [78]. Oligodendrocytes do not have a basal lamina, although there exists some evidence that developing oligodendrocyte precursor cells can secrete low levels of laminin [79], which suggests oligodendrocytes may interact with outside sources of laminin. Oligodendrocytes myelinate axons through extending multiple cell processes capable of ensheathing numerous axons [80, 81]. Expression of laminins during development correlates with the onset of CNS myelination [80, 82], and varied degrees of defects have been found in white matter tracts of patients suffering from congenital muscle dystrophy [83, 84]. Mice lacking laminin $\alpha 2$ have a developmental delay in oligodendrocyte maturation, resulting in hypomyelination [85, 86]. The degree of developmental delay is region-specific, which may reflect different laminin $\alpha 2$ requirements [86].

1.4. Vitronectin

Vitronectin (Vn) is a multifunctional plasma and ECM glycoprotein with multiple domains for interactions with plasma proteins like thrombin, anti-thrombin III, and plasminogen activator inhibitor-1 [87]. Vn is primarily synthesized in the liver [88] and has affinity for different integrins expressed on T-cells, platelets, endothelial cells, and macrophages. Comparatively little is known about the expression patterns of Vn during CNS development; however, a role for Vn in the induction of neurite outgrowth has been shown [89, 90]. In the normal adult CNS, vitronectin is localized mostly to blood vessels, with the exception of capillaries, suggesting that small amounts of vitronectin can be deposited in the CNS under normal conditions [91].

2. Matrix influence on astrocyte differentiation in CNS development

Astrocytes are specialized glial cells that are the major cell Component of the adult CNS, outnumbering neurons by over fivefold and Comprising roughly 20–40% of all glia [92]. They extend numerous processes that locally contact the surrounding neurons, other glial cells, and endothelial cells. Besides their pure barrier function, they also play a vital role in the control of cerebral blood flow and glucose homeostasis in the brain [92]. During the early development of the CNS, the overall expression of ECM molecules is relatively low and subsequently increases toward the end of embyrogenesis and during postnatal development [93, 94]. Despite their prominent expression during neural development, little is known about the functional importance of specific ECM molecules for astrocyte development. While the importance of how chondroitin sulfate proteoglycans (CSPGs) influence the differentiation of NPCs to astrocytes during embryonic development cannot be overlooked, for the purposes of this book chapter we will instead focus on the known role of tenascin-c in NPC differentiation into astrocytes. The tenascin gene family has recently gained increased attention with regard to glial development Owing to their late embryonic and early postnatal expression. Karus et al. have recently shown that TnC is capable of regulating the maturation of astrocytes during embryonic spinal cord development, primarily by orchestrating the responsiveness of NPCs to growth factors [27]. Within the developing brain and spinal cord, NPCs have been observed to initially generate neuronal cells. However, changes in the expression patterns of growth factor receptors result in the specification of astroglial cells. The expression of the epidermal growth factor receptor has been shown to be critical for normal astrocyte development [95]. During early embryonic stages, NPCs expressing Nestin, brain lipid binding protein (BLBP), and fibroblast growth factor receptors (FGFR) primarily generate neurons. Upon sustained fibroblast growth factor (FGF) signaling, these NPCs gain responsiveness towards epidermal growth factor (EGF). The expression of the EGF receptor is stimulated by TnC [27]. As a result, a rapid decline in neuron generation is observed in the embryonic spinal cord. Regardless of their location along the rostro-caudal axis of the developing spinal cord, the NPCs begin to express shared molecular markers with astrocytes, such as glutamate aspartate transporter (GLAST) and TnC [96]. Additionally, these cells begin to express additional markers such as S100β, aquaporin-4 [97], and fibroblast growth factor receptor 3 (FGFR3). Subsequently, the NPCs differentiate into GFAP-positive mature astrocytes, which are then often classified into fibrous white matter and gray matter astrocytes. Moreover, CSPGs and potentially TnC are involved in the maturation toward GFAP-positive astrocytes [98].

Astrocytes have been shown to play a prominent role in the developing central nervous system. Astrocytes contribute significantly in coordinating neuronal migration, axon guidance, and synapse formation [92]. This coordination is directed through deposition of specific extracellular matrix protein in the developing CNS—namely fibronectin and laminin. In an early report, Stewart and Pearlman observed fibronectin-like staining in the developing mouse cerebral cortex [99]. The temporal and spatial expression of fibronectin led them to posit that the transient appearance of fibronectin-like immunostaining in the zones that contain early cortical afferents suggests a role for Fn in forming the migratory pathway for the growth cones of these axons. In this role, it may be acting in concert with other extracellular matrix components such as hyaluronectin [100], glycosaminoglycans [101, 102], and laminin [103], which have been shown to have similar spatial distributions. The decline of fibronectin-like immunostaining that occurs as cortical development progresses may be a part of the change from the immature state, which supports profuse axon elongation in the CNS, to the mature state in which neurite outgrowth is quite limited. In addition to fibronectin deposition, astrocytes produce and secrete laminin, a key extracellular matrix guidance molecule in the developing brain. Laminin is synthesized and secreted by astrocytes both *in vivo* [103–107] and *in vitro* [108–112]. Astrocytic laminin is deposited into the ECM and fixed on the cell surface through binding to specific transmembrane receptors known as integrins [113–115]. The regionalization of laminin on the astrocyte surface is determined by the clustering of integrins, which are bound to the microfilaments, into macromolecular complexes known as focal contacts [116, 117]. It is this organization of laminin into specific patterns on the cell surface that provides directional cues to the elongating neurite [118–120]. Similar to fibronectin and TnC, the

http://dx.doi.org/10.5772/62706

expression of laminin in the brain parenchyma is developmentally regulated and coincides with neuronal migration [119]. Once the wiring of the brain network is firmly established, laminin disappears from the brain parenchyma and is restricted to the basal lamina of the vasculature.

3. ECM regulation of astrocytes

Once established, the composition of the mature extracellular matrix is rather stable with little or no turnover of their components [24]. This stability is lost, however, when tissue damage results from injury, inflammation, or disease. Extracellular matrix degradation is triggered through inflammation, which results in changes to the matrix composition. Under these circumstances, the expression of various extracellular matrix molecules is highly up-regulated and major depositions are observed often marking lesion sites, in particular, in association with glial scar tissue formation. The freshly produced ECM components may be secreted by reactive astrocytes, oligodendrocyte precursors, microglia/macrophages, and eventually by meningeal cells. The lesion and consequent reactive processes induce a matrix accumulation that strongly resembles the "juvenile-type" of meshwork previously observed during early nervous system development. In many CNS diseases, it is becoming increasingly clear that some ECM molecules are aberrantly expressed and others cleaved into bioactive fragments known as damage-associated molecular patterns (DAMPs) or "alarmins" [121]. Through their ability to bind to different types of pattern recognition receptors (PRRs), these ECM molecules can influence the phenotype and magnitude of inflammation. Moreover, the enzymes and inflammatory mediators released by immune cells further degrade or alter the composition of the ECM. For the purposes of this book chapter, we will focus on the role of astrocytes in CNS injury and disease and how the extracellular matrix influences their response. We will highlight how the extracellular matrix proteins mentioned in the introduction could have profound effects on CNS injury and disease recovery by discussing their known roles.

3.1. Tenascin-c influences on astrocytes in diseases of the CNS

Tenascin-c can act as a DAMP, eliciting activation of innate immune cells via binding to a TLR-4 [122]. This was first demonstrated in a model of arthritis where inflammatory disease symptoms in TnC KO mice resolved rapidly; conversely, TnC injection elicited joint inflammation. TLR-4 stimulation up-regulate TnC in macrophages so tenascin-c can act in an autocrine loop to amplify acute inflammation [122]. Although acute TnC expression is required for proper would healing [123], persistent expression can be detrimental; TnC is up-regulated in mice with Alzheimer's disease, and its deletion reduces neuropathology and inflammation [124]. TnC is an important factor in propagating chronic inflammation and could act in a similar manner after spinal cord injury.

After spinal cord injury, *de novo* synthesis of TnC occurs around the site within three days. Expression of TnC has been shown to persist for at least 30 days post injury in this model [125]. TnC is expressed by astrocytes in the lesion border, within the dorsal columns, and within the lesion epicenter. Interestingly, astrocytes cultured on TnC express fewer scar-related markers and proliferate less than astrocytes grown on control substrates [126], implying that TnC may restrict astrogliosis and scar formation after spinal cord injury. Additionally, *in vivo* work on spinal cord injury in tenascin-c knockout (KO) mice have shown that spontaneous recovery of locomotor functions after spinal injury is impaired in these animals when compared to wild-type mice. The impaired recovery was associated with attenuated excitability of the plantar Hoffmann reflex, reduced glutamatergic input, reduced sprouting of monaminergic axons in the lumbar spinal cord, and enhanced post-traumatic degeneration of corticospinal axons [127]. In a follow-up study using a model of lumbar spinal cord hemisection injury, global deletion of TnC was associated with enhanced axonal plasticity and growth into the lesion site. While these recent reports provide contrarian views to the role of TnC in the injured spinal cord, the precise mechanisms responsible for these changes have not been determined. In their review on extracellular matrix regulation in the healthy and injured spinal cord, Gaudet and Popovich suggest performing complementary gain-of-function experiments in wild-type mice and analyses of specific cellular and molecular pathways (e.g., inflammation) in tenascin-c KO mice [121]. Clearly the authors the state, consistent up-regulation of TnC after injury and its ability to bind/activate TLRs suggest that it is a candidate for controlling inflammation after spinal cord injury [121]. Further research will need to be performed in order to tease apart the role of this integral ECM protein in spinal cord injury.

TnC has also been implicated in globoid cell leukodystrophy (GLD), also known as Krabbe disease. GLD is a rare and often-fatal demyelinating disease caused by mutations in the galactocerebrosidase (*galc*) gene that results in the accumulation of galactosylsphingosine ("psychosine") [128]. Aberrant deposition of the extracellular matrix protein TnC has been observed in the brains of GLD patients when compared to age-matched control subjects. Elevated deposition and expression of TnC have also been observed in brain tissues from twitcher mice, an authentic mouse model of GLD [129]. The elevated TnC levels have been implicated in enhancing astrocyte responses to psychosine and astrocytic production of matrix metalloproteinase (MMP)-3, which activates microglial responses, inducing the formation of "globoid-like" cells in culture [129, 130]. This dysregulation of astrocytes, in part mediated by altered ECM, is thought to enhance the demyelination seen in this disease [129].

Expression of TnC is also aberrant in multiple sclerosis (MS). This chronic inflammatory and CNS demyelinating disease involves autoimmunity directed against myelin. A neuropathological hallmark of MS is glial scarring, formed by reactive astrocytes. Multiple sclerosis lesions can be broadly defined as inactive, chronic active, and chronic. Inactive MS lesions and the center of chronic active lesions are characterized by few leukocytes and extensive glial fibrillary acidic protein immunoreactivity, indicative of astrogliosis. Within acute MS plaques, a significant loss of tenascin-c immunoreactivity has been observed, whereas tenascin-c was distributed throughout chronic MS plaques at levels similar to or greater than those seen in normal-appearing white matter. Particularly reactive astrocytes have been shown to strongly express TnC, and several reports have shown a correlation between TnC induction and acute inflammation, suggesting that enhanced tenascin-c expression might function as a defense mechanism to control the inflammatory reaction [22, 131]. However, the loss of TnC seen in acute MS lesions is consistent with inflammatory cell-mediated breakdown of the extracellular matrix, which may be a marker of blood-brain barrier breakdown and leukocyte extravasation. Matrix metalloproteinases, which can degrade tenascins, are probably a factor in this inflammatory-mediated matrix breakdown. Such a breakdown in the TnC matrix might lead to a loss of matrix-cellular interactions, influencing the radial extension of the active lesion. Furthermore, the expression and preservation of TnC in normal-appearing white matter beyond the plaque edge may negatively influence migration. The increase in TnC seen in association with a reactive astrocyte subpopulation in extensively demyelinated and subacute lesions and scar formation in chronic MS lesions might also inhibit repair. It suggests that reactive astrocytes continue to produce TnC, which leads to the eventual increase in levels seen in chronic plaques. This increased production and deposition of TnC would then actively inhibit and prevent the differentiation of oligodendrocyte progenitor cells into oligodendrocytes within the lesion, leading to the failure to remyelinate axons, which is seen in MS pathology.

3.2. Astrocytic fibronectin in CNS disease pathology

Inflammation-mediated loss of myelin and incomplete remyelination are pathological hallmarks of multiple sclerosis (MS). Remyelination is essential for both restoration of saltatory conduction and axonal protection [132]. Although remyelination occurs in the early stages of MS, it declines as the disease progresses, resulting in chronically demyelinated plaques and axonal loss [133]. Oligodendrocyte progenitors, the cells responsible for CNS remyelination [134], are present in most MS lesions, but ultimately fail to differentiate into mature myelinating oligodendrocytes, which results in remyelination failure [132, 135]. One of the many factors regulating the migration, proliferation, and differentiation of oligodendrocyte progenitor cells into mature, myelinating oligodendrocytes is the extracellular matrix [136]. In multiple sclerosis brain tissue, enhanced fibronectin deposition was primarily localized to vessel walls, in particular in perivascular infiltrates, and correlated with the extent of inflammation. Fibronectin accumulation was also detected in the parenchyma of active demyelinating MS lesions, suggesting that, in addition to extravasation from affected blood vessels, fibronectin may be locally produced by endothelial cells, infiltrated macrophages in the CNS [15, 137], as well as astrocytes [14]. Recent data have now demonstrated that fibronectin inhibits the differentiation of oligodendrocyte progenitors and, as a result, remyelination [138]. This finding was furthered when Stoffels et al. observed that the production of fibronectin aggregates inhibited oligodendrocyte progenitor cell differentiation in both an animal model of MS and within chronically demyelinated lesions. When they examined the expression of fibronectin on demyelinating injury, they found that the formation of these inhibitory fibronectin aggregates is mediated by inflammation. In toxin-induced lesions, fibronectin expression was transiently increased within demyelinated areas and declined as remyelination proceeded. However, upon the examination of chronically demyelinated MS lesions, fibronectin expression persisted in the form of insoluble, degradation-resistant aggregates. This finding was also observed in a mouse model of MS, chronic experimental autoimmune encephalomyelitis, wherein fibronectin aggregates were found at the relapse phase but not in a toxin-induced demyelination injury model.

Frost et al. [139] showed that fibronectin promoted the migration of oligodendrocyte precursor cells. Connecting segment-1 fibronectin, an alternative splice variant of fibronectin, localized to astrocytes and astrocyte end-feet at the edge of MS lesions [16]. The CS-1 domain serves as a ligand for a4B1, a leukocyte integrin involved in cell-ECM and cell-cell adhesion. The presence of CS-1 fibronectin in astrocyte end-feet may therefore contribute to entry or retention of a4B1 integrin-bearing leukocytes further into the CNS parenchyma. These data indicate that fibronectin and its splice variants have an active part in MS lesion development and progression.

Fibronectin has also been shown to mediate the inflammatory response in spinal cord injury. After spinal cord injury, both a glial and fibrotic scar forms at the site of injury. An excellent review on the glial scar can be found in Sofroniew and Vinters, 2010. Along with the re-appearance of tenascin-c, fibronectin deposition is also increased following spinal cord injury. While fibronectin has been shown to be a growth-permissive substrate for axons, the fibrotic scar is inhibitory to axon regeneration [140]. In a compression trauma model of spinal cord injury, Farooque et al. found that fibronectin was present within sites of severe spinal cord compression trauma; however, when distal portions of the spinal cord were probed for fibronectin antigen, there were no signs of deposition [141]. This indicates that fibronectin is deposited proximal to the site of injury. Additionally, fibronectin has been shown to stimulate astrocyte proliferation through two means: (1) the $\alpha5\beta1$ integrin found on astrocytes, and (2) the up-regulation of the P2Y1 receptor. The up-regulation of P2Y1 by fibronectin requires [Ca2+]i and the activation of the integrin-linked kinase (ILK) and Akt [142]. The [Ca2+]i stimulated by fibronectin is a5B1 integrin receptor dependent and the phosphorylation of Akt or extracellular signal-regulated protein kinase (ERK1/2) induced by fibronectin mediates the action of cAMP response element-binding protein (CREB) and signal transducer and activator of transcription 3 (Stat3). Through these various pathways, fibronectin release could stimulate the astrocyte proliferation seen after spinal cord injury, that the increased expression of the P2Y1 receptor would provide more sites for ATP to bind, which could further aggravate the proliferation and inflammation of spinal cord astrocytes, thus worsening the recovery of Spinal cord injury patients [142].

3.3. Laminin in CNS disease pathology

Laminins are high-molecular weight (~400 kDa) proteins of the extracellular matrix. They are a major component of the basal lamina (one of the players of the basement membrane), a protein network foundation for most cells and organs. The laminins are an important and biologically active part of the basal lamina, influencing cell differentiation, migration, and adhesion. Laminin is vital for the maintenance and survival of tissues. In the central nervous system, laminins, similar to other extracellular matrix proteins, are broadly expressed during embryonic brain development. In the adult brain, however, the distribution of laminin is restricted to the vascular basal lamina and the ventricular-subventricular zone stem cell niche. We will be covering laminin expression as it pertains to the vascular basal lamina (i.e., basement membrane). There are two distinct continuous basement membranes that can be identified surrounding the cerebrovasculature: the vascular basement membrane and the astroglial basement membrane. Both of the basement membranes are composed of the characteristic sheet-like structures of laminins, heparan sulfate proteoglycans, entactin, and type IV collagen. The only difference observed between the two basement membranes is the source of the structural components: endothelial cells are the predominate source for the vascular basement membrane, and astrocytes (specifically, astrocytic end-feet) are responsible for the formation of the astroglial basement membrane.

In multiple sclerosis lesions, alterations in the basement membrane are observed [16, 91, 143]. Abnormalities of the basal lamina surrounding the brain capillaries and local deposition of matrix molecules may influence blood-brain barrier permeability and thus leukocyte migration and retention. The basement membrane barriers previously discussed—vascular and astroglial—define the inner and outer limits, respectively, of the Virchow-Robin perivascular space where leukocytes accumulate before they migrate into the CNS neuroparenchyma. Recently, the presence of extensive basement membrane alterations in MS brain tissue was described [16, 143]. It was found that inflammatory MS lesions contained irregular and discontinuous basement membranes throughout the lesion area. It was also found that organized deposition of basement membrane proteins occurred within the perivascular infiltrates in MS lesions. This group hypothesized that these structures contributed to the influx of leukocytes by forming a reservoir of chemotactic agents. However, they also posited that the perivascular ECM structures might function as a conduit network, thereby facilitating the transport of myelin-laden macrophages out of the CNS toward cervical lymph nodes [144]. The deposition of such compact parenchymal basement membrane deposits may have further consequences such as hampering axonal regeneration and outgrowth through the formation of an anatomical barrier, which could lead to the persistence of MS lesions.

It has also been demonstrated that the only laminin isoforms present in the vascular basement membranes are $\alpha4$ and $\alpha5$, whereas isoforms $\alpha1$ and $\alpha2$ were restricted to the astroglial basement membrane [145]. When investigating the expression of these laminin isoforms in the brain tissue of experimental autoimmune encephalomyelitis mice, leukocyte infiltration was associated with a pronounced loss of laminin $\alpha5$ immunoreactivity in the vascular basement membrane. However, in regions where laminin $\alpha4$ and $\alpha5$ were detected, no leukocyte infiltration was detected. Interestingly, there was major leukocyte infiltration occurring at sites where the parenchymal basement membrane contained both the laminin $\alpha1$ and $\alpha2$ chains, isoforms produced primarily by the astrocytic basement membrane. This suggests that leukocyte migration across the astroglial basement membrane is markedly different compared to the migration observed across the vascular basement membrane [145]. There was also a recent study looking into the differential distribution of several laminin isoforms in control and MS brain tissue. In this study, the authors confirmed the previous finding that the vascular basement membrane is mainly composed of laminin-5,-8, and -10, whereas the astroglial basement membrane predominantly consists of laminin-1 and -2. However, in active demyelinating MS lesions, they observed leukocytes accumulating around vascular basement membranes containing laminin $\alpha5$. In addition, disruption and loss of vascular laminin expression in active demyelinating lesions have been reported [146].

Laminin also plays a role in maintaining the integrity of the blood-brain barrier (BBB). The BBB is a dynamic network that regulates material exchange between the circulatory system and the brain parenchyma, which aids in homeostatic maintenance of the central nervous system [147]. In the context of central nervous system injury, BBB malfunction has been reported. The BBB is mainly composed of brain microvascular endothelial cells, astrocytic end-feet, pericytes, and the basement membrane, of which laminin is a key component. Astrocytes wrap around endothelial cells using their end-feet, and pericytes, which are sandwiched between endothelial cells and astrocytes, signal to both cell types. Recently, it has been shown that pericytes are necessary for the formation of the BBB during embryogenesis, and loss of pericytes leads to comprised BBB integrity and age-dependent vascular-mediated neurodegeneration in adult mice, which suggests an important role for pericytes in BBB regulation. In a recent report, a group found that astrocyte laminin, by binding to the integrin $\alpha 2$ receptor, prevents pericyte differentiation from the BBB-stabilizing resting stage to the BBB-disrupting contractile stage, which helps to maintain the integrity of the BBB [148]. However, when astrocytic laminin was down-regulated using functional blocking antibodies and RNA interference, there were decreases in aquaporin-4 expression on astrocyte end-feet and decreases in tight junction protein expression. Further, in laminin knockdown animals, the lack of astrocytic laminin induced the differentiation of pericytes from the resting stage to the contractile stage. This loss of astrocytic laminins could be one of the major driving forces behind the leakiness of the BBB seen in many neurodegenerative diseases and CNS injuries.

3.4. Vitronectin in CNS disease and injury

Unlike the preceding extracellular matrix proteins, vitronectin has remained elusive in its functional role in central nervous system inflammation and injury. The earliest reports observed an enhancement of vitronectin expression in the blood vessel walls of active demyelinating lesions, in demyelinated axons, and on a small number of hypertrophic astrocytes. However, a negative role for vitronectin has not been found. In contrast, vitronectin has been shown to promote neurite outgrowth [149] and enhance astrocyte migration [150]. As vitronectin mRNA is almost undetectable in the normal adult brain, it might be synthesized by infiltrating leukocytes or derived from the plasma as a result of blood-brain barrier breakdown. In the EAE model of multiple sclerosis, vitronectin expression was shown to be enhanced, as well as contribute to the up-regulation of matrix metalloproteinases and activation of microglia [151]. Increasing research into the role of this under-studied extracellular matrix protein could provide clues as to its functional role in CNS inflammation.

4. Concluding remarks

In this chapter, we have outlined many lines of evidence linking the activity of astrocytes to changes in the ECM associated with neuropathology. We postulate that establishing the nature of astrocyte heterogeneity will be important for understanding the growing number of diseases in which astrocytes have been identified as having a primary or causal role. The growing awareness that astrocyte dysfunction, not just reactivity, contributes to neuropathology as a concept we have called "gliodystrophy" [152]. This term reflects more than the presence of astrocytes in pathology, but denotes the loss or gain of astrocyte functions as a result of astrocyte plasticity and disease-associated heterogeneity. Therefore, understanding the basis for astrocyte heterogeneity, as a component of astrocyte dysfunction, is of increasing importance as astrocytes are recognized to play a central role in a wide range of neurological diseases.

How might we define astrocyte heterogeneity and is astrocyte reactivity a form of heterogeneity? To begin, we would propose that the heterogeneity of astrocytes is divergence in functions between ontogenically identical cells. By this definition, we would suggest that there would exist homeostatic heterogeneity among astrocytes related to their anatomical location. One could argue that the metabolic and physiological demands on an astrocyte within the cortical gray matter would be different from an astrocyte located within the heavily myelinated tracts of the corpus callosum. Indeed, astrocytes in these two locations are easily discernible by their overt appearance as either protoplasmic or fibrous [92]. Then, reactivity would add another dimension to this heterogeneity as each would, in theory, have potentially unique starting states from which reactivity could also be distinctive. If we depict this idea in a (perhaps overly) simplified manner (**Figure 2**), we could envision naive astrocytes within the CNS lying along an X-axis at different points. In response to a stimulus, you might then predict that each cell would in response to that trigger rise up the Y-axis to different points. When considered in terms of neurological diseases, identical acute injuries or trauma to different anatomical loci may be expected to evoke different responses from astrocytes in terms of their proliferation, gene response, and secretome contribution to the immediate environment. When considered in terms of chronic neurodegenerative conditions, where time plays a crucial role (as conveyed along the Z-axis in **Figure 2**), heterogeneous astrocytes may be expected to develop adaptations to disease in different ways. Where one activated cell may become quiescent, interacting less with its immediate environment and failing to sustain homeostatic functions, as one might envision occurring in a glial scar. Another astrocyte may instead adopt a gain of function with an enhanced or altered secretome that modifies its immediate environment and interacts with adjacent cells in a pathological way. This concept may contribute to how we might explain the role of astrocytes in dementia where dysregulation of synaptic connectivity and impaired cognitive function may relate to astrocyte senescence in Alzheimer's disease [153].

From all of this discussion, we should also consider the potential utility of the information gleaned from what could be considered the basic biology of the astrocyte. How could we apply our present and future understanding of astrocyte heterogeneity to developing new, or possibly enhancing current treatments for neurological disease? One possible approach of harnessing the potential of heterogenous astrocytes has already been applied to models of spinal cord injury and Parkinson's disease. In these studies, Proschel and colleagues have determined that astrocyte transplants can dramatically improve the outcomes in these degenerative conditions. For instance, in 6-hydroxydopamine lesioned rats, the behavioral deficit and dopaminergic denervation of the striatum were attenuated when these animals received transplants of astrocytes pre-exposed to bone morphogeneic protein [154]. In a previous study, this group also demonstrated enhanced axon growth in a spinal cord injury model with these astrocyte transplants [155]. These data show that astrocytes possess therapeutic potential to address important neurological diseases. To build upon the ideas set forth by these transplant studies, one could ask how could we target the endogenous astrocytes to achieve similar outcomes? While the answer to this important question is likely complex, if our own ideas on the origins of astrocyte heterogeneity are valid, then select targeting of ECM-Astrocyte communication may be one approach to try. For instance, targeting of the β1 integrin using the RGD peptide has been shown to prevent astrogliosis in the injured spinal cord and improve functional recovery [156]. With an advanced understanding on how the ECM controls, or at the very least influences, the function of astrocytes *in situ* during brain injury or disease, we may be able to target and promote brain recovery and repair.

In the future, we suggest that technical approaches are now available to advance this line of investigation in ways not previously feasible. For instance, cataloging astrocyte diversity using single-cell laser-capture sequencing may be expected to identify unique markers to distinguish different subtypes of astrocytes from tissues. This approach would also allow for the important distinction of acquiring astrocytes that are spatially and temporally associated with specific types of neural injury [157]. A similar approach has recently been used to identify markers of reactive astrocytes. Results from these types of future investigations should enable us to delve deeper into the complexity of astrocyte biology and better understand the nature and function of these cells as they maintain the CNS and react and participate in neurological disease states.

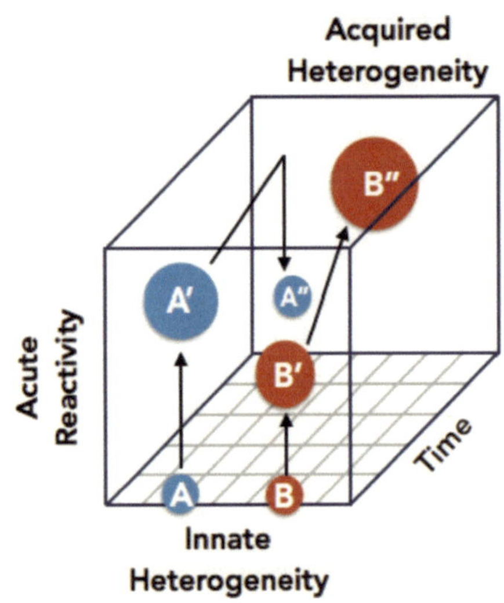

Figure 2. Hypothesized model of influence of ECM on innate and acquired astrocyte heterogeneity. Depicted are two different astrocytes, labeled A and B which have been positioned along the X-axis to reflect innate heterogeneity on the basis of their location within the central nervous system (CNS). In response to a stimulus (labeled A′ and B′, respectively), the innate heterogeneity impacts the reactivity, as depicted as different locations along the Y-axis. Lastly, with chronic stimulation, these two distinct cells develop distinct long-term phenotypes, labeled A″ and B″, where the innate heterogeneity results in different outcomes to the long-term stimulation. Whereas A depicted as a smaller sphere lower on the Y axis may become chronically less active, perhaps related to development of an astrocyte scar, the other astrocyte labeled B″ with chronic stimulation adapts to become a more active, perhaps physiologically adapted, phenotype contributing to neuropathology in disease.

http://dx.doi.org/10.5772/62706

Acknowledgements

We thank Alexandra Nicaise for her helpful feedback on this manuscript. CW and SJC were supported in part by grants from the National Multiple Sclerosis Society (USA) (Grant# RG5001-A-3, to SJC) and the National Institutes of Health (Grant# NS087578, to SJC).

Author details

Cory M. Willis and Stephen J. Crocker*

*Address all correspondence to: crocker@uchc.edu

Department of Neuroscience, University of Connecticut School of Medicine, Farmington, CT, USA

References

[1] Cragg B. Brain extracellular space fixed for electron microscopy. Neuroscience Letters. 1979;15:301-306.

[2] Bignami A, Hosley M, Dahl D. Hyaluronic acid and hyaluronic acid-binding proteins in brain extracellular matrix. Anatomy and Embryology. 1993;188:419-433.

[3] Hynes RO, Naba A. Overview of the matrisome—an inventory of extracellular matrix constituents and functions. Cold Spring Harbor Perspectives in Biology. 2012;4:a004903. 10.1101/cshperspect.a004903

[4] Rauch U. Brain matrix: structure, turnover and necessity. Biochemical Society Transactions. 2007;35:656-660. 10.1042/bst0350656

[5] Lau LW, Cua R, Keough MB, Haylock-Jacobs S, Yong VW. Pathophysiology of the brain extracellular matrix: a new target for remyelination. Nature Reviews Neuroscience. 2013;14:722-729. 10.1038/nrn3550

[6] Colognato H, Tzvetanova ID. Glia unglued: how signals from the extracellular matrix regulate the development of myelinating glia. Developmental Neurobiology. 2011;71:924-955. 10.1002/dneu.20966

[7] Berezin V, Walmod PS, Filippov M, Dityatev A. Targeting of ECM molecules and their metabolizing enzymes and receptors for the treatment of CNS diseases. Progress in Brain Research. 2014;214:353-388. 10.1016/b978-0-444-63486-3.00015-3

[8] Dzyubenko E, Gottschling C, Faissner A. Neuron-glia interactions in neural plasticity: contributions of neural extracellular matrix and perineuronal nets. Neural Plasticity. 2016;2016:5214961. 10.1155/2016/5214961

[9] Pankov R, Yamada KM. Fibronectin at a glance. Journal of Cell Science. 2002;115:3861-3863.

[10] Nishinarita S, Yamamoto M, Takizawa T, Hayakawa J, Karasaki M, Sawada S. Increased plasma fibronectin in patients with systemic lupus erythematosus. Clinical Rheumatology. 1990;9:214-219.

[11] Goos M, Lange P, Hanisch UK, Prinz M, Scheffel J, Bergmann R, et al. Fibronectin is elevated in the cerebrospinal fluid of patients suffering from bacterial meningitis and enhances inflammation caused by bacterial products in primary mouse microglial cell cultures. Journal of Neurochemistry. 2007;102:2049-2060. 10.1111/j.1471-4159.2007.04683.x

[12] George EL, Georges-Labouesse EN, Patel-King RS, Rayburn H, Hynes RO. Defects in mesoderm, neural tube and vascular development in mouse embryos lacking fibronectin. Development (Cambridge, England). 1993;119:1079-1091.

[13] Scanzello CR, Plaas A, Crow MK. Innate immune system activation in osteoarthritis: is osteoarthritis a chronic wound? Current Opinion in Rheumatology. 2008;20:565-572. 10.1097/BOR.0b013e32830aba34

[14] Stoffels JM, de Jonge JC, Stancic M, Nomden A, van Strien ME, Ma D, et al. Fibronectin aggregation in multiple sclerosis lesions impairs remyelination. Brain: A Journal of Neurology. 2013;136:116-131. 10.1093/brain/aws313

[15] Sobel RA, Mitchell ME. Fibronectin in multiple sclerosis lesions. American Journal of Pathology. 1989;135:161-168.

[16] van Horssen J, Bo L, Vos CM, Virtanen I, de Vries HE. Basement membrane proteins in multiple sclerosis-associated inflammatory cuffs: potential role in influx and transport of leukocytes. Journal of Neuropathology and Experimental Neurology. 2005;64:722-729.

[17] Satoh JI, Tabunoki H, Yamamura T. Molecular network of the comprehensive multiple sclerosis brain-lesion proteome. Multiple Sclerosis (Houndmills, Basingstoke, England). 2009;15:531-541. 10.1177/1352458508101943

[18] Hibbits N, Yoshino J, Le TQ, Armstrong RC. Astrogliosis during acute and chronic cuprizone demyelination and implications for remyelination. ASN Neuro. 2012;4:393-408. 10.1042/an20120062

[19] Faissner A. The tenascin gene family in axon growth and guidance. Cell and Tissue Research. 1997;290:331-341.

[20] Jones FS, Jones PL. The tenascin family of ECM glycoproteins: structure, function, and regulation during embryonic development and tissue remodeling. Developmental

Dynamics: An Official Publication of the American Association of Anatomists. 2000;218:235-259. 10.1002/(sici)1097-0177(200006)218:2<235::aid-dvdy2>3.0.co;2-g

[21] Joester A, Faissner A. The structure and function of tenascins in the nervous system. Matrix Biology: Journal of the International Society for Matrix Biology. 2001;20:13-22.

[22] Chiquet-Ehrismann R, Chiquet M. Tenascins: regulation and putative functions during pathological stress. The Journal of Pathology. 2003;200:488-499. 10.1002/path.1415

[23] Wiese S, Karus M, Faissner A. Astrocytes as a source for extracellular matrix molecules and cytokines. Frontiers in Pharmacology. 2012;3:120. 10.3389/fphar.2012.00120

[24] Zimmermann DR, Dours-Zimmermann MT. Extracellular matrix of the central nervous system: from neglect to challenge. Histochemistry and Cell Biology. 2008;130:635-653. 10.1007/s00418-008-0485-9

[25] Crossin KL, Hoffman S, Grumet M, Thiery JP, Edelman GM. Site-restricted expression of cytotactin during development of the chicken embryo. The Journal of Cell Biology. 1986;102:1917-1930.

[26] Prieto AL, Jones FS, Cunningham BA, Crossin KL, Edelman GM. Localization during development of alternatively spliced forms of cytotactin mRNA by in situ hybridization. The Journal of Cell Biology. 1990;111:685-698.

[27] Karus M, Denecke B, ffrench-Constant C, Wiese S, Faissner A. The extracellular matrix molecule tenascin C modulates expression levels and territories of key patterning genes during spinal cord astrocyte specification. Development (Cambridge, England). 2011;138:5321-5331. 10.1242/dev.067413

[28] von Holst A, Egbers U, Prochiantz A, Faissner A. Neural stem/progenitor cells express 20 tenascin C isoforms that are differentially regulated by Pax6. The Journal of Biological Chemistry. 2007;282:9172-9181. 10.1074/jbc.M608067200

[29] Garcion E, Halilagic A, Faissner A, ffrench-Constant C. Generation of an environmental niche for neural stem cell development by the extracellular matrix molecule tenascin C. Development (Cambridge, England). 2004;131:3423-3432. 10.1242/dev.01202

[30] Garcion E, Faissner A, ffrench-Constant C. Knockout mice reveal a contribution of the extracellular matrix molecule tenascin-C to neural precursor proliferation and migration. Development (Cambridge, England). 2001;128:2485-2496.

[31] Garwood J, Garcion E, Dobbertin A, Heck N, Calco V, ffrench-Constant C, et al. The extracellular matrix glycoprotein Tenascin-C is expressed by oligodendrocyte precursor cells and required for the regulation of maturation rate, survival and responsiveness to platelet-derived growth factor. The European Journal of Neuroscience. 2004;20:2524-2540. 10.1111/j.1460-9568.2004.03727.x

[32] Czopka T, Von Holst A, Schmidt G, Ffrench-Constant C, Faissner A. Tenascin C and tenascin R similarly prevent the formation of myelin membranes in a RhoA-dependent

manner, but antagonistically regulate the expression of myelin basic protein via a separate pathway. Glia. 2009;57:1790-1801. 10.1002/glia.20891

[33] Faissner A, Kruse J. J1/tenascin is a repulsive substrate for central nervous system neurons. Neuron. 1990;5:627-637.

[34] Lochter A, Vaughan L, Kaplony A, Prochiantz A, Schachner M, Faissner A. J1/tenascin in substrate-bound and soluble form displays contrary effects on neurite outgrowth. The Journal of Cell Biology. 1991;113:1159-1171.

[35] Husmann K, Faissner A, Schachner M. Tenascin promotes cerebellar granule cell migration and neurite outgrowth by different domains in the fibronectin type III repeats. The Journal of Cell Biology. 1992;116:1475-1486.

[36] Gotz B, Scholze A, Clement A, Joester A, Schutte K, Wigger F, et al. Tenascin-C contains distinct adhesive, anti-adhesive, and neurite outgrowth promoting sites for neurons. The Journal of Cell Biology. 1996;132:681-699.

[37] Gotz M, Bolz J, Joester A, Faissner A. Tenascin-C synthesis and influence on axonal growth during rat cortical development. The European Journal of Neuroscience. 1997;9:496-506.

[38] Meiners S, Geller HM. Long and short splice variants of human tenascin differentially regulate neurite outgrowth. Molecular and Cellular Neurosciences. 1997;10:100-116. 10.1006/mcne.1997.0643

[39] Meiners S, Mercado ML, Nur-e-Kamal MS, Geller HM. Tenascin-C contains domains that independently regulate neurite outgrowth and neurite guidance. The Journal of Neuroscience: The Official Journal of the Society for Neuroscience. 1999;19:8443-8453.

[40] Bartsch U, Bartsch S, Dorries U, Schachner M. Immunohistological localization of tenascin in the developing and lesioned adult mouse optic nerve. The European Journal of Neuroscience. 1992;4:338-352.

[41] Dorries U, Schachner M. Tenascin mRNA isoforms in the developing mouse brain. Journal of Neuroscience Research. 1994;37:336-347. 10.1002/jnr.490370306

[42] Gates MA, Thomas LB, Howard EM, Laywell ED, Sajin B, Faissner A, et al. Cell and molecular analysis of the developing and adult mouse subventricular zone of the cerebral hemispheres. The Journal of Comparative Neurology. 1995;361:249-266. 10.1002/cne.903610205

[43] Thomas LB, Gates MA, Steindler DA. Young neurons from the adult subependymal zone proliferate and migrate along an astrocyte, extracellular matrix-rich pathway. Glia. 1996;17:1-14. 10.1002/(SICI)1098-1136(199605)17:1<1::AID-GLIA1>3.0.CO;2-7

[44] Theodosis DT, Pierre K, Cadoret MA, Allard M, Faissner A, Poulain DA. Expression of high levels of the extracellular matrix glycoprotein, tenascin-C, in the normal adult

hypothalamoneurohypophysial system. The Journal of Comparative Neurology. 1997;379:386-398.

[45] Timpl R, Rohde H, Robey PG, Rennard SI, Foidart JM, Martin GR. Laminin—a glycoprotein from basement membranes. The Journal of Biological Chemistry. 1979;254:9933-9937.

[46] Campos LS, Leone DP, Relvas JB, Brakebusch C, Fassler R, Suter U, et al. Beta1 integrins activate a MAPK signalling pathway in neural stem cells that contributes to their maintenance. Development (Cambridge, England). 2004;131:3433-3444. 10.1242/dev.01199

[47] Lathia JD, Patton B, Eckley DM, Magnus T, Mughal MR, Sasaki T, et al. Patterns of laminins and integrins in the embryonic ventricular zone of the CNS. The Journal of Comparative Neurology. 2007;505:630-643. 10.1002/cne.21520

[48] Calof AL, Reichardt LF. Response of purified chick motoneurons to myotube conditioned medium: laminin is essential for the substratum-binding, neurite outgrowth-promoting activity. Neuroscience Letters. 1985;59:183-189.

[49] Lander AD, Fujii DK, Reichardt LF. Laminin is associated with the "neurite outgrowth-promoting factors" found in conditioned media. Proceedings of the National Academy of Sciences of the United States of America. 1985;82:2183-2187.

[50] Lander AD, Fujii DK, Reichardt LF. Purification of a factor that promotes neurite outgrowth: isolation of laminin and associated molecules. The Journal of Cell Biology. 1985;101:898-913.

[51] Hall DE, Neugebauer KM, Reichardt LF. Embryonic neural retinal cell response to extracellular matrix proteins: developmental changes and effects of the cell substratum attachment antibody (CSAT). The Journal of Cell Biology. 1987;104:623-634.

[52] Tomaselli KJ, Reichardt LF. Peripheral motoneuron interactions with laminin and Schwann cell-derived neurite-promoting molecules: developmental regulation of laminin receptor function. Journal of Neuroscience Research. 1988;21:275-285. 10.1002/jnr.490210220

[53] Timpl R, Brown JC. Supramolecular assembly of basement membranes. BioEssays: News and Reviews in Molecular, Cellular and Developmental Biology. 1996;18:123-132. 10.1002/bies.950180208

[54] Erickson AC, Couchman JR. Still more complexity in mammalian basement membranes. The Journal of Histochemistry and Cytochemistry: Official Journal of the Histochemistry Society. 2000;48:1291-1306.

[55] Sievers J, von Knebel Doeberitz C, Pehlemann FW, Berry M. Meningeal cells influence cerebellar development over a critical period. Anatomy and Embryology. 1986;175:91-100.

[56] von Knebel Doeberitz C, Sievers J, Sadler M, Pehlemann FW, Berry M, Halliwell P. Destruction of meningeal cells over the newborn hamster cerebellum with 6-hydroxydopamine prevents foliation and lamination in the rostral cerebellum. Neuroscience. 1986;17:409-426.

[57] Sievers J, Pehlemann FW, Gude S, Berry M. A time course study of the alterations in the development of the hamster cerebellar cortex after destruction of the overlying meningeal cells with 6-hydroxydopamine on the day of birth. Journal of Neurocytology. 1994;23:117-134.

[58] Radakovits R, Barros CS, Belvindrah R, Patton B, Muller U. Regulation of radial glial survival by signals from the meninges. The Journal of Neuroscience: the Official Journal of the Society for Neuroscience. 2009;29:7694-7705. 10.1523/jneurosci.5537-08.2009

[59] Drago J, Nurcombe V, Bartlett PF. Laminin through its long arm E8 fragment promotes the proliferation and differentiation of murine neuroepithelial cells in vitro. Experimental Cell Research. 1991;192:256-265.

[60] Liesi P. Neuronal migration on laminin involves neuronal contact formation followed by nuclear movement inside a preformed process. Experimental Neurology. 1992;117:103-113.

[61] Liesi P, Seppala I, Trenkner E. Neuronal migration in cerebellar microcultures is inhibited by antibodies against a neurite outgrowth domain of laminin. Journal of Neuroscience Research. 1992;33:170-176. 10.1002/jnr.490330122

[62] Kearns SM, Laywell ED, Kukekov VK, Steindler DA. Extracellular matrix effects on neurosphere cell motility. Experimental Neurology. 2003;182:240-244.

[63] Flanagan LA, Rebaza LM, Derzic S, Schwartz PH, Monuki ES. Regulation of human neural precursor cells by laminin and integrins. Journal of Neuroscience Research. 2006;83:845-856. 10.1002/jnr.20778

[64] Hall PE, Lathia JD, Caldwell MA, Ffrench-Constant C. Laminin enhances the growth of human neural stem cells in defined culture media. BMC Neuroscience. 2008;9:71. 10.1186/1471-2202-9-71

[65] Ma W, Tavakoli T, Derby E, Serebryakova Y, Rao MS, Mattson MP. Cell-extracellular matrix interactions regulate neural differentiation of human embryonic stem cells. BMC Developmental Biology. 2008;8:90. 10.1186/1471-213x-8-90

[66] Silva A, Pereira J, Oliveira CR, Relvas JB, Rego AC. BDNF and extracellular matrix regulate differentiation of mice neurosphere-derived cells into a GABAergic neuronal phenotype. Journal of Neuroscience Research. 2009;87:1986-1996. 10.1002/jnr.22041

[67] Pierret C, Morrison JA, Rath P, Zigler RE, Engel LA, Fairchild CL, et al. Developmental cues and persistent neurogenic potential within an in vitro neural niche. BMC Developmental Biology. 2010;10:5. 10.1186/1471-213x-10-5

[68] Smyth N, Vatansever HS, Murray P, Meyer M, Frie C, Paulsson M, et al. Absence of basement membranes after targeting the LAMC1 gene results in embryonic lethality due to failure of endoderm differentiation. The Journal of Cell Biology. 1999;144:151-160.

[69] Halfter W, Dong S, Yip YP, Willem M, Mayer U. A critical function of the pial basement membrane in cortical histogenesis. The Journal of Neuroscience: The Official Journal of the Society for Neuroscience. 2002;22:6029-6040. 20026580

[70] Chen ZL, Haegeli V, Yu H, Strickland S. Cortical deficiency of laminin gamma1 impairs the AKT/GSK-3beta signaling pathway and leads to defects in neurite outgrowth and neuronal migration. Developmental Biology. 2009;327:158-168. 10.1016/j.ydbio.2008.12.006

[71] Yoshida A, Kobayashi K, Manya H, Taniguchi K, Kano H, Mizuno M, et al. Muscular dystrophy and neuronal migration disorder caused by mutations in a glycosyltransferase, POMGnT1. Developmental Cell. 2001;1:717-724.

[72] Beltran-Valero de Bernabe D, Currier S, Steinbrecher A, Celli J, van Beusekom E, van der Zwaag B, et al. Mutations in the O-mannosyltransferase gene POMT1 give rise to the severe neuronal migration disorder Walker-Warburg syndrome. American Journal of Human Genetics. 2002;71:1033-1043. 10.1086/342975

[73] Grewal PK, Holzfeind PJ, Bittner RE, Hewitt JE. Mutant glycosyltransferase and altered glycosylation of alpha-dystroglycan in the myodystrophy mouse. Nature Genetics. 2001;28:151-154. 10.1038/88865

[74] Michele DE, Barresi R, Kanagawa M, Saito F, Cohn RD, Satz JS, et al. Post-translational disruption of dystroglycan-ligand interactions in congenital muscular dystrophies. Nature. 2002;418:417-422. 10.1038/nature00837

[75] Moore SA, Saito F, Chen J, Michele DE, Henry MD, Messing A, et al. Deletion of brain dystroglycan recapitulates aspects of congenital muscular dystrophy. Nature. 2002;418:422-425. 10.1038/nature00838

[76] Garcia-Alonso L, Fetter RD, Goodman CS. Genetic analysis of Laminin A in Drosophila: extracellular matrix containing laminin A is required for ocellar axon pathfinding. Development (Cambridge, England). 1996;122:2611-2621.

[77] Yoshimura T, Kawano Y, Arimura N, Kawabata S, Kikuchi A, Kaibuchi K. GSK-3beta regulates phosphorylation of CRMP-2 and neuronal polarity. Cell. 2005;120:137-149. 10.1016/j.cell.2004.11.012

[78] Bradl M, Lassmann H. Oligodendrocytes: biology and pathology. Acta Neuropathologica. 2010;119:37-53. 10.1007/s00401-009-0601-5

[79] Yang D, Bierman J, Tarumi YS, Zhong YP, Rangwala R, Proctor TM, et al. Coordinate control of axon defasciculation and myelination by laminin-2 and -8. The Journal of Cell Biology. 2005;168:655-666. 10.1083/jcb.200411158

[80] Colognato H, ffrench-Constant C, Feltri ML. Human diseases reveal novel roles for neural laminins. Trends in Neurosciences. 2005;28:480-486. 10.1016/j.tins.2005.07.004

[81] Simons M, Trotter J. Wrapping it up: the cell biology of myelination. Current Opinion in Neurobiology. 2007;17:533-540. 10.1016/j.conb.2007.08.003

[82] Colognato H, Baron W, Avellana-Adalid V, Relvas JB, Baron-Van Evercooren A, Georges-Labouesse E, et al. CNS integrins switch growth factor signalling to promote target-dependent survival. Nature Cell Biology. 2002;4:833-841. 10.1038/ncb865

[83] Caro PA, Scavina M, Hoffman E, Pegoraro E, Marks HG. MR imaging findings in children with merosin-deficient congenital muscular dystrophy. AJNR American Journal of Neuroradiology. 1999;20:324-326.

[84] Leite CC, Lucato LT, Martin MG, Ferreira LG, Resende MB, Carvalho MS, et al. Merosin-deficient congenital muscular dystrophy (CMD): a study of 25 Brazilian patients using MRI. Pediatric Radiology. 2005;35:572-579. 10.1007/s00247-004-1398-y

[85] Chun SJ, Rasband MN, Sidman RL, Habib AA, Vartanian T. Integrin-linked kinase is required for laminin-2-induced oligodendrocyte cell spreading and CNS myelination. The Journal of Cell Biology. 2003;163:397-408. 10.1083/jcb.200304154

[86] Relucio J, Tzvetanova ID, Ao W, Lindquist S, Colognato H. Laminin alters fyn regulatory mechanisms and promotes oligodendrocyte development. The Journal of Neuroscience: The Official Journal of the Society for Neuroscience. 2009;29:11794-11806. 10.1523/jneurosci.0888-09.2009

[87] Preissner KT. Structure and biological role of vitronectin. Annual Review of Cell Biology. 1991;7:275-310. 10.1146/annurev.cb.07.110191.001423

[88] Tomasini BR, Mosher DF. Vitronectin. Progress in Hemostasis and Thrombosis. 1991;10:269-305.

[89] Grabham PW, Gallimore PH, Grand RJ. Vitronectin is the major serum protein essential for NGF-mediated neurite outgrowth from PC12 cells. Experimental Cell Research. 1992;202:337-344.

[90] Martinez-Morales JR, Marti E, Frade JM, Rodriguez-Tebar A. Developmentally regulated vitronectin influences cell differentiation, neuron survival and process outgrowth in the developing chicken retina. Neuroscience. 1995;68:245-253.

[91] van Horssen J, Dijkstra CD, de Vries HE. The extracellular matrix in multiple sclerosis pathology. Journal of Neurochemistry. 2007;103:1293-1301. 10.1111/j.1471-4159.2007.04897.x

[92] Sofroniew MV, Vinters HV. Astrocytes: biology and pathology. Acta Neuropathologica. 2010;119:7-35. 10.1007/s00401-009-0619-8

[93] Milev P, Maurel P, Chiba A, Mevissen M, Popp S, Yamaguchi Y, et al. Differential regulation of expression of hyaluronan-binding proteoglycans in developing brain:

aggrecan, versican, neurocan, and brevican. Biochemical and Biophysical Research Communications. 1998;247:207-212. 10.1006/bbrc.1998.8759

[94] Rauch U. Extracellular matrix components associated with remodeling processes in brain. Cellular and Molecular Life Sciences: CMLS. 2004;61:2031-2045. 10.1007/s00018-004-4043-x

[95] Kornblum HI, Hussain R, Wiesen J, Miettinen P, Zurcher SD, Chow K, et al. Abnormal astrocyte development and neuronal death in mice lacking the epidermal growth factor receptor. Journal of Neuroscience Research. 1998;53:697-717.

[96] Lehre KP, Levy LM, Ottersen OP, Storm-Mathisen J, Danbolt NC. Differential expression of two glial glutamate transporters in the rat brain: quantitative and immunocytochemical observations. The Journal of Neuroscience: The Official Journal of the Society for Neuroscience. 1995;15:1835-1853.

[97] Nagelhus EA, Mathiisen TM, Ottersen OP. Aquaporin-4 in the central nervous system: cellular and subcellular distribution and coexpression with KIR4.1. Neuroscience. 2004;129:905-913. 10.1016/j.neuroscience.2004.08.053

[98] Sirko S, von Holst A, Wizenmann A, Gotz M, Faissner A. Chondroitin sulfate glycosaminoglycans control proliferation, radial glia cell differentiation and neurogenesis in neural stem/progenitor cells. Development (Cambridge, England). 2007;134:2727-2738. 10.1242/dev.02871

[99] Stewart GR, Pearlman AL. Fibronectin-like immunoreactivity in the developing cerebral cortex. The Journal of Neuroscience: The Official Journal of the Society for Neuroscience. 1987;7:3325-3333.

[100] Bignami A, Delpech B. Extracellular matrix glycoprotein (hyaluronectin) in early cerebral development Immunofluorescence study of the rat embryo. International Journal of Developmental Neuroscience: The Official Journal of the International Society for Developmental Neuroscience. 1985;3:301-307. 10.1016/0736-5748(85)90035-8

[101] Derer P, Nakanishi S. Extracellular matrix distribution during neocortical wall ontogenesis in "normal" and "Reeler" mice. Journal fur Hirnforschung. 1983;24:209-224.

[102] Nakanishi S. Extracellular matrix during laminar pattern formation of neocortex in normal and reeler mutant mice. Developmental Biology. 1983;95:305-316.

[103] Liesi P. Do neurons in the vertebrate CNS migrate on laminin? The EMBO Journal. 1985;4:1163-1170.

[104] Liesi P. Laminin and fibronectin in normal and malignant neuroectodermal cells. Medical Biology. 1984;62:163-180.

[105] Liesi P, Kaakkola S, Dahl D, Vaheri A. Laminin is induced in astrocytes of adult brain by injury. The EMBO Journal. 1984;3:683-686.

[106] Cohen J, Burne JF, McKinlay C, Winter J. The role of laminin and the laminin/fibronectin receptor complex in the outgrowth of retinal ganglion cell axons. Developmental Biology. 1987;122:407-418.

[107] Liesi P, Silver J. Is astrocyte laminin involved in axon guidance in the mammalian CNS? Developmental Biology. 1988;130:774-785.

[108] Liesi P, Dahl D, Vaheri A. Laminin is produced by early rat astrocytes in primary culture. The Journal of Cell Biology. 1983;96:920-924.

[109] Liesi P, Risteli L. Glial cells of mammalian brain produce a variant form of laminin. Experimental Neurology. 1989;105:86-92.

[110] Chiu AY, Espinosa de los Monteros A, Cole RA, Loera S, de Vellis J. Laminin and s-laminin are produced and released by astrocytes, Schwann cells, and schwannomas in culture. Glia. 1991;4:11-24. 10.1002/glia.440040103

[111] Baghdassarian D, Toru-Delbauffe D, Gavaret JM, Pierre M. Effects of transforming growth factor-beta 1 on the extracellular matrix and cytoskeleton of cultured astrocytes. Glia. 1993;7:193-202. 10.1002/glia.440070302

[112] Trentin AG, Moura Neto V. T3 affects cerebellar astrocyte proliferation, GFAP and fibronectin organization. Neuroreport. 1995;6:293-296.

[113] Ruoslahti E, Pierschbacher MD. New perspectives in cell adhesion: RGD and integrins. Science (New York, NY). 1987;238:491-497.

[114] Ruoslahti E. Integrins. The Journal of Clinical Investigation. 1991;87:1-5. 10.1172/jci114957

[115] Hynes RO. Integrins: versatility, modulation, and signaling in cell adhesion. Cell. 1992;69:11-25.

[116] Reichardt LF, Tomaselli KJ. Extracellular matrix molecules and their receptors: functions in neural development. Annual Review of Neuroscience. 1991;14:531-570. 10.1146/annurev.ne.14.030191.002531

[117] Tawil N, Wilson P, Carbonetto S. Integrins in point contacts mediate cell spreading: factors that regulate integrin accumulation in point contacts vs. focal contacts. The Journal of Cell Biology. 1993;120:261-271.

[118] Dodd J, Jessell TM. Axon guidance and the patterning of neuronal projections in vertebrates. Science (New York, NY). 1988;242:692-699.

[119] Liesi P. Extracellular matrix and neuronal movement. Experientia. 1990;46:900-907.

[120] Rakic P. Principles of neural cell migration. Experientia. 1990;46:882-891.

[121] Gaudet AD, Popovich PG. Extracellular matrix regulation of inflammation in the healthy and injured spinal cord. Experimental neurology. 2014;258:24-34. 10.1016/j.expneurol.2013.11.020

[122] Goh FG, Piccinini AM, Krausgruber T, Udalova IA, Midwood KS. Transcriptional regulation of the endogenous danger signal tenascin-C: a novel autocrine loop in inflammation. Journal of immunology. 2010;184:2655-2662. 10.4049/jimmunol.0903359

[123] Sumioka T, Kitano A, Flanders KC, Okada Y, Yamanaka O, Fujita N, et al. Impaired cornea wound healing in a tenascin C-deficient mouse model. Laboratory Investigation. 2013;93:207-217. 10.1038/labinvest.2012.157

[124] Xie K, Liu Y, Hao W, Walter S, Penke B, Hartmann T, et al. Tenascin-C deficiency ameliorates Alzheimer's disease-related pathology in mice. Neurobiol Aging. 2013;34:2389-2398. 10.1016/j.neurobiolaging.2013.04.013

[125] Zhang Y, Winterbottom JK, Schachner M, Lieberman AR, Anderson PN. Tenascin-C expression and axonal sprouting following injury to the spinal dorsal columns in the adult rat. Journal of Neuroscience Research. 1997;49:433-450.

[126] Holley JE, Gveric D, Whatmore JL, Gutowski NJ. Tenascin C induces a quiescent phenotype in cultured adult human astrocytes. Glia. 2005;52:53-58. 10.1002/glia.20231

[127] Chen J, Joon Lee H, Jakovcevski I, Shah R, Bhagat N, Loers G, et al. The extracellular matrix glycoprotein tenascin-C is beneficial for spinal cord regeneration. Molecular Therapy. 2010;18:1769-1777. 10.1038/mt.2010.133

[128] Suzuki K, Suzuki Y. Globoid cell leucodystrophy (Krabbe's disease): deficiency of galactocerebroside beta-galactosidase. Proceedings of the National Academy of Sciences of the United States of America. 1970;66:302-309.

[129] Claycomb KI, Winokur PN, Johnson KM, Nicaise AM, Giampetruzzi AW, Sacino AV, et al. Aberrant production of tenascin-C in globoid cell leukodystrophy alters psychosine-induced microglial functions. Journal of Neuropathology and Experimental Neurology. 2014;73:964-974. 10.1097/NEN.0000000000000117

[130] Ijichi K, Brown GD, Moore CS, Lee JP, Winokur PN, Pagarigan R, et al. MMP-3 mediates psychosine-induced globoid cell formation: implications for leukodystrophy pathology. Glia. 2013;61:765-777. 10.1002/glia.22471

[131] Chiquet M, Sarasa-Renedo A, Tunc-Civelek V. Induction of tenascin-C by cyclic tensile strain versus growth factors: distinct contributions by Rho/ROCK and MAPK signaling pathways. Biochimica et Biophysica Acta. 2004;1693:193-204. 10.1016/j.bbamcr.2004.08.001

[132] Franklin RJ, Ffrench-Constant C. Remyelination in the CNS: from biology to therapy. Nature Reviews in Neuroscience. 2008;9:839-855. 10.1038/nrn2480

[133] Goldschmidt T, Antel J, Konig FB, Bruck W, Kuhlmann T. Remyelination capacity of the MS brain decreases with disease chronicity. Neurology. 2009;72:1914-1921. 10.1212/WNL.0b013e3181a8260a

[134] Zawadzka M, Rivers LE, Fancy SP, Zhao C, Tripathi R, Jamen F, et al. CNS-resident glial progenitor/stem cells produce Schwann cells as well as oligodendrocytes during repair of CNS demyelination. Cell Stem Cell. 2010;6:578-590. 10.1016/j.stem.2010.04.002

[135] Kuhlmann T, Miron V, Cui Q, Wegner C, Antel J, Bruck W. Differentiation block of oligodendroglial progenitor cells as a cause for remyelination failure in chronic multiple sclerosis. Brain. 2008;131:1749-1758. 10.1093/brain/awn096

[136] Maier O, van der Heide T, van Dam AM, Baron W, de Vries H, Hoekstra D. Alteration of the extracellular matrix interferes with raft association of neurofascin in oligodendrocytes. Potential significance for multiple sclerosis? Molecular and Cellular Neuroscience. 2005;28:390-401. 10.1016/j.mcn.2004.09.012

[137] Esiri MM, Morris CS. Immunocytochemical study of macrophages and microglial cells and extracellular matrix components in human CNS disease. 2. Non-neoplastic diseases. Journal of Neurology Science. 1991;101:59-72.

[138] Siskova Z, Baron W, de Vries H, Hoekstra D. Fibronectin impedes "myelin" sheet-directed flow in oligodendrocytes: a role for a beta 1 integrin-mediated PKC signaling pathway in vesicular trafficking. Molecular and Cellular Neuroscience. 2006;33:150-159. 10.1016/j.mcn.2006.07.001

[139] Frost E, Kiernan BW, Faissner A, ffrench-Constant C. Regulation of oligodendrocyte precursor migration by extracellular matrix: evidence for substrate-specific inhibition of migration by tenascin-C. Developmental Neuroscience. 1996;18:266-273.

[140] Zhu Y, Soderblom C, Trojanowsky M, Lee DH, Lee JK. Fibronectin matrix assembly after spinal cord injury. Journal of Neurotrauma. 2015;32:1158-1167. 10.1089/neu.2014.3703

[141] Farooque M, Zhang Y, Holtz A, Olsson Y. Exudation of fibronectin and albumin after spinal cord injury in rats. Acta Neuropathologica. 1992;84:613-620.

[142] Xia M, Zhu Y. Fibronectin enhances spinal cord astrocyte proliferation by elevating P2Y1 receptor expression. Journal of Neuroscience Research. 2014;92:1078-1090. 10.1002/jnr.23384

[143] van Horssen J, Bo L, Dijkstra CD, de Vries HE. Extensive extracellular matrix depositions in active multiple sclerosis lesions. Neurobiology of Disease. 2006;24:484-491. 10.1016/j.nbd.2006.08.005

[144] Fabriek BO, Zwemmer JN, Teunissen CE, Dijkstra CD, Polman CH, Laman JD, et al. In vivo detection of myelin proteins in cervical lymph nodes of MS patients using ultrasound-guided fine-needle aspiration cytology. Journal of Neuroimmunology. 2005;161:190-194. 10.1016/j.jneuroim.2004.12.018

[145] Sixt M, Engelhardt B, Pausch F, Hallmann R, Wendler O, Sorokin LM. Endothelial cell laminin isoforms, laminins 8 and 10, play decisive roles in T cell recruitment across the

blood-brain barrier in experimental autoimmune encephalomyelitis. Journal of Cellular Biology. 2001;153:933-946.

[146] Sobel RA, Hinojoza JR, Maeda A, Chen M. Endothelial cell integrin laminin receptor expression in multiple sclerosis lesions. American Journal of Pathology. 1998;153:405-415. 10.1016/S0002-9440(10)65584-8

[147] Persidsky Y, Ramirez SH, Haorah J, Kanmogne GD. Blood-brain barrier: structural components and function under physiologic and pathologic conditions. Journal of Neuroimmune Pharmacology. 2006;1:223-236. 10.1007/s11481-006-9025-3

[148] Yao Y, Chen ZL, Norris EH, Strickland S. Astrocytic laminin regulates pericyte differentiation and maintains blood brain barrier integrity. Nature Communications. 2014;5:3413. 10.1038/ncomms4413

[149] Neugebauer KM, Emmett CJ, Venstrom KA, Reichardt LF. Vitronectin and thrombospondin promote retinal neurite outgrowth: developmental regulation and role of integrins. Neuron. 1991;6:345-358.

[150] Milner R, Huang X, Wu J, Nishimura S, Pytela R, Sheppard D, et al. Distinct roles for astrocyte alphavbeta5 and alphavbeta8 integrins in adhesion and migration. Journal of Cell Science. 1999;112 (Pt 23):4271-4279.

[151] Milner R, Crocker SJ, Hung S, Wang X, Frausto RF, del Zoppo GJ. Fibronectin- and vitronectin-induced microglial activation and matrix metalloproteinase-9 expression is mediated by integrins alpha5beta1 and alphavbeta5. Journal of Immunology (Baltimore, MD: 1950). 2007;178:8158-8167.

[152] Claycomb KI, Johnson KM, Winokur PN, Sacino AV, Crocker SJ. Astrocyte regulation of CNS inflammation and remyelination. Brain Sciences. 2013;3:1109-1127. 10.3390/brainsci3031109

[153] Bhat R, Crowe EP, Bitto A, Moh M, Katsetos CD, Garcia FU, et al. Astrocyte senescence as a component of Alzheimer's disease. PloS One. 2012;7:e45069. 10.1371/journal.pone.0045069

[154] Proschel C, Stripay JL, Shih CH, Munger JC, Noble MD. Delayed transplantation of precursor cell-derived astrocytes provides multiple benefits in a rat model of Parkinsons. EMBO Molecular Medicine. 2014;6:504-518. 10.1002/emmm.201302878

[155] Davies SJ, Shih CH, Noble M, Mayer-Proschel M, Davies JE, Proschel C. Transplantation of specific human astrocytes promotes functional recovery after spinal cord injury. PLoS One. 2011;6:e17328. 10.1371/journal.pone.0017328

[156] Pan L, North HA, Sahni V, Jeong SJ, McGuire TL, Berns EJ, et al. beta1-Integrin and integrin linked kinase regulate astrocytic differentiation of neural stem cells. PLoS One. 2014;9:e104335. 10.1371/journal.pone.0104335

[157] Zamanian JL, Xu L, Foo LC, Nouri N, Zhou L, Giffard RG, et al. Genomic analysis of reactive astrogliosis. Journal of Neuroscience. 2012;32:6391-6410. 10.1523/JNEUROSCI.6221-11.2012

Chapter 7

Neuronal Plasticity in the Juvenile and Adult Brain Regulated by the Extracellular Matrix

Max F.K. Happel and Renato Frischknecht

Additional information is available at the end of the chapter

http://dx.doi.org/10.5772/62452

Abstract

In brains of higher vertebrates, the delicate balance of structural remodeling and stabilization of neuronal networks changes over the life-span. While the juvenile brain is characterized by high structural plasticity, it is more restricted in the adult. During brain maturation, the occurrence of the extracellular matrix (ECM) is a critical step to restrict the potential for neuronal remodeling and regeneration, but providing structural tenacity. How this putative limitation of adult neuronal plasticity might impact on learning-related plasticity, lifelong memory reformation, and higher cognitive functions is subject of current research. Here, we summarize recent evidence that recognizes the ECM and its activity-dependent modulation as a key regulator of learning-related plasticity in the adult brain. We will first outline molecular concepts of enzymatic ECM modulation and its impact on synaptic plasticity mechanisms. Thereafter, the ECM's role in converting juvenile to adult plasticity will be explained by several key studies in wild-type and genetic knockout animals. Finally, current research evidences the impact of ECM dynamics in different brain areas including neocortex on learning-related plasticity in the adult brain impacting on lifelong learning and memory. Experimental modulation of the ECM in local neuronal circuits further opens short-term windows of activity-dependent reorganization. Malfunctions of the ECM might contribute to a variety of neurological disorders. Therefore, experimental ECM modulation might not only promote complex forms of learning and cognitive flexible adaptation of valuable behavioral options, but has further implications for guided neuroplasticity with regenerative and therapeutic potential.

Keywords: Learning, Plasticity, Memory, Cortex, Protein turnover

1. Introduction

In the brains of higher vertebrates, both neurons and glial cells produce and secrete the molecules that accumulate and form the extracellular matrix (ECM). In the nineteenth century, pioneers of brain cell biology including Camillo Golgi and Santiago Ramon y Cajal have discovered a mesh-like structure surrounding the most neurons and synapses in the adult brain [1]. This extracellular scaffold has been seen initially as a key component for ensuring the structural stability of the respective tissue [2]. Later research evidenced that the ECM has multiple functions including regulation of cell adhesion, cell-to-cell communication, cell differentiation, cellular compartmentalization, and several forms of neuronal plasticity [3]. During brain maturation, the ECM undergoes profound changes. In late embryonic phases, the ECM is composed of particular proteoglycans and glycoproteins like neurocan and tenascin-C, which are downregulated during adulthood. Other components like the glycoprotein tenascin-R and chondroitin sulfate (CS) proteoglycans (CSPGs), such as brevican and aggrecan, are the main molecular components of the adult brain ECM. One of the most important structural components in the adult ECM is the unbranched polysaccharide hyaluronic acid (HA). HA forms the backbone that structurally orchestrates the enmeshment of all other ECM components. Interestingly, it has been shown that this developmental shift of the juvenile and adult forms of the ECM coincides with the closure of the so-called critical periods during brain maturation of respective brain regions. This led to the hypothesis that the brain's ECM is involved in regulating the switch from juvenile to adult plasticity by structural tenacity restricting the potential for neuronal reorganization [4]. Thereby, the brain has evolved mechanisms that guarantee structural stability of the neuronal networks established during experience-dependent learning. This brain function is fundamental for strengthening neuronal connections and their respective forms of processing impacting on long-term memory storage and recall. Nevertheless, current research has shown that dynamic adaptations of the ECM can alter several forms of synaptic plasticity and thereby regulate flexible learning and memory organization in the adult brain.

2. The structural foundation of the ECM in the vertebrate's brain

Components of the ECM in the mature brain are interlinked in a complex netlike architecture [5]. The linear HA backbone binds and coordinates proteoglycans especially of the lectican family that are cross-linked by glycoproteins such as the tenascins. Thus, this form of ECM is referred to as HA-based ECM. It is rich in the glycosaminoglycan CS attached to CSPGs of the lectican family, as for instance brevican and aggrecan [6]. The main carrier of CS within the ECM is aggrecan with its multiple attachment sites, while Brevican is a part-time proteoglycan existing glycoprotein and proteoglycan [7,8]. Brevican and aggrecan both bind to the ECM glycoprotein tenascin-R (**Figure 1A**). Other so-called cartilage link proteins form a complex with N-terminal domains of the lecticans and HA and thereby contribute to the ECM stability [6]. A large variety of other components including reelin, laminins, thrombospondins, heparin-sulfate proteoglycans, guidance molecules, and even transcription factors are incorporated in the complex ECM structure (**Figure 1A**). This form of the homogenous HA-based ECM loosely enwraps cell bodies, dendritic, and synaptic structures of most neurons. Further, the mature brain contains a specialized glycosaminoglycan-rich ECM structure around synapses and somata of a small proportion of neurons. This more rigid form is called the perineuronal nets (PNNs), which are especially rich in aggrecan and CS. Such PNNs are most abundant on

http://dx.doi.org/10.5772/62452

GABAergic interneurons expressing the calcium buffer protein parvalbumin (PV; **Figure 1B**). Recent evidence shows that PNNs are highly heterogeneous and can be found on various types of neurons throughout the CNS. For the formation of WFA-positive PNNs, the cartilage link protein Crtl1/Hapln1 has been identified as a key regulator [9].

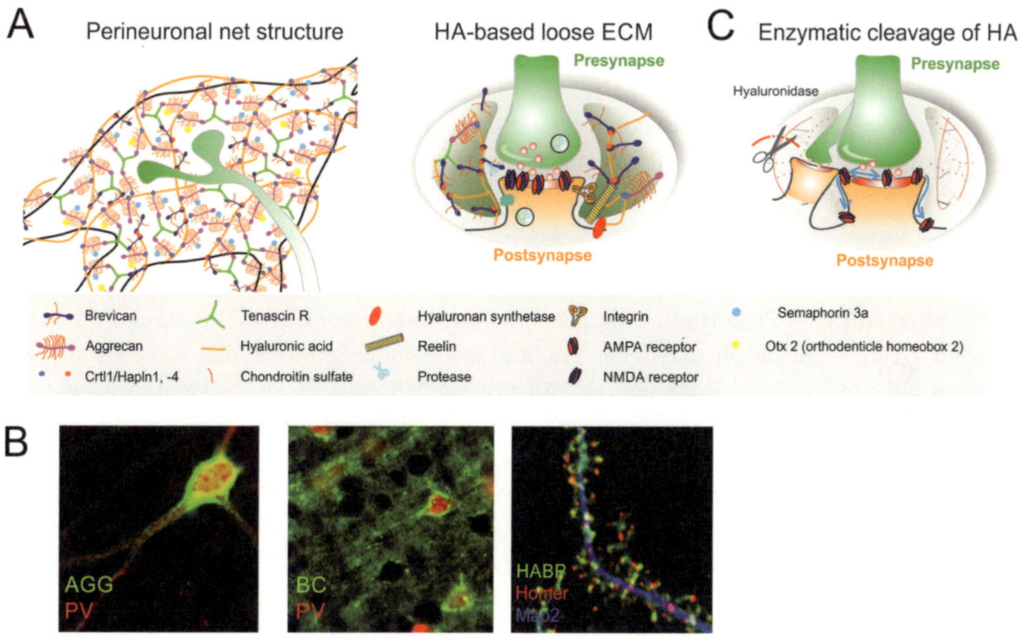

Figure 1. Cell types and their specific forms of the hyaluronan-based ECM. (**A**) The mature ECM is based on the backbone of hyaluronic acid (orange). Schematized are the specialized form of the PNN (*left*) and the loose ECM (*right*) around synaptic contacts. Densely packed PNNs are rich in CS and aggrecan. Small signaling molecules such as semaphorin 3a or Oxt2 are bound to CS within the PNN and mediate several functions or regulate gene expression. The loose ECM around excitatory, spiny synapses is rich in brevican and contains only little aggrecan and is thus less rich in CS. ECM proteins (e.g., reelin) signal through their receptors (e.g., integrins) and regulate several cellular processes including trafficking of glutamate receptors (see C). ECM function is modulated by proteases (scissors) that may free synapses by removing ECM or unmask signaling molecules. (**B**) *Left*, Parvalbumine positive interneuron (PV, red) with typical PNN stained for aggrecan (AGG, green). *Middle*, Dendritic spines and synapses of excitatory neurons (red) are also surrounded by brevican (BC, green), although it is less specific for PNN's. *Right*, Dissociated cortical neuron stained with the dendritic marker Map2 (blue) and Homer (red) to stain excitatory synapses and hyaluronic acid binding protein (HABP, green) to label extracellular matrix. Note the loose appearance around dendrites, spines, and synapses. (**C**) Enzymatic weakening of the ECM by glycosidases (e.g., hyaluronidase, see scissors) changes the micromillieu around synapses by for instance increased lateral diffusion and synaptic exchange of AMPA receptors (blue arrows). Further, removal of stabilizing cues facilitates structural plasticity such as *de-novo* formation of new synapses and synaptic scaling. Modified from Ref. [22].

The structural foundation of the mature ECM allows regulating several functions beyond mechanical stability of neuronal networks. In the juvenile brain, the ECM regulates neuro- and gliogenesis, cell migration, axonal pathfinding, and synaptogenesis. Adult forms of the HA-based ECM are in the service of multiple functions including regulation of synaptogenesis and synaptic plasticity, compartmentalization of the neuronal surface, neuroprotection, regulation of ion homeostasis, and neuron–glia interactions. The remainder of this review will consider several ECM-based mechanisms for regulating synaptic plasticity and its effects on brain development and adult learning behavior.

3. ECM-guided switch from juvenile to adult synaptic plasticity

Early in life, high structural plasticity allows profound shaping of brain circuits by experience. Such critical periods in the juvenile brain are limited by the occurrence of the ECM implementing adult brain plasticity modes. For instance, in wild-type animals, dark rearing delays not only the critical periods of developmental plasticity in visual cortex of rodents, but also the formation of PNNs. During the development, the cartilage link protein Crtl1/Hapln1 is organizing the formation of PNNs. Crtl1/Hapln1 k.o. mice do not develop normal PNN structures in the visual cortex. These mice show juvenile forms of ocular dominance (OD) plasticity and sensitivity of the visual system to deprivation throughout the life-span [9].

The seminal study by Pizzorusso et al. [10] has elucidated the ECM as the regulatory switch between juvenile and adult plasticity. The authors combined monocular deprivation with injection of the ECM-cleaving enzyme chondroitinase ABC (chABC) into visual cortex of adult rats (see also **Figure 1C**). The local weakening of the ECM "re-juvenated" the visual cortex and restored the critical period form of OD plasticity (**Figure 2**). By the same manipulation Pizzorusso and colleagues restored in a follow-up study, the visual acuity in adult animals grown up with long-term monocular deprivation [11]. Similarly, application of the serine protease tissue-type plasminogen activator (tPA) into the visual cortex can prolong or reactivate critical periods of OD plasticity in visual cortex [12] based on increased structural remodeling [13].

Later studies identified the regulatory role of the ECM in other forms of developmental plasticity during brain maturation of different vertebrate species. For instance, birdsong learning in the zebra finch occurs during a sensitive period similar to the language development in humans. It has been shown that with the end of this critical period PNNs around PV-positive neurons emerge in brain areas that are dedicated to singing [14]. In another set of experiments, Gogolla et al. [15] shown that the maturing ECM in the amygdala essentially makes fear memories erasure resistant in adult animals. In rats not older than 3–4 weeks, a conditioned fear memory trace can be erased permanently by extinction, that is, the presentation of the conditioned stimulus without the aversive stimulus. However, after this period extinction only attenuates the fear response, but it reinstates instantaneously if the aversive stimulus is presented again. Hence, a permanent loss of the fear memory is only found before the ECM in the amygdala is formed and is preserving established fear memories. Gogolla and colleagues now attenuated the ECM in the amygdala by chABC injections in adult rats. This led to a complete erasure of the fear response after an extinction phase even if the aversive stimulus is presented to these animals again. In addition, the early preweaning environment impacts on rodent ECM maturation in a functional manner. Improved performance in water maze learning in the adult age after early postnatal-enriched housing has been correlated with increased PNN formation in the striatum reflecting functional shaping of neuronal circuits involved in motor learning [16].

http://dx.doi.org/10.5772/62452

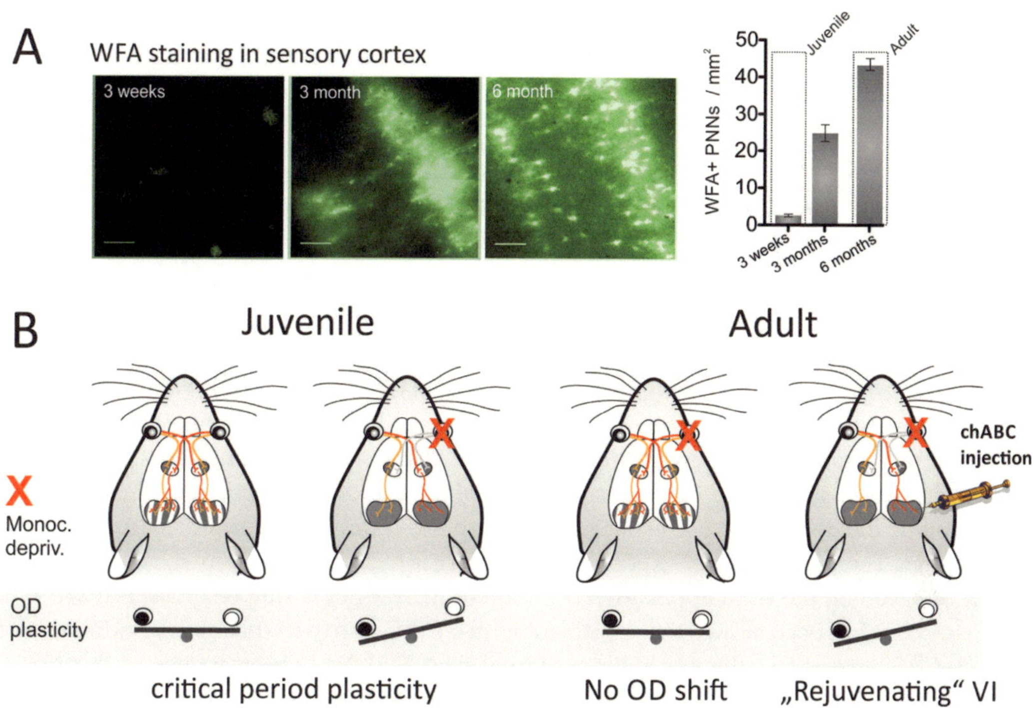

Figure 2. The emerging ECM during brain maturation and its regulatory role for juvenile brain plasticity (**A**) *Wisteria floribunda* (WFA) fluorescein staining against sugar chains of the CSPGs in rodent sensory cortex shows that the diffuse HA-based ECM and perineuronal nets around cells emerge with brain maturation (≥3 month), but were abundant in juvenile state (3 weeks). Modified from Ref. [42]. (**B**) Pizzorusso et al. [10] demonstrated the inhibitory role of the developmental maturation of the ECM in visual cortex of rats for early experience-dependent plasticity. In juvenile rats, monocular deprivation (MD) leads to an ocular dominance shift (left). After the critical period, MD alone did not cause such a shift in the adult. In this study, the authors found that weakening of the ECM by chABC treatment in visual cortex reactivated OD plasticity shifting toward the nondeprived eye.

The emerging ECM in the developing brain seems to be involved in a complex molecular machinery organizing the maturation of neuronal networks. For instance, in mouse visual cortex, it has been shown that critical periods are initiated, maintained, and closed by the action of the orthodenticle homeobox 2 homeoprotein (Otx2) on PV+-GABAergic interneurons [17]. Interestingly, Otx2 contains a glycosaminoglycan binding sequence that mediates its allocation to the PNNs by specific binding to CS D and E. Thereby, a constant level of Otx2 in PV+-neurons keeps a mature, consolidated, and persistent PNN-state in the adult brain. Hydrolysis of the PNNs by chABC reduces the amount of endogenous Otx2 in PV cells mediating the reopening of OD plasticity in adult mice [18].

The reviewed studies in this paragraph have shown how the change of the juvenile form of the ECM to a more rigid adult ECM mediates the different modes of neuronal plasticity during brain development. Why has this restriction of adult reorganizational and regenerative plasticity only evolved in higher vertebrates? The evolutionary benefit may be to preserve the costly acquired hardwired connections during early life experience, which are fundamental for rapid experience-based behavioral adaptations of higher vertebrates [4]. Nonetheless, the adult, healthy brain retains a remarkable capability of plastic reorganization that is essential for constantly adapting to our ever-changing environment.

4. The hyaluronic acid-based ECM regulates adult synaptic plasticity

The functional mechanisms by which the HA-based ECM implements adult forms of brain plasticity including classical (Hebbian) and homeostatic plasticity are still elusive (for an overview see [3]). Research investigating knockout models lacking particular components of the ECM have provided major insights into the impact of the ECM on adult synaptic plasticity. For instance, different mouse models deficient in specific matrix components, as tenascin-R, brevican, or neurocan, all showed impaired forms of hippocampal long-term potentiation (LTP). However, during the brain development of k.o.-models compensatory mechanisms might mimic the deficit of a particular matrix component and hence limit the significance of the findings. Therefore, another experimental strategy is to use acute enzymatic weakening of the ECM based on the local application of different matrix-degrading enzymes (**Figure 1C**). For instance, it has been shown that treatment with chABC impaired theta-burst-induced LTP in CA1–CA3 pyramidal synapses in hippocampal slices [19]. It has been suggested that this is due to an increased excitability of GABAergic perisomatic interneurons [3]. Similar findings have been found with Injection of the hyaluronidase (HYase) from *Streptomyces hyalurolyticus* [20]. In contrast to other hyaluronidases, this enzyme is highly specific to HA and does not digest CS. The phenotype in these experiments could be rescued by perfusion with HA. It has been suggested that this is due to the fact that HA directly regulates L-type voltage-gated calcium channels (L-VDCC; Cav1.2) of CA1 neurons and thus postsynaptic Ca^{2+} entry and hippocampal-dependent forms of learning [3,20]. In a recent study, we have described a molecular mechanism by which acute ECM removal was altering short-term-dependent forms of synaptic plasticity [21]. We measured paired pulse ratios in dissociated hippocampal cultures by cell-attached recordings. We found the typical robust paired pulse depression (PPD) under control conditions. Interestingly, digestion of the ECM by infusion of HYase prevented cells from expressing PPD. We further found increased lateral diffusion of extrasynaptic AMPA receptors after ECM digestion as key correlate of this effect [21]. This results in a higher exchange between synaptic desensitized receptors with extrasynaptic naïve ones, which quickly replenishes the pool of excitable receptors in the active zone of a synapse (**Figure 1C**). In such conditions, synapses are able to follow higher firing frequencies. Blockade of lateral diffusion of AMPA receptors by cross-linking with antibodies, on the other hand, led to stronger PPD caused by accumulated desensitized synaptic AMPA receptors [22]. Similarly, mobility of other receptor types can be modulated by specific proteolytic enzymes. For instance, MMP9 increases the mobility of NMDA-type, but not AMPA receptors. These results have shown that the perisynaptic ECM forms surface compartments that act as diffusion barrier for membrane proteins such as AMPA receptors. Modulation of lateral receptor diffusion provides a novel mechanism of short-term plasticity due to a changed extracellular micromillieu at individual synapses. Interestingly, lack of the major hyaluronan synthetase

http://dx.doi.org/10.5772/62452

HAS3 in the hippocampus did not lead to a striking morphological change in the ECM. However, the extracellular space was reduced, which resulted in a more dense packing of cells and lower diffusion of soluble molecules in the CA1 *stratum pyramidale* [23]. Further, these animals were prone to epileptic seizures which underlines the importance of the regulatory function of the HA-based ECM that is required for instance for volume transmission and ion homeostasis.

Weakening of the ECM and the corresponding synaptic networks might further promote architectural remodeling and even increased spine motility. In *in vitro* hippocampal cultures, microinjections of chABC treatment increased the motility of local spines and induced spine remodeling in a β1-integrin-dependent manner [24].

5. ECM proteolysis and the generation of synaptic signaling molecules

Endogenous ECM-modulating enzymes regulate synaptic function in the juvenile and adult brain [25,26]. Such enzymes can exert their function by either altering the extracellular milieu via digestion of the ECM or by generating proteolytic fragments that may act as signaling molecules. An important group of such enzymes are the metalloproteases of the ADAMTS-family (a disintegrin and metalloproteinase with thrombospondin motifs). Within this family, ADAMTS-4/-5 are particularly interesting, as they are known for their ability to digest aggrecan and brevican. Therefore, they have been termed previously aggrecanase-1/-2. The current terminology, however, better reflects the ability of these enzymes to digest all members of the lectican family. Interestingly, their activity is increased after epileptic seizures and regulates homeostatic plasticity [27]. However, their impact on synaptic plasticity remains elusive and is subject of current research. The best-studied extracellular protease is the matrix metalloprotease 9 (MMP9). The activity-dependent expression of MMP9 influences synaptic plasticity by regulating spine enlargement and synaptic potentiation [25]. Recently, a molecular signaling cascade regulating synaptic plasticity has been identified based on the MMP-9-dependent cleavage of neuroligin-1 [28]. This study has demonstrated that focal activation of a single spine by glutamate uncaging is sufficient to cleave neuroligin-1. Moreover, the activity of MMP9 has been shown to be NMDA-receptor dependent and hence implemented locally input-specific forms of synaptic plasticity. Thereby, extracellular MMP-9 triggers a specific retrograde regulation of presynaptic efficacy by targeting postsynaptic neuroligin-1 [26,28]. Similarly, the brain-specific serine protease neurotrypsin is regulated in an activity-dependent manner and requires concomitant activation of the postsynaptic neuron [29]. Proteolytic cleavage of agrin by neurotrypsin unmasks a signaling molecule harboring a single laminin G3 domain. This 22 kDa molecule can further regulate spine morphology and de-novo synapse generation. Together, this suggests that proteolysis of components of the ECM by exoenzymes not only modify the structural rigidity, but also activates instructive signal molecules that locally modulate synaptic functions [25]. This may temporally restore local divisions of "juvenile" environments as a major constituent of the balance between plasticity and tenacity in the mature brain.

We have shown that the ECM in the adult brain is a plastic structural scaffold shaped by network activity. Depending on the current activity level, the ECM can incorporate secreted components or release signaling messengers by proteolytic cleavage. Cleaved products can trigger signaling through diverse ECM receptors and modulate the activities of transmitter receptor, ion channels, or integrin signaling impacting on plastic shaping of individual synapses.

6. Role of the ECM in control of adult learning behavior and cognitive flexibility

Experimental weakening of the ECM by local injection of matrix-digesting enzymes can promote functional neurorehabilitation in the injured brain. This has been related mostly to injuries on the level of the peripheral nervous system and spinal cord [4,30,31]. Experience-driven plasticity does, however, not only lead sensory development or neuronal rehabilitation, but is also indispensable during learning, memory formation, and re-consolidation throughout life. The question now arises how forms of ECM-dependent plasticity in the adult brain might govern learning-related plasticity, lifelong memory reformation, and the organization of cognitively flexible behavior. In this respect, several studies have investigated the involvement of ECM functions in memory storage in adult animals. This has been characterized the best for long-term plasticity in the hippocampus and fear memory in the amygdala. However, available evidence is controversial about how ECM functions may impact on learning and memory processes.

For instance, it has been reported that tenascin-R knockout mice show normal hippocampus-dependent spatial memory acquisition in a Water maze. In subsequent reversal learning though animals showed more vulnerable spatial long-term memory yielding enhanced relearning performance due to less conflicting past and actual learning contingencies [32]. Another study found an already impaired acquisition of hippocampus-dependent contextual memory in same knockout mice [33]. Injection of chABC in the bilateral striatum, however, has been related to an improvement of water maze acquisition learning, while the recall of the learned values was unaffected [34].

In addition to deficits in matrix components, studies also found effects of deficits in exoproteases modeling the ECM. Loss of MMP9 activity has been associated with impaired hippocampal-dependent learning and amygdala-dependent learning. This is in line with findings of wild-type mice trained in an inhibitory avoidance (IA) learning paradigm [35,36]. Hippocampal LTP has been related to increased levels of MMP3 and MMP9. Both proteases were upregulated for at least ~48 h promoting local plastic synaptic environments underlying the learning performance. Intra-hippocampal injections of MMP9 blockers completely abrogated memory for the IA response when tested days later. Comparably, hippocampal MMP3 and MMP9 were found to be increased during water maze acquisition learning in a NMDA-dependent manner. Hippocampal injection of the broad-spectrum MMP9 inhibitor FN-439 also prevented elevated MMP9 levels, altered hippocampal LTP, and prevented spatial acquisition learning [35].

Similarly, spatial training in a water maze in wild-type rats has been found to correlate with increased levels of hippocampal brevican and versican in the membrane fraction [37]. These findings indicate that hippocampal-dependent learning induced a period of intrinsic activity-induced focal MMP-mediated proteolysis driving long-lasting synaptic modifications underlying learning and memory consolidation. Effects of changes in the ECM on initial learning are, however, still unclear [25,33].

More recently, insights into the impact of the ECM onto behavior came from studies using experimental, enzymatic, and local weakening of the ECM. The study of Gogolla et al. manipulating adult fear extinction suggested that memory acquisition differs in juvenile and adult brains due to changes of the mature ECM functions. The authors further argued that intra-amygdala injections of chABC in adult rats had no effect on acquisition learning of fear, but only on extinction, reinstatement, and renewal of the fear memory [15]. A further study showed that intra-hippocampal and prefrontal injection of chABC and HYase in mice impair long-term trace contextual fear conditioning [38]. This finding has been related to the impairment in the L-VDCC-dependent component of hippocampal LTP by cleaved extracellular HA [20].

In addition to spatial memory, another set of studies examined the function of the ECM in memory consolidation of drug seeking. A recent study showed that intracerebral injection of FN-439 impaired the acquisition of a cocaine-induced conditioned place preference (CPP) of rats. FN-439 injection 30 min prior to cocaine memory re-activation further attenuated the reinstatement of CPP in extinguished animals. The study further showed that intra-amygdala injections of chABC during active extinction of cocaine-induced CPP prevented its subsequent priming-induced reinstatement. ChABC injections alone had no effect on the retention, retrieval, or relearning of CPP [39]. Similarly, enzymatic weakening of PNNs in the prelimbic cortex or in the amygdala of adult rats impaired the acquisition and reconsolidation of drug-induced memories [40,41].

With respect to cognitively flexible adaptation of behavior, we have recently shown that weakening of the ECM in auditory cortex promotes complex forms of cortex-dependent relearning in the Mongolian gerbil [42]. In our experiments, we trained animals on frequency-modulated tone discrimination based on the rising or falling modulation direction in a go-/nogo-task. Such auditory learning is known to depend on learning-induced plastic reorganization of neuronal circuits in the auditory cortex. After acquiring robust discrimination of the stimulus contingencies, the animals were trained to reverse their choice. We found that ECM weakening by local HYase injection in bilateral auditory cortex accelerated the demanding relearning performance (**Figure 3**). Specifically, animals had to inhibit the obsolete initial behavioral strategy and then establish its successful reversal. Importantly, attenuation of the ECM did neither affect the acquisition learning nor erased already established, learned memory traces (**Figure 3B**). That means attenuation of the ECM in sensory cortex of these animals promoted the flexible adaptation of the effectively appropriate strategy during cortex-dependent learning behavior that bases on "reprogramming" previously acquired auditory memories. The ECM reconstitutes after several days to weeks limiting again the promoting effects onto cognitive flexibility (**Figure 3A**). A comparable finding investigated log-term object recognition memory in knockout mice of the link protein Crtl1/Hapln1—a key molecule for stabilization of PNNs. The Ctrl1/Hapln1 knockout mice have attenuated PNNs in the perirhinal cortex. Long-term object recognition memory, a task depending on perirhinal cortex, was enhanced in these mice. Local injection of chABC in wild-type mice had the same memory-prolonging effect in the object recognition task, but also attenuated over time [43]. In this study, the attenuation of the PNNs was accompanied by enhanced perirhinal LTD, which is thought to be the major synaptic mechanism underlying object recognition memory.

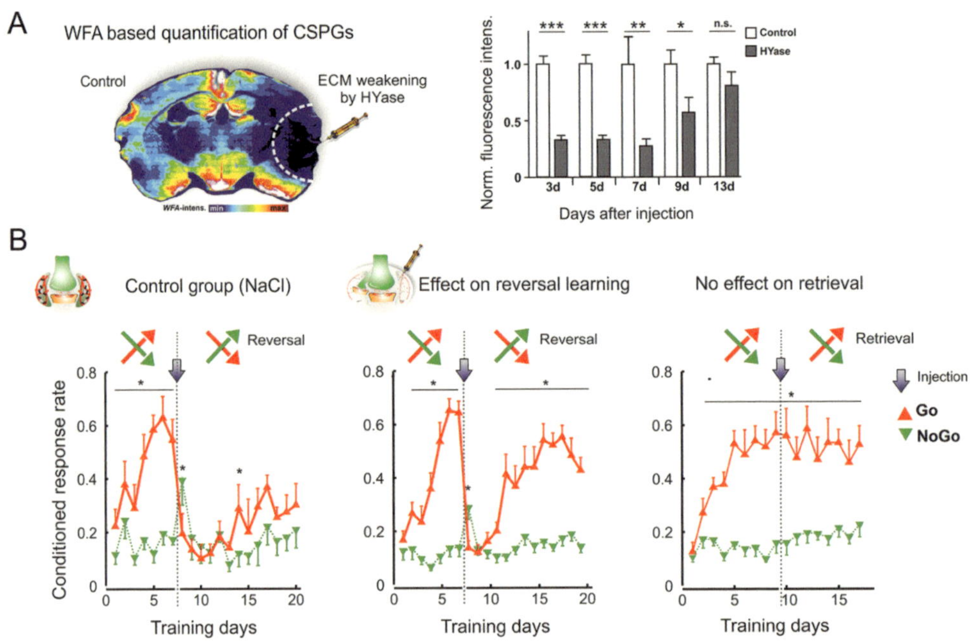

Figure 3. Local enzymatic weakening of the ECM in auditory cortex of Mongolian gerbils enhanced the cognitive flexibility in a relearning paradigm. (**A**) *Right*, Quantification of ECM weakening after local injection of HYase in unilateral auditory cortex of Mongolian gerbils (right) compared to control (left) based on WFA staining. *Left*, HYase injection significantly weakened the ECM for about 1 week and reconstituted fully after 2 weeks. (**B**) Mongolian gerbils were trained in a two-compartmental go/no-go Shuttle-box in order to discriminate two frequency-modulated sounds (modulation direction indicated by rising and falling arrows). Gerbils showed successful acquisition depending on the contingency of the stimuli as a go-stimulus (red) or Nogo-stimulus (green). In two groups, the contingency was reversed after seven training days (*left and middle*). Conditioned response rates were strongly reduced in both experimental groups indicating the active inhibition of the previously established discrimination strategy. HYase-treated animals were significantly better in correcting the behavioral strategy and successfully relearn the task (middle). Interestingly, HYase treatment did not interfere with the recall of already established cortex-dependent auditory memories (*right*). Modified from Ref. [42].

The both last-mentioned studies therefore promote the view that the perineuronal ECM in the adult brain actively organizes the balance between memory stability and flexibility. Cortical attenuation of the ECM in the mature brain might hence promote the cognitive flexibility that can build on learned behaviors and allows for an enhanced activity-dependent memory re-

http://dx.doi.org/10.5772/62452

organization (see also Ref. [44]). And regeneration of the ECM gradually restores normal, restrictive adult plasticity levels. Generally, all studies summarized in this review emphasized that the increased experience-based plasticity by acute, enzymatic PNN diminution is activity-dependent and the rather inconspicuous effect of mere ECM attenuation in general. Mechanistically, enzymatic ECM degradation might facilitate the rearrangement of functional network connectivity by a shift in the balance between excitation and inhibition leading to destabilized existing patterns of neuronal network interactions [45].

7. Impact of the ECM in old age on memory function and cognitive integrity

With respect to age, the importance to provide profound tenacity to conserve experience-based memories might increase over the life span. Deficits or malfunctions of several ECM molecules or ECM-chopping enzymes can affect cognitive and psychological conditions. For instance, in humans neurotrypsin has been identified as essential component for cognitive functions. Deficits in the neurotrypsin genotype have been correlated with severe mental retardation [46]. Further, Cichon et al. [47] reported a genetic variation of neurocan as susceptibility factor for bipolar disorders. With relation to ageing, hippocampal ECM levels have been suggested to show an age-dependent increase conquering age-related cognitive decline. In this line, the Alzheimer's disease (AD) mouse model APP/PS1 showed a significant upregulation of several matrix components correlating with impairments in hippocampal LTP and contextual memory [48]. Intra-hippocampal injections of chABC restored both [48] suggesting an important, but yet elusive role for the ECM in early memory impairment in AD, as the mere correlative findings about ECM alterations in dementia are highly controversial and are far from conclusive [33]. That these data might have relevant impacts for human AD is indicated by findings of correlating HA levels in the cerebrospinal fluid of female AD patients and particular AD-related biomarkers [49]. Further, MMP9 levels have been found to be increased in Alzheimer patients [50] and to cleave the amyloid beta peptide leading to AD-typical neuritic plaques [51]. Its role in Aβ-induced cognitive decline is however elusive [52].

8. Outlook

We have summarized recent evidence showing that experimental modulation of the ECM promotes "windows of opportunities" with an increase in learning-related plasticity yielding cognitively flexible adaptation of learned behaviors and the underlying memories. How the ECM, in addition, impacts onto several mental disorders that generally develop after the closure of major critical periods for higher brain functions, as for instance affective disorders or schizophrenia, are exciting new research directions. We are envisaging future challenges in developing new tools for guided neuroplasticity with therapeutic potential for memory disorders, stroke, or neuroprosthetic applications based on ECM manipulations.

Author details

Max F.K. Happel[1,2*] and Renato Frischknecht[3*]

*Address all correspondence to: mhappel@lin-magdeburg.de and rfrischk@lin-magdeburg.de

1 Department Systems Physiology of Learning, Leibniz-Institute for Neurobiology, Magdeburg, Germany

2 Institute for Biology, Otto-von-Guericke University Magdeburg, Magdeburg, Germany

3 AG Perisynaptic Extrazellular Matrix, Department Neurochemistry and Molecular Biology, Leibniz-Institute for Neurobiology, Magdeburg, Germany

References

[1] Celio MR, Spreafico R, De Biasi S, Vitellaro-Zuccarello L: Perineuronal nets: past and present. *Trends Neurosci* 1998, 21:510–5.

[2] Berardi N, Pizzorusso T, Maffei L: Extracellular matrix and visual cortical plasticity: freeing the synapse. *Neuron* 2004, 44:905–8.

[3] Dityatev A, Schachner M, Sonderegger P: The dual role of the extracellular matrix in synaptic plasticity and homeostasis. *Nat Rev Neurosci* 2010, 11:735–46.

[4] Gundelfinger ED, Frischknecht R, Choquet D, Heine M: Converting juvenile into adult plasticity: a role for the brain's extracellular matrix. *Eur J Neurosci* 2010, 31:2156–65.

[5] Berardi N, Pizzorusso T, Ratto GM, Maffei L: Molecular basis of plasticity in the visual cortex. *Trends Neurosci* 2003, 26:369–78.

[6] Yamaguchi Y: Lecticans: organizers of the brain extracellular matrix. *Cell Mol Life Sci* 2000, 57:276–89.

[7] Giamanco KA, Morawski M, Matthews RT: Perineuronal net formation and structure in aggrecan knockout mice. *Neuroscience* 2010, 170:1314–27.

[8] Frischknecht R, Seidenbecher CI: Brevican: a key proteoglycan in the perisynaptic extracellular matrix of the brain. *Int J Biochem Cell Biol* 2012, 44:1051–54.

[9] Carulli D, Pizzorusso T, Kwok JCF, Putignano E, Poli A, Forostyak S, Andrews MR, Deepa SS, Glant TT, Fawcett JW: Animals lacking link protein have attenuated perineuronal nets and persistent plasticity. *Brain* 2010, 133:2331–47.

[10] Pizzorusso T, Medini P, Berardi N, Chierzi S, Fawcett JW, Maffei L: Reactivation of ocular dominance plasticity in the adult visual cortex. *Science* 2002, 298:1248–51.

[11] Pizzorusso T, Medini P, Landi S, Baldini S, Berardi N, Maffei L: Structural and functional recovery from early monocular deprivation in adult rats. *Proc Natl Acad Sci USA* 2006, 103:8517–22.

[12] Oray S, Majewska A, Sur M: Dendritic spine dynamics are regulated by monocular deprivation and extracellular matrix degradation. *Neuron* 2004, 44:1021–30.

[13] Hensch TK: Critical period plasticity in local cortical circuits. *Nat Rev Neurosci* 2005, 6:877–88.

[14] Balmer TS, Carels VM, Frisch JL, Nick TA: Modulation of perineuronal nets and parvalbumin with developmental song learning. *J Neurosci* 2009, 29:12878–85.

[15] Gogolla N, Caroni P, Lüthi A, Herry C: Perineuronal nets protect fear memories from erasure. *Science* 2009, 325:1258–61.

[16] Simonetti T, Lee H, Bourke M, Leamey CA, Sawatari A: Enrichment from birth accelerates the functional and cellular development of a motor control area in the mouse. *PLoS One* 2009, 4:e6780.

[17] Sugiyama S, Di Nardo AA, Aizawa S, Matsuo I, Volovitch M, Prochiantz A, Hensch TK: Experience-dependent transfer of Otx2 homeoprotein into the visual cortex activates postnatal plasticity. *Cell* 2008, 134:508–20.

[18] Beurdeley M, Spatazza J, Lee HHC, Sugiyama S, Bernard C, Di Nardo AA, Hensch TK, Prochiantz A: Otx2 binding to perineuronal nets persistently regulates plasticity in the mature visual cortex. *J Neurosci* 2012, 32:9429–37.

[19] Bukalo O, Schachner M, Dityatev A: Modification of extracellular matrix by enzymatic removal of chondroitin sulfate and by lack of tenascin-R differentially affects several forms of synaptic plasticity in the hippocampus. *Neuroscience* 2001, 104:359–69.

[20] Kochlamazashvili G, Henneberger C, Bukalo O, Dvoretskova E, Senkov O, Lievens PM-J, Westenbroek R, Engel AK, Catterall WA, Rusakov DA, Schachner M, Dityatev A: The extracellular matrix molecule hyaluronic acid regulates hippocampal synaptic plasticity by modulating postsynaptic L-type Ca(2+) channels. *Neuron* 2010, 67:116–28.

[21] Frischknecht R, Heine M, Perrais D, Seidenbecher CI, Choquet D, Gundelfinger ED: Brain extracellular matrix affects AMPA receptor lateral mobility and short-term synaptic plasticity. *Nat Neurosci* 2009, 12:897–904.

[22] Heine M, Groc L, Frischknecht R, Béïque J-C, Lounis B, Rumbaugh G, Huganir RL, Cognet L, Choquet D: Surface mobility of postsynaptic AMPARs tunes synaptic transmission. *Science* 2008, 320:201–5.

[23] Arranz AM, Perkins KL, Irie F, Lewis DP, Hrabe J, Xiao F, Itano N, Kimata K, Hrabetova S, Yamaguchi Y: Hyaluronan deficiency due to Has3 knock-out causes altered neuronal

activity and seizures via reduction in brain extracellular space. *J Neurosci* 2014, 34:6164–76.

[24] Orlando C, Ster J, Gerber U, Fawcett JW, Raineteau O: Perisynaptic chondroitin sulfate proteoglycans restrict structural plasticity in an integrin-dependent manner. *J Neurosci* 2012, 32:18009–17, 18017a.

[25] Huntley GW: Synaptic circuit remodelling by matrix metalloproteinases in health and disease. Nat Rev Neurosci. 2012, 13(11):743–57

[26] Shinoe T, Goda Y: Tuning synapses by proteolytic remodeling of the adhesive surface. *Curr Opin Neurobiol* 2015, 35:148–55.

[27] Juan RV, Christopher H, Gilbert F, Jeet S, Barbara S, Constanze S, Frischknecht R: Hyaluronan-based extracellular matrix under conditions of homeostatic plasticity. *Philos Trans R Soc B Biol Sci* 2014. 19;369(1654):20130606

[28] Peixoto RT, Kunz PA, Kwon H, Mabb AM, Sabatini BL, Philpot BD, Ehlers MD: Transsynaptic signaling by activity-dependent cleavage of neuroligin-1. *Neuron* 2012, 76:396–409.

[29] Frischknecht R, Fejtova A, Viesti M, Stephan A, Sonderegger P: Activity-induced synaptic capture and exocytosis of the neuronal serine protease neurotrypsin. *J Neurosci* 2008, 28:1568–79.

[30] Bowes C, Massey JM, Burish M, Cerkevich CM, Kaas JH: Chondroitinase ABC promotes selective reactivation of somatosensory cortex in squirrel monkeys after a cervical dorsal column lesion. *Proc Natl Acad Sci USA* 2012, 109:2595–600.

[31] Fawcett JW: The extracellular matrix in plasticity and regeneration after CNS injury and neurodegenerative disease. *Prog Brain Res* 2015, 218:213–26.

[32] Morellini F, Sivukhina E, Stoenica L, Oulianova E, Bukalo O, Jakovcevski I, Dityatev A, Irintchev A, Schachner M: Improved reversal learning and working memory and enhanced reactivity to novelty in mice with enhanced GABAergic innervation in the dentate gyrus. *Cereb Cortex* 2010, 20:2712–27.

[33] Dityatev A, Wehrle-Haller B, Pitkänen A: *Brain extracellular matrix in health and disease.* Progress in Brain Research Vol. 214; 2014 - Elsevier.

[34] Lee H, Leamey CA, Sawatari A: Perineuronal nets play a role in regulating striatal function in the mouse. *PLoS One* 2012, 7:e32747.

[35] Meighan SE, Meighan PC, Choudhury P, Davis CJ, Olson ML, Zornes PA, Wright JW, Harding JW: Effects of extracellular matrix-degrading proteases matrix metalloproteinases 3 and 9 on spatial learning and synaptic plasticity. *J Neurochem* 2006, 96:1227–41.

[36] Nagy V, Bozdagi O, Huntley GW: The extracellular protease matrix metalloproteinase-9 is activated by inhibitory avoidance learning and required for long-term memory. *Learn Mem* 2007, 14:655–64.

[37] Saroja SR, Sase A, Kircher SG, Wan J, Berger J, Höger H, Pollak A, Lubec G: Hippocampal proteoglycans brevican and versican are linked to spatial memory of Sprague–Dawley rats in the morris water maze. *J Neurochem* 2014, 130:797–804.

[38] Hylin MJ, Orsi SA, Moore AN, Dash PK: Disruption of the perineuronal net in the hippocampus or medial prefrontal cortex impairs fear conditioning. *Learn Mem* 2013, 20:267–73.

[39] Brown TE, Forquer MR, Cocking DL, Jansen HT, Harding JW, Sorg BA: Role of matrix metalloproteinases in the acquisition and reconsolidation of cocaine-induced conditioned place preference. *Learn Mem* 2007, 14:214–23.

[40] Xue Y-X, Xue L-F, Liu J-F, He J, Deng J-H, Sun S-C, Han H-B, Luo Y-X, Xu L-Z, Wu P, Lu L: Depletion of perineuronal nets in the amygdala to enhance the erasure of drug memories. *J Neurosci* 2014, 34:6647–58.

[41] Slaker M, Churchill L, Todd RP, Blacktop JM, Zuloaga DG, Raber J, Darling RA, Brown TE, Sorg BA: Removal of perineuronal nets in the medial prefrontal cortex impairs the acquisition and reconsolidation of a cocaine-induced conditioned place preference memory. *J Neurosci Off J Soc Neurosci* 2015, 35:4190–202.

[42] Happel MFK, Niekisch H, Castiblanco Rivera LL, Ohl FW, Deliano M, Frischknecht R: Enhanced cognitive flexibility in reversal learning induced by removal of the extracellular matrix in auditory cortex. *Proc Natl Acad Sci USA* 2014, 111:2800–5.

[43] Romberg C, Yang S, Melani R, Andrews MR, Horner AE, Spillantini MG, Bussey TJ, Fawcett JW, Pizzorusso T, Saksida LM: Depletion of perineuronal nets enhances recognition memory and long-term depression in the perirhinal cortex. *J Neurosci* 2013, 33:7057–65.

[44] Tsien RY: Very long-term memories may be stored in the pattern of holes in the perineuronal net. *Proc Natl Acad Sci USA* 2013, 110:12456–61.

[45] Bikbaev A, Frischknecht R, Heine M: Brain extracellular matrix retains connectivity in neuronal networks. *Sci Rep* 2015, 5:14527.

[46] Senkov O, Andjus P, Radenovic L, Soriano E, Dityatev A: Neural ECM molecules in synaptic plasticity, learning, and memory. *Prog Brain Res* 2014, 214(January 2016):53–80.

[47] Cichon S, Mühleisen TW, Degenhardt FA, Mattheisen M, Miró X, Strohmaier J, Steffens M, Meesters C, Herms S, Weingarten M, Priebe L, Haenisch B, Alexander M, Vollmer J, Breuer R, Schmäl C, Tessmann P, Moebus S, Wichmann HE, Schreiber S, Müller-Myhsok B, Lucae S, Jamain S, Leboyer M, Bellivier F, Etain B, Henry C, Kahn JP, Heath S, Hamshere M, et al.: Genome-wide association study identifies genetic variation in neurocan as a susceptibility factor for bipolar disorder. *Am J Hum Genet* 2011, 88:372–81.

[48] Végh MJ, Heldring CM, Kamphuis W, Hijazi S, Timmerman AJ, Li K, van Nierop P, Mansvelder HD, Hol EM, Smit AB, van Kesteren RE: Reducing hippocampal extracel-

lular matrix reverses early memory deficits in a mouse model of Alzheimer's disease. *Acta Neuropathol Commun* 2014, 2:76.

[49] Nielsen HM, Palmqvist S, Minthon L, Londos E, Wennstrom M: Gender-dependent levels of hyaluronic acid in cerebrospinal fluid of patients with neurodegenerative dementia. *Curr Alzheimer Res* 2012, 9:257–66.

[50] Asahina M, Yoshiyama Y, Hattori T: Expression of matrix metalloproteinase-9 and urinary-type plasminogen activator in Alzheimer's disease brain. *Clin Neuropathol* 2001, 20:60–3.

[51] Backstrom JR, Lim GP, Cullen MJ, Tökés ZA: Matrix metalloproteinase-9 (MMP-9) is synthesized in neurons of the human hippocampus and is capable of degrading the amyloid-beta peptide (1–40). *J Neurosci* 1996, 16:7910–9.

[52] Mizoguchi H, Yamada K, Nabeshima T: Matrix metalloproteinases contribute to neuronal dysfunction in animal models of drug dependence, Alzheimer's disease, and epilepsy. *Biochem Res Int* 2011, 2011:681385.

Section 3

ECM of Connective Tissues and Organs

Chapter 8

CCN Family: Matricellular Proteins in Cartilage and Bone Development

John A. Arnott, Kathleen Doane and
Sonia Lobo Planey

Additional information is available at the end of the chapter

http://dx.doi.org/10.5772/63199

Abstract

The extracellular matrix is the intricate scaffolding which surrounds and supports cells and helps to organize them into tissues and organs. The CCN family of matricellular proteins helps to regulate and modulate production, degradation, and remodeling of the extracellular matrix. In this chapter, we review the extracellular matrix of cartilage and bone, including an overview of chondrogenesis and skeletogenesis, and summarize the importance of the CCN proteins in establishment of the skeletal system. CCN proteins have both positive and negative regulatory roles in skeletal development, and their abnormal expression is related to the pathogenesis of several diseases observed in cartilage and bone that arise when inflammation or tissue injury becomes chronic, including fibrosis, arthritis, and cancer. Understanding the biological functions of the CCN proteins within this context offers opportunities for developing therapeutics by targeting CCN functions.

Keywords: matricellular proteins, CCN family, chondrogenesis; osteogenesis, extracellular matrix

1. Introduction

The extracellular matrix (ECM) is the intricate scaffolding, which surrounds and supports cells, and helps to organize them into tissues and organs. The composition of the extracellular matrix varies based on tissue type and organ function, and this matrix affects many cellular behaviors including proliferation, differentiation, and wound healing. Matricellular proteins, present in the immediate environment of the cell, help to regulate and modulate these functions. We present a discussion of the important matrix elements of cartilage and bone, and the supporting connective tissues of the body, with regard to their function in health and disease. The major aim of this book chapter is to summarize the role of extracellular matrix proteins and their roles in chondrogenesis and osteogenesis. Additionally, we will discuss the importance of the CCN proteins—a family of secreted extracellular matrix-associated proteins—in functional pathways of chondrogenesis and osteogenesis, and include a discussion of the role of CCNs in pathophysiology of these tissues. The reader will gain an increased awareness of the importance of CCN proteins in a wide range of important functional pathways, providing in-depth focus on their role in bone and cartilage development and repair. These proteins have also been implicated in many human diseases and thus are important targets for drug discovery and development.

2. Overview of chondrogenesis

Cartilage is a specialized type of supporting connective tissue that contains cartilage cells, or chondrocytes, and a specialized extracellular matrix. The type of matrix is specific to the type of cartilage. Depending on the type of cartilage, there may be a perichondrium present, which is a connective tissue layer that contains mesenchymal cells or fibroblasts, as well as chondroblasts. Cartilage develops from mesenchyme cells that commit to become chondrocytes. Cartilage is first seen in the developing embryo in the areas that serve as templates for endochondral ossification, as well as in Meckel's cartilage in the head area [1]. Differentiated chondrocytes are surrounded by matrix and can appear to have space around them and they retain their proliferative capacity within their matrix. Development of cartilage involves several different transcription factors, such as Sox 9, that control this process, as well as growth factors that interact with cellular receptors to facilitate various cellular functions [2]. Interaction with the extracellular matrix has been implicated in development as well. Various epigenetic mechanisms and several different microRNAs also are important in the development of cartilage [3,4]. Chondrocytes may remain as cartilage in specific regions of the body or may be used as a template for the formation of many of the bones in the body during endochondral ossification. Cartilage remains in the adult in several specific places including the bridge of the nose, the trachea and bronchi, the ear, the intervertebral discs of the spinal column, and on the ends of bones as articular cartilage [5]. Chondrocytes are usually considered quiescent in the adult, although this can change during pathological processes such as osteoarthritis. Cartilage has very little capacity to regenerate, even though progenitor cells are present in the adult [6]. This lack of repair ability causes a number of disease issues as humans age, including the most prevalent form of degenerative joint disease, osteoarthritis [7].

2.1. Types of cartilage

Three kinds of cartilage have been identified: hyaline, elastic, and fibrous based on differences in matrix and cellular arrangement and function [5,8]. Hyaline is present in the bronchi, parts of the trachea, and also as articular cartilage. It also forms the model for bone formation during endochondral ossification. Another elastic cartilage is present in the oropharynx and ear, among other regions of the body. Both of these types of cartilage are typically surrounded by a connective tissue layer and a chondrogenic layer of the perichondrium. Cells from this layer can undergo proliferation and can differentiate into chondrocytes for growth or renewal of cartilage, a type of growth that is appositional in that the newly formed chondrocytes come from outside the cartilage matrix. In addition, cells within the cartilage can proliferate, forming isogenous groups of similar chondrocytes, which is termed interstitial growth. By contrast, articular cartilage, at the surface, is not covered by a perichondrium. This lack of a perichondrium results in the proliferation of articular cartilage occurring only from within the matrix itself. An unfortunate consequence of this lack of a perichondrium is that the articular cartilage, continually exposed to stress from the movement of the joint, is less able to repair itself since no proliferation can occur from the perichondrium. Aging also leads to decreased healing ability and thus damage to cartilage may not repair sufficiently to prevent disease [9]. Damage to the articular cartilage leads to the most common [10] affliction of the aging population, osteoarthritis.

The third type of cartilage, fibrocartilage, is present as a transitional tissue generally connecting two other structures or tissues. Fibrocartilage does not have a perichondrium and thus only grows interstitially. Due to stress on this cartilage, the chondrocytes are typically arranged in a linear fashion [5]. One prominent place where fibrocartilage is present is in the annulus fibrosus of intervertebral discs. It is present in other areas such as part of the knee joint as well. Damage to this tissue may lead to herniation of the intervertebral disc, another major health issue as this may cause back pain [11].

Chondrocytes are surrounded by a thin pericellular matrix containing specific proteins, termed matricellular proteins, which have significantly different properties than the majority of the cartilage matrix [12]. This matrix and the associated chondrocyte are termed a chondron [13]. The matricellular proteins may confer the ability to withstand degradation by matrix metalloproteinases (MMPs), and this may be due to the localization of type VI collagen in this region [14,15]. This collagen is not broken down by many matrix metalloproteinases. Surrounding the pericellular matrix is the matrix termed histologically the territorial matrix and in between groups of chondrocytes is the interterritorial matrix [5].

3. Overview of skeletogenesis

The skeletal system is primarily composed of bone and cartilage, both mesodermal-derived tissues formed through the differentiation and cellular function of osteoblasts and chondrocytes. Skeletogenesis is a complex, multistep process involving the coordinated actions of osteoblast gene expression and cellular activity that are regulated by a multitude of systemic and locally produced growth factors, as well as complex extracellular matrix interaction [16]. Osteoblasts are highly differentiated bone lining cells responsible for the production of bone through secretion and mineralization of ECM constituents. Osteoblasts are protein-producing cells characterized morphologically by a round nucleus, an intensely basophilic cytoplasm, a prominent Golgi complex, and a well-developed rough endoplasmic reticulum. Osteoblast precursors originate from local pluripotent mesenchymal stem cells from the bone marrow stromal mesenchymal stem cells. Under specific conditions, including ECM interactions, osteoblast precursors will proliferate and differentiate into pre-osteoblasts cells that are committed to differentiate into mature osteoblasts [16–18]. The majority of these mature synthetic osteoblasts will become osteocytes as they become embedded in their secreted ECM. Osteocytes are the most abundant cell type in the bone and have the ability to resorb and synthesize bone matrix but in a more limited capacity compared to osteoclasts and osteoblasts, respectively [19]. These osteocytes form an interconnected network called the lacunocanalicular system that is embedded within the mineralized tissue. The matrix directly surrounding the lacunocanalicular system is not mineralized and has fluid-like mechanical properties and contains ECM proteins secreted by osteocytes [20]. This system is thought to play an active role in bone remodeling and adaptation by sensing and responding to biomechanical and other systemic stimuli. In addition, the ECM proteins produced here also play a role in mineralization [21,22]. Osteoclasts are bone-lining cells that are responsible for bone resorption. Osteoclasts are characterized morphologically as very large, multinucleated cell with abundant Golgi complexes and numerous transport vesicles loaded with lysosomal enzymes. The zone of contact with bone is characterized by the presence of a ruffled border where bone resorption takes place [16,18]. Osteoclasts originate from hematopoietic precursors of the mononuclear/ phagocytic lineage that are capable of multilineage differentiation as well as self-renewal. Thus, skeletogenesis results from the total contribution of all of these cell types and occurs as the result of two distinct processes in vivo: endochondral and intramembranous ossification.

3.1. Intramembranous ossification

Intramembranous ossification occurs in the flat skull bones and the clavicle and involves the condensation and direct differentiation of osteochondral progenitors into bone-forming osteoblasts [23]. The process begins with mesenchymal cell proliferation and condensation into compact nodules at the future sites of bone formation. As these condensations begin to form, mesenchymal cells within the interior of these nodules stop replicating and undergo changes in gene expression (i.e., Runx2/Cfba 1; Osx) [24,25] and associated morphological changes that are characteristic of osteoprogenitor cells. These osteoprogenitor cells begin to secrete a collagen-proteoglycan, pre-bone matrix termed osteoid that eventually becomes mineralized through deposits of minerals such as calcium and phosphates. Typically, osteoblasts remain separated from this mineralized matrix by a layer of the osteoid matrix they secrete; however, some osteoblasts remain trapped in this mineral matrix and become osteocytes. As this process proceeds further, bony spicules are formed and begin to radiate from the region where ossification began. Compacted mesenchymal cells on the exterior of these nascent bony spicules also begin to form a dense layer of vascularized connective tissue that surrounds the mineralized tissue to form the periosteum. The periosteum consists of two primary histological layers: a fibrous layer that has dense collagenous tissue and a cambium layer that is cellular and functions to provide osteoblast cells during bone expansion, regeneration, and fracture repair [26–28]. The cells that remain on the inner surface of the periosteum also differentiate into osteoblasts and deposit osteoid matrix parallel to that of the existing spicules forming bone layers.

3.2. Endochondral ossification

Endochondral ossification occurs for most axial and appendicular bones (e.g., long bones) and involves the differentiation of osteochondral progenitors into chondrocytes that form a cartilage template (anlagen) that will eventually be transformed into bone tissue [23]. Endochondral ossification also begins with condensations of mesenchymal stem cells that migrate to the site of future bone formation and resemble the eventual shape of the bone that will be formed [29]. The distinction between intramembranous and endochondral bone formation lies in the fact that the mesenchymal cells found within the interior of these condensations express cartilage-specific genes [30,31] and differentiate into chondrocytes to form cartilaginous models. N-cadherin appears to be important in the initiation of these condensations, and N-CAM also seems to be critical for their maintenance [32,33]. These chondrocytes also secrete collagen II, aggrecan, and other matrix molecules that constitute the ECM of hyaline cartilage [29]. As development proceeds, further chondrocyte differentiation in the center of these anlagen results in chondrocyte hypertrophy, expression of collagen type X and fibronectin, and subsequent mineralization of the surrounding matrix. Exteriorly located perichondrial cells differentiate into osteoblasts and stimulate the invasion of blood vessels that initiate the conversion of the perichondrial layer into the periosteum where osteoblasts differentiate and secrete collagen type I and other bone matrix-specific proteins. This collar will eventually mineralize by means of intramembranous bone formation forming a bone collar of cortical bone around the periphery of the tissue [29]. The combination of matrix mineralization and vascular invasion promotes the invasion of osteoblast precursor cells, osteoclasts, blood vessel endothelial cells, and hematopoietic cells from the perichondrium into the hypertrophic cartilage. This results in the resorption of the hypertrophic cartilage, osteoblast differentiation, and bone formation, and the hematopoietic and endothelial cells establish the bone marrow stroma in what becomes the primary ossification center [34]. Eventually, this primary ossification center expands and secondary ossification centers form at the distal ends of the developing bone. This results in the development of epiphyseal growth-plate cartilage, responsible for the longitudinal growth of bones [35,36]. The epiphyseal growth plate contains chondrocytes that are organized into structural and functional zones, each with distinct gene expression patterns [37]. In the reserve zone, chondrocytes are spherical with large amounts of matrix consisting of collagen type II and proteoglycans. This zone transitions into a zone of chondrocyte proliferation, where chondrocytes appear discoid and columnar. Elongation of the cartilaginous anlagen mainly occurs from the proliferation of this zone. This zone transitions into the zone of maturation where chondrocytes become prehypertrophic. Here proliferation ceases; however, cell size increases as a result of growth. As this zone progresses, chondrocytes continue to hypertrophy and secrete a collagen X-rich matrix until eventually they undergo apoptosis leaving behind spaces for subsequent vascular invasion and osteoblast differentiation and bone formation [29].

4. Cartilage and bone extracellular matrix

4.1. Cartilage ECM

The extracellular matrix of cartilage varies based on the type of cartilage. All cartilage types have the aggregated proteoglycan formed by the glycosaminoglycan hyaluronic acid, the chondroitin sulfate/keratan sulfate proteoglycan aggrecan, and link protein, as well as other types of proteoglycans. Cartilage tissues vary in types of collagen present, and may contain elastic fibers. The predominant fibrillar collagen in hyaline and elastic cartilage is type II collagen, while in fibrocartilage the predominant fibrillar collagen is type I. Elastic cartilage has elastic fibers comprising fibrillin and elastin [5]. Differences in the distribution of these matrix molecules are observed in cartilage, with the pericellular matrix, territorial matrix, and interterritorial matrix having distinct matrix compositions [8,38]. In addition, zones in articular cartilage (superficial, intermediate, deep, and calcified) have a distinct organization of matrix. The pericellular matrix of a chondrocyte contains proteins termed "matricellular proteins" to distinguish them from matrix proteins present in other locations. A chondrocyte with its pericellular matrix is termed a chondron [13] and the type of matrix determines the mechanophysical properties of the cartilage [12,39]. These major differences in the most prevalent matrix molecules of the cartilage are not the only distinctions between the various matrices of cartilages. Other fibrillar collagens present in cartilage include types III [40,41], V, and XI [41, 42]. Collagens IX, XII, and XIV (members of the FACIT group of collagens, or Fibril Associated Collagens with Interrupted Triple helices) are present in cartilage matrix [42,43], as is type X collagen during hypertrophy of cartilage in endochondral ossification [42]. Type IV, present in basal laminae, is found in cartilage [44], as are type VI [8,42], type VII, type VIII, and type XVIII collagen. Other proteins present in cartilage include fibronectin [45], thrombospondin or COMP (cartilage oligomeric matrix protein) [46], SPARC or osteonectin, tenascin-C, laminin [44], and the subject of this chapter, CCNs [47]. Proteoglycans in cartilage consist predominantly of the aggregated proteoglycan, with hyaluronic acid and aggrecan associated through link protein. Other proteoglycans are observed including perlecan, decorin, lumican, fibromodulin, syndecan, glypican, biglycan, and epiphycan; all may be present in cartilage in addition to aggrecan [48]. The interaction of chondrocytes with their environment, and particularly the matricellular molecules, can regulate proliferation, differentiation, shape changes, apoptosis, and motility [38,47]. In addition to direct interaction of chondrocyte cellular receptors, such as integrins, with matricellular proteins, the matrix of cartilage can bind and release soluble molecules such as growth factors that regulate many cellular functions [49].

4.2. Bone ECM

Bone ECM is a composite material that provides the bulk, shape, and strength of bone tissue through a combination of mineral, collagen, and non-collagenous protein components. However, bone ECM provides much more than just mechanical and structural support, due to its role in the developmental patterning of bone and by providing a spatial context necessary for regulating cellular behavior important for bone development and maintenance. ECM proteins can signal through cell surface receptors on bone cells to regulate cell functions, and factors contained within the ECM including growth factors, cytokines, chemokines, and extracellular enzymes can also modulate the activity of bone cells and/or affect the ECM itself [50]. Bone ECM contains numerous components (over 100 proteins) and various bone tissue compartments (e.g., periosteum, marrow stroma, epiphyseal growth plate, etc.) have unique ECM environments that play diverse roles in directing bone development through regulating the differentiation process of mesenchymal stem cells and remodeling of bone through the coupled activity of osteoclasts and osteoblasts [20]. Bone ECM can be generally broken down into collagenous components that represent the majority of the ECM proteins and non-collagenous components including proteoglycans, glycosylated proteins, small-integrin-binding, N-glycosylated proteins (SIBLINGs), Gla-containing proteins, numerous MMPs, matricellular proteins, and cell-associated proteins such as integrins and cadherins. Non-collagenous components have multifactorial roles in organizing the ECM, coordinating cell-cell and mineral-matrix interactions, and regulating the mineralization process [51]. Here, we summarize some of the key components of bone and cartilage ECM, describe their role in various tissue compartments and their contributions for bone and cartilage development, maintenance, and disease.

4.2.1. Collagenous proteins

Collagen type I—Collagen type I is by far the most abundant ECM protein found in the organic component of the bone matrix accounting for roughly 90% and is the basic building block of the bone matrix fiber network. It serves as a scaffolding substrate for mineralization and also binds and orients other matrix proteins that nucleate the mineral depositions [51]. Collagen type I is secreted by committed pre-osteoblast cells and primarily determines the material strength attributes of the skeleton, but is also involved in osteoblast lineage progression. Collage type I can bind with integrins on pre-osteoblasts to initiate signaling cascades that activate Runx2 (a master transcriptional activator of osteoblast differentiation), which controls the differentiation of osteoblasts and expression of other important bone-specific ECM proteins [e.g., osteopontin (OPN), bone sialoprotein (BSP), etc.] [52–54]. Human mutations in collagen type I result in phenotypic features of osteogenesis imperfecta (OI) [55]. OI is characterized by bone brittleness leading to a higher rate of fracture in patients with the disorder potentially from thinner or osteoporotic-like bone mass; however, the exact cause of these symptoms remains an active area of investigation [20,56]. Animal models of the disease display brittle/mechanically weak bones that possibly result from impaired or improper collagen mineralization [57] and defects in the microarchitecture of the bone structure [58,59]. More recently, recessive forms of OI have confirmed that while the quantity and structure of type I collagen is critical to maintaining proper bone strength, posttranslational modification and assembly of type I collagen into its normal lattice structure are key regulators of bone strength as well [60].

Collagen type II—Collagen type II is the major structural protein in cartilage ECM (~85%) and the major collagen found in the growth plate during endochondral ossification [61]. It is primarily found in the matrix secreted by the reserve zone chondrocytes. More than a hundred human mutations of collagen type II have been identified; however, only heterozygous mutations causing autosomal-dominant phenotypes have been described to date due to phenotypic variations and age-dependent phenotypic transitions [62]. Type II collagen mutations give rise to a spectrum of phenotypes predominantly affecting cartilage and bone that range from severe disorders that are perinatally lethal to the milder conditions that are recognized in the postnatal period and childhood. These can include skeletal abnormalities (e.g., Stickler syndrome a.k.a. hereditary progressive arthro-opthalmopathy) or chondrodysplasias that are characterized by disproportionate short stature, eye abnormalities, cleft palate, and hearing loss [62]. Mouse models where collagen II is deleted phenotypically resemble human achondrogenesis type II and die immediately before or at birth and are smaller than their littermates [63]. Long bones from these animals lack endochondral bone and the epiphyseal growth plate and intervertebral discs are not developed [63,64]. Other mouse models with mutations of collagen type II also display growth-plate anomalies and chondrodysplasia [65]; some of these mice have similar phenotypes found in the human forms of the disease including short bones, cleft palate, and respiratory failure [62,65].

Collagen type III and collagen type V—Collagen type III is a fibrillar collagen that is found in extensible connective tissues such as vascular system, skin, gut, and lung, frequently in association with type I collagen. Collagen type V is a minor fibrillar collagen found in skin, tendons, and ligaments [66]. Both collagen type III and type V are found in bone and in trace amounts and may play a role in regulating collagen diameter. Collagen type III is expressed particularly during bone healing [51]. Human mutations in collagen type III and V are associated with Ehlers-Danlos syndrome, a disease characterized by defects in the structure, production, or processing of collagen that leads to wide-ranging symptoms in the digestive, excretory, and particularly the cardiovascular systems [66]. Collagen type III mutations are also associated with aortic and intracranial arterial aneurysms. Collagen type III null mice are embryonic lethal, but analysis of null cells in culture or heterozygous mice suggests that type III collagen may promote bone differentiation [67].

Collagen type VI is a major matricellular protein present in cartilage and most connective tissues [8, 42, 68–70]. In cartilage, collagen type VI is present in the pericellular matrix and its localization determines the boundary of this region [14]. Both integrins and the integral membrane proteoglycan NG2 can bind to this collagen in cartilage [71–74]. Differentiated chondrocytes express collagen type VI, while this expression is lost when chondrocytes undergo dedifferentiation [75]. Human mutations of collagen type VI lead to a range of disorders from a milder Bethlem myopathy to a more severe Ullrich muscular dystrophy associated with muscle weakness [76]. Knockout mice lacking type VI collagen show alterations in the skeleton, including less density of bone, delayed secondary ossification, and faster development of osteoarthritis [70].

Collagen type X—type X collagen is found in cartilage, but due to its expression in hypertrophic chondrocytes found in bone at the epiphyseal growth plate, it is important for endochondral ossification [77–80]. Human mutations in COL10A1 lead to autosomal-dominant Schmid metaphyseal dysplasia [81]. Schmid metaphyseal dysplasia is phenotypically characterized by long bones that are short and curved, with widened growth plates and metaphyses [82]. Interestingly, initial studies of collagen type X null mice showed no obvious abnormalities in the development and growth of long bones [38]; however, phenotypic changes that in part mimicked Schmid metaphyseal dysplasia were observed in a second, separate collagen X null mouse line [83]. In these mice, the height of the resting zone of growth-plate chondrocytes was reduced and trabecular bone architecture was altered in the chondro-osseous junction as well as differences in the distribution and organization of growth-plate ECM components [83]. In transgenic mice models expressing dominant negative collagen type X, endochondral ossification was also effected displaying variable phenotypes [84,85]. In all of these models, the hypertrophic zone of the growth plate was compressed and the degree of compression correlated with phenotype severity [84,85]. Additionally, two other collagen X models the first where mice express collagen X containing a deletion similar to one found in human Schmid metaphyseal dysplasia patients and a second knock-in mouse with a collagen X Asn617Lys mutation displayed shortened limbs, consistent with a role in the epiphyseal growth plate [86–88].

4.2.2. Non-collagenous proteins

4.2.2.1. Proteoglycans

Bone ECM contains several proteoglycans that, other than collagens, represent the major constituents of the bone ECM. Proteoglycans are macromolecules that contain a central core protein and one or more acidic polysaccharide side chains (glycosaminoglycans) [51,89]. Proteoglycans exhibit diverse biological functions including cell proliferation, adhesion, migration, and differentiation and act as structural components in tissue organization. They can also interact with growth factors and cytokines, as well as with growth factor receptors, and are implicated in cell signaling [90]. Bone ECM contains several classes of proteoglycans that include small leucine-rich proteoglycans (SLRP), aggrecan, heparin sulfate proteoglycans (HSPGs), and hyaluronic acid [90].

Members of the SLRP family are composed of core proteins of leucine-rich repeats that are approximately 25 amino acids in length and represent the most abundant type of proteoglycans in bone. Some of the most studied in bone include biglycan decorin, fibromodulin, and lumican and these studies have demonstrated that SLRPs are involved in the structural organization of the bone ECM and regulation of growth factor activity [90]. Biglycan is distributed evenly throughout bone ECM [91] and can bind collagen type I and several important bone growth factors such as transforming growth factor-beta 1 (TGF-β1) [89]. Biglycan null mice have reduced biomechanical bone strength [92] and fail to achieve peak bone mass due to a decrease in bone formation due to low osteoblast numbers and activity [92]. These mice also have an age-related reduction in capacity to produce bone marrow stromal cells (MSCs), reduced responses to TGF-β1, reduced collagen synthesis, and relatively more cellular apoptosis [93]. Additionally, quantitative variations in the range, mean, and distribution profiles of the collagen fibril diameters were detected [93]. Decorin null mice have no skeletal defect, with no major phenotypic changes in bone at macroscopic or histological levels; however, changes in collagen fibril size and shape in bone have been observed. However, decorin/biglycan double null mice have a more severe osteopenia than in biglycan-deficient mice, with earlier onset and severely reduced cortical and trabecular bone mass [93]. TGF-β1-binding experiments demonstrated that it bounds to decorin with high affinity and that this interaction may increase TGF-β1-receptor interaction to enhance its bioactivity [94]. Fibromodulin is a small keratan sulfate proteoglycan that is found in bone in a pericellular fashion near late-hypertrophic chondrocytes of the secondary ossification centers and in the growth plate suggesting a role during endochondral ossification. Fibromodulin null mice have no apparent skeletal phenotype but abnormal and fewer collagen fibril bundles in the tail than in wild-type animals [95,96]. Fibromodulin possesses the capacity for TGF-β1 binding [97] and levels have been correlated with decreased TGF-β1 expression in multiple fetal and adult rodent models. Recent studies suggest that fibromodulin coordinates temporospatial distribution of TGF-β ligands and receptors to modulate TGF-β bioactivity [98]. Lumican is secreted specifically by differentiating and mature osteoblasts and is a significant proteoglycan component of the bone matrix, playing an essential role in the regulation of collagen fibril formation [99]. Lumican null mice display altered collagen fibril structure [100]; however, no alteration in bone structure was reported in these mice. Interestingly, double null lumican/fibromodulin mice are smaller than their wild-type littermates and display age-dependent osteoarthritis [100].

Aggrecan is a large chondroitin sulfate proteoglycan and is the major proteoglycan component of cartilage [101,102], but it is also expressed in developing bone tissue [103]. Human mutations in aggrecan cause two types of spondyloepiphyseal dysplasia, an autosomal-dominant Kimberley type and autosomal-recessive Aggrecan type, resulting in dwarfism, skeletal abnormalities, and problems with vision and hearing, and an autosomal-dominant familial osteochondritis dissecans, which displays abnormal cartilage formation and joint issues. In mice, a natural truncating mutation of aggrecan leads to an autosomal-recessive cartilage matrix-deficiency syndrome with abnormal craniofacial structures, shortened limbs and tail, and perinatal lethality [88]. In these mice, chondrocytes are disorganized and the amount of hypertrophic chondrocytes is significantly reduced and expression of other ECM genes is altered in the growth plates of these mutant mice [104].

HSPGs act as regulators of skeletal patterning, differentiation, growth, and homeostasis and are a critical component of the hematopoietic stem cell niche within the growth plate and bone marrow [105]. Studies suggest that osteoblast precursors and osteoblasts synthesize HSPGs that both membrane- and matrix-associated HSPG are found in bone tissue and may play an important role in cell-cell interactions between fibroblast-like cells and osteoclast-lineage cells by interacting with heparin-binding growth factors, growth factor receptors, and/or other heparin-binding adhesion molecules, such as fibronectin [90,106]. HSPGs act as co-receptors for numerous signaling molecules, such as fibroblast growth factors (FGFs), vascular endothelial growth factor (VEGF), TGF-β1, and TGF-β2 in addition to various other cytokines [107,108]. The binding of these signaling molecules to HSPGs can serve a variety of functional purposes including immobilization of these factors or cytokines and/or protection against proteolytic degradation, thereby affecting their biological availability and/or modulation of their biological activity [90].

One example of the role that HSPGs can have on skeletal development can be seen with perlecan. Perlecan is a large heparan sulfate/chondroitin sulfate proteoglycan and is a ubiquitous component of basement membranes and articular cartilage, and in bone it is present in the extracellular matrix of the growth plate where it plays an important role in bone structure [109–112]. Mutations in the human perlecan gene cause autosomal-recessive skeletal disorders including the severe and lethal Silverman-Handmaker type characterized by a flat face, disorganized growth plate, cleft palate, and death at birth to the milder Rolland-Desbuquois type dyssegmental dwarfism skeletal dysplasia [81]. Schwartz-Jampel syndrome (myotonic chondrodystrophy) is a milder, progressive disease as a result of reduced perlecan levels characterized by abnormalities of the skeletal muscles (myotonic myopathy), bone dysplasia, joint contractures, and/or growth delays resulting in dwarfism [81,113]. While most perlecan null mice are embryonic lethal, surviving embryos have defects in skeletal development starting at E14.5 with disorganized growth plates, reduced proliferation of chondrocytes, reduced endochondral ossification, and reduced numbers of type II collagen fibrils [114,115]. The mice also die shortly after birth [114,115]. Additionally, it has been shown that perlecan mediates binding and delivery of FGF-2 to FGF receptors [116] and that it regulates VEGF signaling and is essential for vascularization during endochondral bone formation [117].

Hyaluronic acid (hyaluronan) is a non-sulfated linear polysaccharide present in the extracellular matrix of every vertebrate's tissue important for bone regeneration [118]. Large amounts of hyaluronan are synthesized during bone formation and it plays a role in enhancing chondrogenic and osteogenic differentiation potentials of mesenchymal stem cells [118] by regulating expression of chondrogenic markers including sulfated glycosaminoglycans, SOX-9, aggrecan, and collagen type II and osteogenic markers alkaline phosphatase (ALP), osterix, runx2, and collagen type I [119–121]. Additionally, hyaluronan synthase-2 (Has2) knockout mouse model demonstrates that it is important for spine development [122].

4.2.2.2. Glycosylated proteins (glycoproteins)

The majority of bone ECM proteins are modified posttranslationally with either N- or O-linked oligosaccharides. In addition, many glycoproteins found in bone ECM contain an arginine-glycine-aspartic acid (RGD) sequence that facilitates integrin binding important for mechanotransduction [89]. Thus, glycoprotein proteins represent an ever-growing group of bone ECM proteins with diverse functions and thus we will only focus on some key proteins in this chapter.

Osteonectin (SPARC, BM-40) is a secreted, multifunctional calcium-binding glycoprotein that participates in tissue remodeling, morphogenesis, and bone mineralization and is secreted by many different types of cells including osteoblasts [123,124]. Osteonectin can initiate mineralization by binding to type I collagen and synthetic hydroxyl apatite and mediating mineralization of the type I collagen [123]. It has also been demonstrated that osteonectin can bind growth factors and regulate their activity of growth factors including platelet-derived growth factor (PDGF), FGF, or (VEGF) [125,126]. It also regulates ECM and matrix metalloprotease production [127,128]. Osteonectin also regulates cell proliferation, promotes osteoblastogenesis [129], and it can stimulate angiogenesis [89]. Osteonectin-deficient mice display decreased bone remodeling with a marked negative bone balance that leads to osteopenia in older animals [20]. Bone turnover is decreased as a result of both reduced osteoclast and osteoblast surface with consequential development of low bone mass [20]. Expression of osteonectin in the intervertebral disc decreases with age [130]; mice lacking osteonectin have disc herniations [131] and may exhibit increased pain [132]. Osteonectin immunostaining is increased in osteoarthritis in cartilage as compared to age-matched controls [133], indicating a role for this protein in the pathogenesis of osteoarthritis. Pseudoachondroplasia, which is related to mutations in osteonectin, is caused by retention of this protein within the endoplasmic reticulum. This disease is an autosomal-dominant disorder that causes dwarfism [134].

Fibronectin is a high-molecular weight glycoprotein dimer that is synthesized by numerous connective tissues throughout the body and contains three alternative spliced domains. Fibronectin is produced from a single gene but as a result of alternative splicing exists in vivo in two forms: a soluble, circulating form known as plasma fibronectin that is synthesized in hepatocytes and a cellular form produced by a number of cell types including osteoblasts and gets incorporated into the bone matrix [135,136]. Fibronectin has been shown to bind to extracellular matrix components including collagen, fibrin, and HSPGs [137]. Fibronectin can also bind to 11 different integrins, six of which are expressed by osteoblasts; however, the primary adhesion integrin that fibronectin binds to on osteoblasts remains an active area of investigation [138]. It appears that fibronectin's key function is in the assembly of collagen as fibronectin is critical for collagen polymerization and matrix integrity [139,140]. Osteoblasts express fibronectin in multiple stages of their differentiation including during proliferation and differentiation concurrently with collagen type I expression. FN null mice die in utero at embryonic day 8.5, prior to skeletal development [141]. Studies using conditional null animals found that while conditional deletion of fibronectin in differentiating osteoblasts failed to show a decrease in fibronectin in the bone ECM, conditional deletion of fibronectin in the liver showed a marked decrease in fibronectin content in bone ECM associated with a decreased mineral-to-matrix ratio and changed biomechanical properties [138]. These studies suggest that while osteoblast-derived fibronectin affected osteoblast differentiation and function, fibronectin found in bone ECM originates from the liver [138].

Thrombospondins are a family of ECM glycoproteins that consist of thrombospondins 1–5 and can be divided into two subgroups: A, which contains thrombospondins 1 and 2, and B, which contains thrombospondins 3–5 (a.k.a. cartilage oligomeric protein or COMP) [142]. We will focus on thrombospondins 1, 2, and 5 (COMP) in this section. Thrombospondin 1 functions in a wide variety of physiological functions including platelet aggregation, inflammatory responses, and the regulation of angiogenesis during wound repair and tumor growth. Thrombospondin 1 binds a variety of cell receptors including CD36, CD47 (integrin-associated protein), numerous integrins, proteoglycans, and calcium [142–145]. Thrombospondin 2 has similar physiological as thrombospondin 1, but it also plays a role in the assembly of connective tissue ECM [142], while COMP is primarily expressed in cartilage and certain other connective tissues and has roles in chondrocyte attachment, differentiation, and cartilage ECM assembly [146]. Thrombospondins 1 and 2 can each bind latent TGF-β, and unlike thrombospondin 2, thrombospondin 1 can control TGF-β bioactivity by releasing it from its latency complex [20]. Thrombospondin 1 null mice are reported to have minor trabecular bone abnormalities [147], mild spine deformation [148], and mild growth-plate cartilage disorganization [149] in addition to enhanced angiogenesis [150]. Additionally, these models suggest that thrombospondin 1 plays a role in bone homeostasis, both in mediating bone matrix integrity and by regulating OC formation as a matrix-derived paracrine signaling molecules [151]. Conversely, thrombospondin 2 null mice have an increase in MSC number, suggesting that it serves as an inhibitor of MSC proliferation [152,153]. Human mutations in thrombospondin cause a pseudoachondroplasia and a type of epiphyseal dysplasia [81]. COMP mutations cause a type of pseudochondrodysplasia and an epiphyseal dysplasia [81,154,155]. Mice lacking this protein tend to appear normal with regard to their skeleton [156], but do have issues with wound healing [157,158]. COMP may affect cellular functions by affecting the interaction of matrix and growth factors with chondrocytes [46]. Thrombospondins function primarily as anti-angiogenic proteins [159]. Thrombospondin is present in fibrocartilage of the intervertebral disc, and lack of this protein causes disorganization of this region [160]. No disc herniation was noted in these mice. Thrombospondins help to organize the growth plate [149]. In osteoarthritis, inappropriate angiogenesis within the cartilage may cause changes in the articular cartilage that lead to degradation of the matrix. Gene transfer of thrombospondin 1 using an adenoviral vector into knee joints suppressed osteoarthritis disease progression, with concomitant reductions in new vessel growth and inflammatory cell infiltration [161].

Periostin is a disulfide-linked, heparin-binding *N* terminus-glycosylated secreted protein that appears to be essential for proper ECM synthesis, particularly with respect to collagen I fibrillogenesis [162,163]. Expression of periostin occurs in many tissues including bone and cartilage and under many pathologic states. It is expressed in osteoblasts and is present in the intervertebral disc and its expression increases in degenerated human discs [130,164]. Overall, the function of periostin in cartilage is not well understood [165]. Periostin expression is also known to be prominent in fibrotic conditions, including subepithelial fibrosis in bronchial asthma [166] and in bone marrow fibrosis [167]. Periostin null mice display severe growth retardation, suggesting that periostin is essential for postnatal development. Histological analysis of the periostin knockout mice demonstrated severe incisor enamel defects, periodontal disease, a lack of trabecular bone, cartilage, and cardiac valve defects; however, some of these phenotypes might be secondary due to eating difficulties as a result of the lesions in the periodontium [168].

Alkaline phosphatase is a glycoprotein enzyme that is found both bound to cell surfaces and also within the mineralized matrix [51]. Human ALP is classified into four isoenzyme types, tissue-nonspecific (TNAP), intestinal, placenta, and germ cell of which the TNAP type is ubiquitously expressed in many tissues, including bone and is the form implicated in its biomineralization [169]. Mineralization of cartilage, bone, and teeth occurs by a series of physicochemical and biochemical processes that facilitate hydroxyapatite deposition (mineralization) into collagen fibrils of the matrix [170] and/or within the lumen of chondrocyte and osteoblast-derived matrix vesicles [171,172] through hydroxylation of pyrophosphate providing inorganic phosphates to promote mineralization [169]. Deactivating mutations in the TNAP gene causes the inborn error of metabolism known as hypophosphatasia [173], characterized by poorly mineralized cartilage (rickets) and bones (osteomalacia), spontaneous bone fractures, and elevated extra-cellular inorganic pyrophosphate (PPi) concentrations [174,175]. TNAP null mice also display skeletal disease at approximately 10 days of age and featured worsening rachitic changes, osteopenia, and fracture [176]. Histologically, these mice displayed developmental arrest of chondrocyte differentiation in epiphyses and in growth plates with diminished or absence of hypertrophic zones with progressive osteoidosis from defective skeletal matrix mineralization [176].

4.2.2.3. Small-integrin-binding, N-glycosylated proteins (SIBLINGs)

SIBLINGs are classified by the presence of the RGD sequence and large amounts of sialic acid known as small-integrin-binding, N-glycosylated proteins (SIBLINGs) [29]. SIBLINGs, along with other matrix proteins, are thought to play a role in cell attachment facilitated by transient or stable focal adhesions to ECM molecules mediated by cell surface receptors. They also play a role in intracellular signaling. Although these SIBLINGs were initially only found in mineralized tissue, many of them can now be found in other tissues [177,178]. Bone cells synthesize at least five SIBLING members including OPN, BSP, dentin matrix protein-1, dentin sialoprotein, and matrix extracellular phosphoprotein. Here, we discuss on OPN and BSP.

Osteopontin (OPN; BSP-1) is a highly negatively charged bone ECM protein that can undergo extensive posttranslational modification mainly phosphorylation [179]. OPN is one of the most extensively studied SIBLING proteins and has broad physiological and pathological functions including development [180], bone remodeling [181], immune function [182], fibrosis [183], and cancer [184]. It is expressed by a wide range of cells and promotes attachment, proliferation, migration, chemotaxis, and apoptosis of macrophages, lymphocytes, osteoblasts, and a range of tumor cells [185–187]. OPN binds calcium and is a key regulator of hydroxyapatite nucleation and is produced by osteoblasts during their terminal differentiation prior to matrix mineralization. OPN regulates bone mass and overall bone quality by minimizing strain-induced fatigue damage and microcrack propagation in bone [188]. In addition, it can also mediate the attachment of osteoclasts and can affect the shape and size of hydroxyapatite crystals in the bone ECM [189–191]. OPN can mediate signaling through integrins, for example, it can bind to $\alpha 4\beta 1$ integrin and trigger a cell type-specific integrin-mediated signaling cascade [192]. OPN null mice showed no bone phenotype [193], but in stress situations, such as oophorectomy, the mice do not develop osteoporosis [91]. Recent data confirm and extend this observation that the skeleton of OPN null mice does not respond properly under stress underlining the importance of OPN in bone metabolism [194,195]. However, other studies have demonstrated that OPN null mice do have a bone phenotype also under physiological conditions [196–198]. More recent OPN null models demonstrate that while bone volume was normal in young null animals, the volume and number of osteoclasts were increased, but that osteoclasts from these mice have a lower resorptive capacity providing evidence for a role of osteopontin in osteoclast activity [199].

BSP-2 is a SIBLING glycoprotein and is expressed in chondrocytes, hypertrophic cartilage, and in osteoblasts at the onset of mineralization and in osteoclasts [200]. BSP is highly expressed at sites of primary bone formation [201], and it coincides with the initial formation of membranous and endochondral bone, the maximal level being reached during the formation of embryonic bone [202]. BSP-2 binds calcium; however, it does not nucleate the hydroxyl apatite found in the bone ECM and BSP-2 mediates cell attachment through interaction with vitronectin receptor [29]. BSP-2 null mice are shorter than their wild type counterparts and display a low level of bone remodeling, with both bone formation and mineralization severely impairing in vivo and in vitro models [203]. These mice also have lower osteoclast numbers and surfaces in vivo; osteoclast recruitment and activity in vitro were impaired and impairment of chondrocyte proliferation was suspected [203,204].

4.2.2.4. Gla-containing proteins

Gla-containing proteins refer to a group of endogenously made bone ECM proteins that undergo vitamin K-dependent gamma-carboxylase modification. The dicarboxylic glutamyl (gla) residues enhance calcium binding. The formation of gamma-carboxy-glutamic acid (Gla) occurs via a unique posttranslational modification of specific peptide-bound glutamate residues, which is required for the biological activities of these proteins.

Osteocalcin (OCN) is one of the most abundant non-collagenous proteins of the bone ECM. It is produced by differentiated osteoblasts [205,206] and once transcribed undergoes posttranslational modifications within osteoblasts that include the carboxylation of three glutamic residues [207]. Vitamin D stimulates osteocalcin transcription and vitamin K regulates the carboxylation processes. Various growth factors, hormones, or cytokines can also modulate osteocalcin production [206]. Osteocalcin is secreted by osteoblasts during active bone formation and can bind with the mineralized bone ECM [207]; however, its exact role in bone physiology remains an active area of investigation. Osteocalcin promotes the recruitment and differentiation of circulating monocytes and osteoclast precursors, suggesting its role on osteoblast-osteoclast interaction and bone resorption [206–208] and other studies have shown that osteoclast resorption is impaired in OCN null bone [208]. However, OCN null mice have a higher bone mineral density without any change in bone resorption and mineralization [209]. Recently, tissue-specific transgenic mice with osteoblast-specific overexpression or reduced OCN production suggested that OCN might have an important endocrine function for glucose metabolism and lipid homeostasis [210,211].

5. The CCN family

The CCN family makes up a group of six highly conserved, secreted, extracellular matrix-associated proteins that regulate diverse cellular functions, including skeletal development, wound healing, fibrosis, and cancer. Originally named after the three identified members—cysteine-rich 61 (Cyr61, CCN1), connective tissue growth factor (CTGF, CCN2), and nephroblastoma overexpressed (Nov, CCN3)—this family also includes the Wnt-induced secreted proteins 1–3 (i.e., WISP1/CCN4, WISP2/CCN5, and WISP3/CCN6). Members of the CCN family share a unique and conserved modular structure and interact with and orchestrate cellular responses to extracellular factors via direct binding to cell surface receptors, including integrins, Notch1, neurotrophic tyrosine kinase receptor type 1 (TrkA), low-density lipoprotein receptor-related proteins (LRPs), and HSPGs. CCN proteins can also mediate biological functions by interacting with growth factors such as tumor growth factor beta (TGF-β), vascular endothelial growth factor, and bone morphogenetic proteins (BMPs) and by associating with other ECM proteins including fibronectin and fibulin 1C. Through these interactions, CCN proteins serve both distinct and overlapping biological roles. Consequently, deregulation of their expression or activities contributes to the pathobiology of several diseases, many of which may arise when inflammation or tissue injury becomes chronic, including vascular diseases, fibrosis, arthritis, and cancer.

5.1. Structure and function of CCN family members

CCN proteins are cysteine-rich and share a modular structure (Modules I–IV), with an N-terminal secretory peptide followed by four conserved domains with sequence homologies to insulin-like growth factor-binding proteins (IGFBPs), von Willebrand factor type C repeat (VWC), thrombospondin type I repeat (TSP1), and a carboxyl-terminal domain (CT) that contains a cysteine-knot motif [212]. The order of these modules has been strictly conserved during evolution, suggesting that it is critically important for these proteins. Each module is involved in protein binding and contains conserved hydrophobic, polar, and cysteine residues. Module I shares 32% sequence homology with the N-terminal cysteine-rich regions of the IGF-binding proteins and contains a GCGC-CXXC motif that is involved in IGF binding. Module II includes a stretch of 70 amino acids with sequence identity to von Willebrand factor as well as various thrombospondins, collagens, and mucins [212]. This domain has been shown to mediate protein oligomerization [213]. Module III is a TSP1 repeat that contains the WSXCSXXCG motif, which is thought to be implicated in the binding of sulfated glycoconjugates and to be important for cell attachment [212,214]. The last module, Module IV, occurs at the carboxy-terminus of various extracellular proteins and is the least conserved of the four domains at the nucleotide sequence level. It consists of several cysteine residues that adopt a cysteine-knot motif. This motif comprises a complex structure of two-stranded β-sheets that lie face to face and are linked by three interlocking disulfide bridges [215] and occurs in TGF-β, PDGF, and nerve growth factor (NGF). It is critical for several of the biological functions of CCN proteins and is thought to mediate dimerization and binding to cell surface receptors. CCN5 is the only family member that lacks the CT domain [216]. A variable, central hinge region that is susceptible to proteolytic processing by MMPs and other proteases links the amino-terminus and carboxy-terminus of these proteins, yielding two halves that bind distinct cell surface receptors [217]. It is not clear whether the individual properties of each of the four modules govern the biological properties of the CCN protein or if it is the combination of the modules and other sequences within the protein that do so. However, all of the modules are highly interactive with a number of other molecules, which include cell surface receptors, ECM components, growth factors, and structural proteins [218].

5.2. Cell surface receptors mediating CCN functions in cartilage and bone

Despite the structural similarities that CCN proteins have to other protein domains as described above, their interactions are unique because of their ability to bind extracellular factors via their modular domains. CCN proteins have been shown to interact specifically with cell surface receptors such as HSPGs, integrins, and LRPs, accounting for their ability to regulate numerous cellular functions. CCN2, the most studied of the CCN family members, shares common functionality with CCN1 with respect to interaction with integrins and HSPGs, causing comparable biological effects. LRP1 is another common receptor shared by CCN2 and CCN1; however, the target cell and biological consequence differ between the two [219,220].

Direct binding of CCN proteins to integrins present on cellular surface drives many of their effects on cartilage and bone. Integrins comprise a large family of cell-cell and cell-matrix receptors that signal both from the ECM to the cytoplasm and from the cell to the matrix (inside-out and outside-in) [221]. Integrins are $\alpha\beta$ heterodimers and can be present in a number of configurations. The different combinations of α and β receptors define what ECM molecules a cell interacts with. These receptors regulate many cellular functions such as proliferation, differentiation, motility, and developmental processes among others [221]. A decrease in the interaction of β integrins with matrix molecules is observed during the pathogenesis of osteoarthritis, and thus the disruption of integrin-matrix interaction causes cellular dysfunction in cartilage [222]. Integrins that have been identified in cartilage include $\alpha_5\beta_1$, $\alpha_v\beta_3$, $\alpha_v\beta_5$, $\alpha_6\beta_1$, $\alpha_1\beta_1$, $\alpha_2\beta_1$, $\alpha_{10}\beta_1$, and $\alpha_3\beta_1$. As is the case with matrix molecules, integrins are expressed differentially in specific regions of cartilage and during development and pathogenesis.

Other matrix receptors are present in cartilage. One example is NG2, a transmembrane proteoglycan in cartilage with the matricellular protein type VI collagen as its ligand [72–74]. NG2 interaction with type VI collagen may be an important interaction in determining the progression of sarcomas [223]. Annexin V, or anchorin II, binds to type II collagen and is mainly expressed in the superficial zone of articular cartilage [224]. CD44 binds to hyaluronic acid; blocking of this receptor causes a loss of the matrix of cartilage [225].

As a matricellular protein, CCN2 also binds the fibronectin receptor ($\alpha_5\beta_1$) and aggrecan, which are major components of the ECM [226]. The interaction of $\alpha_5\beta_1$ with fibronectin in cartilage causes loss of the differentiated state [227]. Blocking of this integrin causes a loss of the differentiation of pre-hypertrophic chondrocytes [228]. A knockout mouse that has loss of β_1 specifically in chondrocytes exhibits a disease similar to chondrodysplasia [222]. β? is the most expressed integrin β subunit in osteoarthritic chondrocytes [229].

CCN2 is induced by signaling from retinoids in cartilage in the growth plate [230]. CCN2 causes differentiation of the chondrocytes and is important for matrix deposition and signaling from cellular receptors [226,231]. Retinoids are important signaling molecules involved in the cartilage becoming hypertrophic during endochondral ossification.

5.3. CCNs in chondrogenesis and osteogenesis

The CCN proteins appear to play a critical role in the establishment of the skeletal system and have been shown to have positive and negative regulatory roles in skeletal development, as demonstrated through cell culture experiments and animal models. Collectively, these studies demonstrate that CCN proteins promote differentiation and proliferation of chondrocytes, osteoblasts, and vascular endothelial cells, which are important in both endochondral and intramembranous ossification.

Cartilage is first seen in the developing embryo in the areas that serve as templates for endochondral ossification, as well as in Meckel's cartilage in the head area [1]. Both CCN1 and CCN2 promote chondrogenic and osteoblastic differentiation [232,233]. CCNs can interact with members of the TGF-β and BMP family via the chordin-like homology found in the VWC domain, and modulate their binding affinity for their respective receptors [234]. CCN2 is important during embryogenesis and bone formation, including during proliferation of mesenchymal cells, their differentiation into chondrocytes, and condensation of these cells into the cartilage model that will form bone [1,235]. CCN2 expression is highest in the vascular tissue and the maturing chondrocytes of the embryo. TGF-β, which induces CCN2, directs the condensation of these cells and regulates secretion of other matricellular proteins such as fibronectin [235]. Cells from CCN2 knockout mice do not undergo this condensation process in vitro and have less synthesis of cartilage proteoglycans [236]. Neutralizing antibodies, siRNA, and antisense oligonucleotides have all been used to demonstrate that reduced or absence of CCN2 prevents condensation from occurring [1,235]. Overexpression of CCN2 causes an increase in the length and density of bones [237]. In vertebral bodies, downregulation of CCN2 is necessary for chondrocyte differentiation [235]. These studies demonstrate that the spatial and temporal regulation of CCNs is an important component of development.

Development of cartilage involves several different transcription factors, such as Sox 9, that control this process, as well as growth factors that interact with cellular receptors to facilitate various cellular functions [2]. Interaction with the extracellular matrix has been implicated in development as well. Various epigenetic mechanisms and several different microRNAs also are important in the development of cartilage [3,4].

BMPs are important during skeletogenesis [238] and are known to be regulated by CCN2 [234]. Expression of CCN2 is observed in both the perichondrium and the chondrocytes during development [239]. BMPs play a role in the repair of cartilage as well as during development [240]. BMP2 causes proliferation of chondrocytes during development, while BMP4 does not [241]. CCN2 and CCN3 can bind BMP2 and abrogate its ability to promote chondrogenic and

http://dx.doi.org/10.5772/63199

osteogenic differentiation [242,243], respectively, whereas CCN4 binding to BMP2 enhances its function in osteogenesis [244]. CCN2 also binds BMP4 and blocks its signaling capabilities [234]. This binding thus modulates the activity of BMP-9. CCN2 also modulates the Wnt pathway in *Xenopus* embryos by binding to LRP6 co-receptor [245]. CCN2 expression in chondrocytes is controlled by both Rac 1 and actin pathways that are mediated by TGF-β/Smad signaling. CCN4 is expressed in developing mesenchymal, pre-osteoblastic, and cartilage cells and is believed to regulate skeletal growth and repair [246]. A truncated form of CCN4, WISP1v, has been shown to regulate the differentiation of chondrocytes toward endochondral ossification [247].

Chondrocytes are usually considered quiescent in the adult, although this can change during pathological processes such as osteoarthritis. Cartilage has very little capacity to regenerate, even though progenitor cells are present in the adult [6]. This lack of repair ability causes a number of disease issues as humans age, including the most prevalent form of degenerative joint disease, osteoarthritis [7]. In adult skeletal systems, CCN2 is highly expressed in the osteoblasts lining metaphyseal trabeculae and in osteogenic surfaces lining fracture calluses, suggesting that its upregulation in these areas may contribute to bone growth and fracture repair [248].

5.4. CCNs in cartilage and bone pathology

Several types of pathologies are observed in cartilage and bone. In cartilage, the most common is osteoarthritis, which significantly increases as the population ages. Disorders of the intervertebral disc are observed as well, and also increase with aging. Wounding and subsequent repair of cartilage occurs and can mimic the progression of osteoarthritis. Various types of chondritis, or inflammation of cartilage, are observed in various regions of the body. Congenital defects in cartilage include conditions such as chondrodysplasia, an example of which would be achondroplasia, causing dwarfism. Finally, chondromas or chondrosarcomas, tumors that can either become or are cancer, can occur within this tissue.

5.4.1. Intervertebral disc

CCN2 is highly expressed in the cartilage present in the intervertebral disc, just as it is in other types of cartilage. The intervertebral disc contains the nucleus pulposus, and this is surrounded by the annulus fibrosus composed of fibrocartilage. The annulus fibrosus contains mainly the fibrillar collagen type I, although type II collagen is present along the inner surface as in reference [249]. Proper function of the annulus fibrosus is important in disc function, and lack of intact matrix can lead to disc instability [249]. Herniation of the nucleus pulposus is sometimes implicated as a cause of back pain [11] and increased expression of CCN2 can be correlated with discs that are painful [250].

5.4.2. Osteoarthritis and cartilage wound healing

Osteoarthritis is the most common form of degenerative joint disease [7]. During the disease process, chondrocytes undergo cell death due to lack of ability to renew, particularly with aging [251]. The initial degenerative changes that occur are termed chondropenia, although this is not often identified as osteoarthritis [252]. Upregulation of the catabolic pathways occurs, causing cartilage degradation, and alteration of the phenotype of the chondrocytes is ultimately responsible for loss of articular cartilage [253]. Much of this change in cartilage structure and function occurs via interactions with matrix and signaling from both matrix and growth factors [254,255]. Initial disease presents with an upregulation of collagen type II synthesis, and a change in collagen type from type II to types I and III [40,256]. A decrease in aggrecan is observed during the pathogenesis of osteoarthritis, while other proteoglycans such as decorin and fibromodulin are increased [257]. Aging causes decreased matrix deposition subsequent to increases in degradation by MMPs as well, causing additional issues with tissue integrity [255]. MMPs and aggrecanase are upregulated during osteoarthritis disease progression [258] causing additional cartilage loss. These processes are very similar to what occurs following an injury to cartilage and its subsequent repair [259].

CCN2 is increased in chondrocytes isolated from human osteoarthritis cartilage [260–262]. Increases in this matricellular protein can lead to fibrosis, and it is possible that mechanisms used to deliver CCN2 to degenerating discs may cause some repair to occur [249] similar to what is seen in an experimental model of osteoarthritis [263]. CCN3 has been shown to be upregulated in an osteoarthritis mouse model [264] and in human osteoarthritis as well [265]. Mutation of CCN3 in mice causes a loss of normal joint function and disease that appears similar to osteoarthritis in humans [266].

5.4.3. Chondromas/chondrosarcomas

A chondroma is a benign tumor of cartilage that may be present within bone and can cause fractures due to its growth pattern. Normally, little is done with these tumors unless there is a danger of fracture. Chondrosarcomas are cancerous tumors that can occur at any age, unlike other sarcomas. These are refractory to treatment and can be highly metastatic. The expression of CCNs can be correlated with chondrosarcoma grades and thus may be useful in clinical identification of these tumors [267]. The same may be true for chondromas as well.

5.4.4. Osteosarcomas

Osteosarcoma is the most common type of cancer that develops in bone, occurring most frequently in children and young adults. Most primary tumors develop in the areas of bone that are growing rapidly such as near the ends of long bones surrounding the knee—the distal femur or the proximal tibia. The proximal humerous is the next most common site, although osteosarcoma can develop in any bone. Evidence suggests that osteosarcoma might originate from mesenchymal cells with osteoblastic features [268,269]. CCN1 expression correlates with poor prognosis of osteosarcoma and overexpression of CCN1 increases cell proliferation and metastatic potential of tumor cell lines [270] and CCN1 knockdown reverses this phenomenon [271]. In humans, CCN3 is associated with increased lung metastasis in osteosarcoma patients [272] and CCN4 expression was shown to be higher in bone from osteosarcoma patients compared to normal tissue [273].

5.4.5. Ewing sarcoma

Ewing sarcoma is the second most common type of bone cancer that predominantly affects children. CCN3 is expressed in approximately 30% of Ewing Sarcoma cases and its expression correlates with lower survival [272]. Elevated CCN3 expression also correlated with recurrences and metastases compared to primary tumors in a study that examined 170 human Ewing sarcoma specimens by immunohistochemistry. In this same study, a low level of CCN3 expression was associated with improved patient prognosis [274].

5.4.6. Chondrodysplasias

Mutations in the major collagen present in hyaline and elastic cartilage, type II, cause several chondrodysplasias [81]. Mutation of type XI collagen also causes skeletal issues due to issues with the structure of the growth plate, in mice with this defect [275]. Mice lacking collagen IX appear normal but have alterations in the cartilage forming the growth plate and ultimately these mice develop osteoarthritis when older [276]. Mutations cause skeletal issues in humans as well, as do mutations in type X collagen. Mice lacking collagen X have several phenotypes, including no obvious changes, metaphyseal dysplasia, and death soon after birth [81,83,277].

5.4.7. CCNs in cartilage and bone pathophysiology

The role of CCNs in cartilage function involves several family members of this class of matricellular proteins. CCN1-6 have all been identified in cartilage matrix, and all CCN genes have an increase in their expression in osteoarthritis or rheumatoid arthritis [265]. CCN3 is integral to the proliferation of chondrocytes, while CCN1, CCN2, and CCN6 are involved in later states of maturation, proliferation, and the calcification of cartilage matrix during endochondral ossification. CCN4 and CCN5 also participate in the differentiation and calcification of cartilage [231]. CCN2 causes proper bone strength, shape, and length, while the counteraction by CCN3 regulates these structural processes.

CCN1 knockout mice die during development, making analysis of their skeleton difficult; however, this protein causes proliferation of chondrocytes and secretion of matrix in vitro [278]. CCN1 is present in chondrocytes in the proliferative and pre-hypertrophic zones during endochondral bone formation [278]. CCN1 signals through the WNT signaling pathway to cause maturation of chondrocytes isolated from sternal cartilage, with overexpression of CCN1 causing damage to chondrocytes [279]. Induction of CCN1 by β-catenin causes maturation of chondrocytes, while overexpression using a cartilage-specific promoter causes chondrodysplasia [279]. CCN1 is decreased in expression during the differentiation of mesenchymal stem cells into chondrocytes and osteoblasts, indicating that it may be important in the maintenance of stem cells [280].

CCN2 promotes differentiation and proliferation of chondrocytes as well as osteoblasts (see reference [47]). The presence of CCN2 is an important modulator of the deposition of cartilage matrix. CCN2 also is a major factor in the induction of fibrosis [281]. This growth factor causes secretion of collagen type II and can induce cells to proliferate and subsequently differentiate [282]. Lack of CCN2 causes issues with bone growth due to lack of hypertrophic zone cartilage growth and loss of angiogenesis [283,284] and causes death immediately after birth due to respiratory failure [239]. CCN2 binds several growth factors, including bone morphogenetic proteins and TGF-β, and can affect cartilage function in this manner [226].

Addition of CCN2 to damaged cartilage in rats can cause enhanced healing [263]. Although exogenous administration of matrix metalloproteases can cause damage to cartilage similar to osteoarthritis [285,286], the use of these inhibitors does not aid in the treatment of the disease [287]. Overexpression of CCN2 using a col2a1 promoter caused reversal of some aging-related changes present in the articular cartilage of aged mice by enhancing matrix deposition and proliferation of chondrocytes. In addition, changes characteristic of cartilage degeneration were reversed, such as the expression of collagens I or X and the presence of MMPs 9 and 13 [288]. These mice also showed greater levels of matrix and faster ossification during endochondral bone growth [237].

CCN2 binds to receptor activator of NF-kappa B (RANK) as seen by plasmon resonance analysis, and enhances RANK signaling. This indicates its importance in the formation of osteoclasts [289].

Fibroblast growth factor causes an increase in BMPs, and this binding causes repair of articular cartilage defects. Different BMPs have differing effects on this repair, but overall BMP3 is an important component during repair of articular cartilage [279].

CCN3 decreases proliferation in several different cell types [290]. CCN3 appears to be downregulated by PTHrP [231], which is involved in the growth of bone. CCN3 may regulate apoptosis under conditions of serum deprivation. The presence of this CCN decreased levels of both proteoglycan and collagen, which may mimic the cartilage environment in which no vascularization occurs and thus conditions are somewhat hypoxic [291]. Bone regeneration in mice that lack CCN3 is enhanced [292]. Loss of CCN2 causes an increase in the expression of CCN3 with a concomitant decrease in proliferation due to the presence of this matricellular protein. This deletion also caused reduced differentiation of chondrocytes due to the upregulation of CCN3 [231]. Loss of CCN2 causes reductions in aggrecan and types II and X collagen during development, a process which mimics the matrix loss that occurs with aging during osteoarthritis development [226]. Overexpression of CCN2 has effects that counter these reductions [288]. Overexpression of CCN4 caused an effect on cartilage differentiation by changing the function of another growth factor, TGF-β3. Mice that completely lacked CCN4 did repair surgical defects well, while mice that expressed CCN4 demonstrated some recovery from this injury [293]. Lack of another member of the CCN family, CCN6, causes a disease in humans that is a form of childhood arthritis, progressive pseudorheumatoid dysplasia [294]. The function of CCN6 in normal cartilage is not well understood, although its expression is high in osteoarthritis [295].

CCN1 and CCN2 expression is elevated during fracture repair in the long bones throughout the reparative phase of the callus, notably in proliferating chondrocytes and osteoblasts [296,297]. Abrogation of CCN1 by antibodies inhibits bone fracture healing in mice [298]. Further, recombinant CCN2 protein promotes the repair of articular cartilage in a rat osteoarthritis model [263]. These studies suggest that CCN proteins may play important roles in the homeostasis of bone and cartilage tissues.

Author details

John A. Arnott, Kathleen Doane and Sonia Lobo Planey*

*Address all correspondence to: splaney@tcmc.edu

The Commonwealth Medical College, Scranton, Pennsylvania

References

[1] Shimo, T., et al., *Expression and roles of connective tissue growth factor in Meckel's cartilage development.* Dev Dyn, 2004. 231(1): p. 136–47.

[2] DeLise, A.M., L. Fischer, and R.S. Tuan, *Cellular interactions and signaling in cartilage development.* Osteoarthr Cartilage, 2000. 8(5): p. 309–34.

[3] Michigami, T., *Regulatory mechanisms for the development of growth plate cartilage.* Cell Mol Life Sci, 2013. 70(22): p. 4213–21.

[4] Mirzamohammadi, F., G. Papaioannou, and T. Kobayashi, *MicroRNAs in cartilage development, homeostasis, and disease.* Curr Osteoporos Rep, 2014. 12(4): p. 410–9.

[5] Junqueira, L.C.U.a., J. Carneiro, and A.N. Contopoulos, *Basic histology*, in *A Concise medical library for practitioner and student*. Los Altos, CA, Norwalk, CT, New York, NY: Lange Medical Publications Appleton & Lange Lange Medical Books/McGraw Hill, p. 11.

[6] Williams, R., et al., *Identification and clonal characterisation of a progenitor cell sub-population in normal human articular cartilage.* PLoS One, 2010. 5(10): p. e13246.

[7] Hayashi, D., F.W. Roemer, and A. Guermazi, *Osteoarthritis year 2011 in review: imaging in OA—a radiologists' perspective.* Osteoarthr Cartilage, 2012. 20(3): p. 207–14.

[8] Naumann, A., et al., *Immunochemical and mechanical characterization of cartilage subtypes in rabbit.* J Histochem Cytochem, 2002. 50(8): p. 1049–58.

[9] Lotz, M. and R.F. Loeser, *Effects of aging on articular cartilage homeostasis.* Bone, 2012. 51(2): p. 241–8.

[10] Arden, N. and M.C. Nevitt, *Osteoarthritis: epidemiology.* Best Pract Res Clin Rheumatol, 2006. 20(1): p. 3–25.

[11] Tschugg, A., et al., *A prospective multicenter phase I/II clinical trial to evaluate safety and efficacy of NOVOCART Disc plus autologous disc chondrocyte transplantation in the treatment*

of nucleotomized and degenerative lumbar disc to avoid secondary disease: study protocol for a randomized controlled trial. Trials, 2016. 17(1): p. 108.

[12] Wilusz, R.E., J. Sanchez-Adams, and F. Guilak, *The structure and function of the pericellular matrix of articular cartilage.* Matrix Biol, 2014. 39: p. 25–32.

[13] Poole, C.A., M.H. Flint, and B.W. Beaumont, *Chondrons in cartilage: ultrastructural analysis of the pericellular microenvironment in adult human articular cartilages.* J Orthop Res, 1987. 5(4): p. 509–22.

[14] Poole, C.A., S. Ayad, and J.R. Schofield, *Chondrons from articular cartilage: I. Immunolocalization of type VI collagen in the pericellular capsule of isolated canine tibial chondrons.* J Cell Sci, 1988. 90 (Pt 4): p. 635–43.

[15] Lee, G.M., et al., *Isolated chondrons: a viable alternative for studies of chondrocyte metabolism in vitro.* Osteoarthr Cartilage, 1997. 5(4): p. 261–74.

[16] Sims, N., Baron R., *Bone cells and their function,* in *Skeletal growth factors,* E. Canalis, Editor. 2000, Lippincott Williams and Wilkins: Philadelphia, PA. p. 1–17.

[17] Katagiri, T. and N. Takahashi, *Regulatory mechanisms of osteoblast and osteoclast differentiation.* Oral Dis, 2002. 8(3): p. 147–59.

[18] Ahdjoudj, S., O. Fromigue, and P.J. Marie, *Plasticity and regulation of human bone marrow stromal osteoprogenitor cells: potential implication in the treatment of age-related bone loss.* Histol Histopathol, 2004. 19(1): p. 151–7.

[19] Marks, S.C., Jr. and S.N. Popoff, *Bone cell biology: the regulation of development, structure, and function in the skeleton.* Am J Anat, 1988. 183(1): p. 1–44.

[20] Alford, A.I., K.M. Kozloff, and K.D. Hankenson, *Extracellular matrix networks in bone remodeling.* Int J Biochem Cell Biol, 2015. 65: p. 20–31.

[21] Bonewald, L.F., *The amazing osteocyte.* J Bone Miner Res, 2011. 26(2): p. 229–38.

[22] Schaffler, M.B., et al., *Osteocytes: master orchestrators of bone.* Calcif Tissue Int, 2014. 94(1): p. 5–24.

[23] Berendsen, A.D. and B.R. Olsen, *Bone development.* Bone, 2015. 80: p. 14–8.

[24] Ducy, P. and G. Karsenty, *Genetic control of cell differentiation in the skeleton.* Curr Opin Cell Biol, 1998. 10(5): p. 614–9.

[25] Nakashima, K., et al., *The novel zinc finger-containing transcription factor osterix is required for osteoblast differentiation and bone formation.* Cell, 2002. 108(1): p. 17–29.

[26] Allen, M.R., J.M. Hock, and D.B. Burr, *Periosteum: biology, regulation, and response to osteoporosis therapies.* Bone, 2004. 35(5): p. 1003–12.

[27] Colnot, C., X. Zhang, and M.L. Knothe Tate, *Current insights on the regenerative potential of the periosteum: molecular, cellular, and endogenous engineering approaches.* J Orthop Res, 2012. 30(12): p. 1869–78.

[28] Grcevic, D., et al., *In vivo fate mapping identifies mesenchymal progenitor cells.* Stem Cells, 2012. 30(2): p. 187–96.

[29] Bilezikian, J.P., L.G. Raisz, and T.J. Martin, *Principles of bone biology.* 3rd ed. 2008, San Diego, CA: Academic Press/Elsevier.

[30] Cserjesi, P., et al., *Scleraxis: a basic helix-loop-helix protein that prefigures skeletal formation during mouse embryogenesis.* Development, 1995. 121(4): p. 1099–110.

[31] Sosic, D., et al., *Regulation of paraxis expression and somite formation by ectoderm- and neural tube-derived signals.* Dev Biol, 1997. 185(2): p. 229–43.

[32] Oberlender, S.A. and R.S. Tuan, *Expression and functional involvement of N-cadherin in embryonic limb chondrogenesis.* Development, 1994. 120(1): p. 177–87.

[33] Hall, B.K. and T. Miyake, *Divide, accumulate, differentiate: cell condensation in skeletal development revisited.* Int J Dev Biol, 1995. 39(6): p. 881–93.

[34] Maes, C., et al., *Osteoblast precursors, but not mature osteoblasts, move into developing and fractured bones along with invading blood vessels.* Dev Cell, 2010. 19(2): p. 329–44.

[35] Karsenty, G. and E.F. Wagner, *Reaching a genetic and molecular understanding of skeletal development.* Dev Cell, 2002. 2(4): p. 389–406.

[36] Kronenberg, H.M., *Developmental regulation of the growth plate.* Nature, 2003. 423(6937): p. 332–6.

[37] Mundlos, S., *Expression patterns of matrix genes during human skeletal development.* Prog Histochem Cytochem, 1994. 28(3): p. 1–47.

[38] Gao, Y., et al., *The ECM-cell interaction of cartilage extracellular matrix on chondrocytes.* Biomed Res Int, 2014. 2014: p. 648459.

[39] Zhang, Z., *Chondrons and the pericellular matrix of chondrocytes.* Tissue Eng Part B Rev, 2015. 21(3): p. 267–77.

[40] Hosseininia, S., et al., *Evidence for enhanced collagen type III deposition focally in the territorial matrix of osteoarthritic hip articular cartilage.* Osteoarthr Cartilage, 2016.

[41] Kadler, K.E., A. Hill, and E.G. Canty-Laird, *Collagen fibrillogenesis: fibronectin, integrins, and minor collagens as organizers and nucleators.* Curr Opin Cell Biol, 2008. 20(5): p. 495–501.

[42] Eyre, D., *Collagen of articular cartilage.* Arthritis Res, 2002. 4(1): p. 30–5.

[43] Wu, J.J., et al., *Type III collagen, a fibril network modifier in articular cartilage.* J Biol Chem, 2010. 285(24): p. 18537–44.

[44] Foldager, C.B., et al., *Distribution of basement membrane molecules, laminin and collagen type IV, in normal and degenerated cartilage tissues.* Cartilage, 2014. 5(2): p. 123–32.

[45] Wurster, N.B. and G. Lust, *Synthesis of fibronectin in normal and osteoarthritic articular cartilage.* Biochim Biophys Acta, 1984. 800(1): p. 52–8.

[46] Acharya, C., et al., *Cartilage oligomeric matrix protein and its binding partners in the cartilage extracellular matrix: interaction, regulation and role in chondrogenesis.* Matrix Biol, 2014. 37: p. 102–11.

[47] Arnott, J.A., et al., *The role of connective tissue growth factor (CTGF/CCN2) in skeletogenesis.* Crit Rev Eukaryot Gene Expr, 2011. 21(1): p. 43–69.

[48] Knudson, C.B. and W. Knudson, *Cartilage proteoglycans.* Semin Cell Dev Biol, 2001. 12(2): p. 69–78.

[49] Kim, S.H., J. Turnbull, and S. Guimond, *Extracellular matrix and cell signalling: the dynamic cooperation of integrin, proteoglycan and growth factor receptor.* J Endocrinol, 2011. 209(2): p. 139–51.

[50] Jun, J.I. and L.F. Lau, *Taking aim at the extracellular matrix: CCN proteins as emerging therapeutic targets.* Nat Rev Drug Discov, 2011. 10(12): p. 945–63.

[51] Rosen, C.J. and American Society for Bone and Mineral Research., *Primer on the metabolic bone diseases and disorders of mineral metabolism.* 8th ed. 2013, Ames, IA: Wiley-Blackwell. xxvi, 1078 p.

[52] Xiao, G., et al., *Ascorbic acid-dependent activation of the osteocalcin promoter in MC3T3-E1 preosteoblasts: requirement for collagen matrix synthesis and the presence of an intact OSE2 sequence.* Mol Endocrinol, 1997. 11(8): p. 1103–13.

[53] Franceschi, R.T. and B.S. Iyer, *Relationship between collagen synthesis and expression of the osteoblast phenotype in MC3T3-E1 cells.* J Bone Miner Res, 1992. 7(2): p. 235–46.

[54] Franceschi, R.T., B.S. Iyer, and Y. Cui, *Effects of ascorbic acid on collagen matrix formation and osteoblast differentiation in murine MC3T3-E1 cells.* J Bone Miner Res, 1994. 9(6): p. 843–54.

[55] Forlino, A., et al., *New perspectives on osteogenesis imperfecta.* Nat Rev Endocrinol, 2011. 7(9): p. 540–57.

[56] Rauch, F. and F.H. Glorieux, *Osteogenesis imperfecta.* Lancet, 2004. 363(9418): p. 1377–85.

[57] Mader, J.T. and W.H. Hulet, *Delayed hyperbaric treatment of cerebral air embolism: report of a case.* Arch Neurol, 1979. 36(8): p. 504–5.

[58] Davis, M.S., et al., *Increased susceptibility to microdamage in Brtl/+ mouse model for osteogenesis imperfecta.* Bone, 2012. 50(3): p. 784–91.

[59] Dong, X.N., et al., *Collagen mutation causes changes of the microdamage morphology in bone of an OI mouse model.* Bone, 2010. 47(6): p. 1071–5.

[60] Morello, R., et al., *CRTAP is required for prolyl 3- hydroxylation and mutations cause recessive osteogenesis imperfecta.* Cell, 2006. 127(2): p. 291–304.

[61] Liu, Y.F., et al., *Type II collagen gene variants and inherited osteonecrosis of the femoral head.* N Engl J Med, 2005. 352(22): p. 2294–301.

[62] Kannu, P., J. Bateman, and R. Savarirayan, *Clinical phenotypes associated with type II collagen mutations.* J Paediatr Child Health, 2012. 48(2): p. E38–43.

[63] Li, S.W., et al., *Transgenic mice with targeted inactivation of the Col2 alpha 1 gene for collagen II develop a skeleton with membranous and periosteal bone but no endochondral bone.* Genes Dev, 1995. 9(22): p. 2821–30.

[64] Aszodi, A., et al., *Collagen II is essential for the removal of the notochord and the formation of intervertebral discs.* J Cell Biol, 1998. 143(5): p. 1399–412.

[65] Myllyharju, J., *Extracellular matrix and developing growth plate.* Curr Osteoporos Rep, 2014. 12(4): p. 439–45.

[66] Czarny-Ratajczak, M. and A. Latos-Bielenska, *Collagens, the basic proteins of the human body.* J Appl Genet, 2000. 41(4): p. 317–30.

[67] Volk, S.W., et al., *Type III collagen regulates osteoblastogenesis and the quantity of trabecular bone.* Calcif Tissue Int, 2014. 94(6): p. 621–31.

[68] Poole, C.A., S. Ayad, and R.T. Gilbert, *Chondrons from articular cartilage. V. Immunohistochemical evaluation of type VI collagen organisation in isolated chondrons by light, confocal and electron microscopy.* J Cell Sci, 1992. 103 (Pt 4): p. 1101–10.

[69] Guilak, F., et al., *The pericellular matrix as a transducer of biomechanical and biochemical signals in articular cartilage.* Ann N Y Acad Sci, 2006. 1068: p. 498–512.

[70] Alexopoulos, L.G., et al., *Developmental and osteoarthritic changes in Col6a1-knockout mice: biomechanics of type VI collagen in the cartilage pericellular matrix.* Arthritis Rheum, 2009. 60(3): p. 771–9.

[71] Salter, D.M., et al., *Integrin expression by human articular chondrocytes.* Br J Rheumatol, 1992. 31(4): p. 231–4.

[72] Burg, M.A., et al., *Binding of the NG2 proteoglycan to type VI collagen and other extracellular matrix molecules.* J Biol Chem, 1996. 271(42): p. 26110–6.

[73] Nishiyama, A. and W.B. Stallcup, *Expression of NG2 proteoglycan causes retention of type VI collagen on the cell surface.* Mol Biol Cell, 1993. 4(11): p. 1097–108.

[74] Stallcup, W.B., K. Dahlin, and P. Healy, *Interaction of the NG2 chondroitin sulfate proteoglycan with type VI collagen.* J Cell Biol, 1990. 111(6 Pt 2): p. 3177–88.

[75] Gouttenoire, J., et al., *Modulation of collagen synthesis in normal and osteoarthritic cartilage.* Biorheology, 2004. 41(3–4): p. 535–42.

[76] Jobsis, G.J., et al., *Bethlem myopathy: a slowly progressive congenital muscular dystrophy with contractures.* Brain, 1999. 122 (Pt 4): p. 649–55.

[77] Myllyharju, J. and K.I. Kivirikko, *Collagens, modifying enzymes and their mutations in humans, flies and worms.* Trends Genet, 2004. 20(1): p. 33–43.

[78] Gordon, M.K. and R.A. Hahn, *Collagens.* Cell Tissue Res, 2010. 339(1): p. 247–57.

[79] Ricard-Blum, S., *The collagen family.* Cold Spring Harb Perspect Biol, 2011. 3(1): p. a004978.

[80] Arnold, W.V. and A. Fertala, *Skeletal diseases caused by mutations that affect collagen structure and function.* Int J Biochem Cell Biol, 2013. 45(8): p. 1556–67.

[81] Warman, M.L., et al., *Nosology and classification of genetic skeletal disorders: 2010 revision.* Am J Med Genet A, 2011. 155A(5): p. 943–68.

[82] Benson, M.K.D.A., *Children's orthopaedics and fractures.* 3rd ed. 2010, London; New York: Springer. xv, 905 p.

[83] Kwan, K.M., et al., *Abnormal compartmentalization of cartilage matrix components in mice lacking collagen X: implications for function.* J Cell Biol, 1997. 136(2): p. 459–71.

[84] Jacenko, O., P.A. LuValle, and B.R. Olsen, *Spondylometaphyseal dysplasia in mice carrying a dominant negative mutation in a matrix protein specific for cartilage-to-bone transition.* Nature, 1993. 365(6441): p. 56–61.

[85] Jacenko, O., et al., *A dominant negative mutation in the alpha 1 (X) collagen gene produces spondylometaphyseal defects in mice.* Prog Clin Biol Res, 1993. 383B: p. 427–36.

[86] Ho, M.S., et al., *COL10A1 nonsense and frame-shift mutations have a gain-of-function effect on the growth plate in human and mouse metaphyseal chondrodysplasia type Schmid.* Hum Mol Genet, 2007. 16(10): p. 1201–15.

[87] Tsang, K.Y., et al., *Surviving endoplasmic reticulum stress is coupled to altered chondrocyte differentiation and function.* PLoS Biol, 2007. 5(3): p. e44.

[88] Watanabe, H., et al., *Mouse cartilage matrix deficiency (cmd) caused by a 7 bp deletion in the aggrecan gene.* Nat Genet, 1994. 7(2): p. 154–7.

[89] Gundberg, C.M., *Matrix proteins.* Osteoporos Int, 2003. 14 Suppl 5: p. S37-40; discussion S40–2.

[90] Lamoureux, F., et al., *Proteoglycans: key partners in bone cell biology.* Bioessays, 2007. 29(8): p. 758–71.

[91] Yoshitake, H., et al., *Osteopontin-deficient mice are resistant to ovariectomy-induced bone resorption.* Proc Natl Acad Sci U S A, 1999. 96(14): p. 8156–60.

[92] Xu, T., et al., *Targeted disruption of the biglycan gene leads to an osteoporosis-like phenotype in mice.* Nat Genet, 1998. 20(1): p. 78–82.

[93] Corsi, A., et al., *Phenotypic effects of biglycan deficiency are linked to collagen fibril abnormalities, are synergized by decorin deficiency, and mimic Ehlers-Danlos-like changes in bone and other connective tissues.* J Bone Miner Res, 2002. 17(7): p. 1180–9.

[94] Takeuchi, Y., Y. Kodama, and T. Matsumoto, *Bone matrix decorin binds transforming growth factor-beta and enhances its bioactivity.* J Biol Chem, 1994. 269(51): p. 32634–8.

[95] Svensson, L., et al., *Fibromodulin-null mice have abnormal collagen fibrils, tissue organization, and altered lumican deposition in tendon.* J Biol Chem, 1999. 274(14): p. 9636–47.

[96] Saamanen, A.M., et al., *Murine fibromodulin: cDNA and genomic structure, and age-related expression and distribution in the knee joint.* Biochem J, 2001. 355(Pt 3): p. 577–85.

[97] Hildebrand, A., et al., *Interaction of the small interstitial proteoglycans biglycan, decorin and fibromodulin with transforming growth factor beta.* Biochem J, 1994. 302 (Pt 2): p. 527–34.

[98] Zheng, Z., et al., *Fibromodulin-deficiency alters temporospatial expression patterns of transforming growth factor-beta ligands and receptors during adult mouse skin wound healing.* PLoS One, 2014. 9(6): p. e90817.

[99] Raouf, A. and A. Seth, *Discovery of osteoblast-associated genes using cDNA microarrays.* Bone, 2002. 30(3): p. 463–71.

[100] Chakravarti, S., *Functions of lumican and fibromodulin: lessons from knockout mice.* Glycoconj J, 2002. 19(4–5): p. 287–93.

[101] Yeung Tsang, K., et al., *The chondrocytic journey in endochondral bone growth and skeletal dysplasia.* Birth Defects Res C Embryo Today, 2014. 102(1): p. 52–73.

[102] Heinegard, D., *Proteoglycans and more--from molecules to biology.* Int J Exp Pathol, 2009. 90(6): p. 575–86.

[103] Wong, M., et al., *Aggrecan core protein is expressed in membranous bone of the chick embryo. Molecular and biomechanical studies of normal and nanomelia embryos.* J Biol Chem, 1992. 267(8): p. 5592–8.

[104] Wai, A.W., et al., *Disrupted expression of matrix genes in the growth plate of the mouse cartilage matrix deficiency (cmd) mutant.* Dev Genet, 1998. 22(4): p. 349–58.

[105] Rodgers, K.D., J.D. San Antonio, and O. Jacenko, *Heparan sulfate proteoglycans: a GAGgle of skeletal-hematopoietic regulators.* Dev Dyn, 2008. 237(10): p. 2622–42.

[106] Nakamura, H. and H. Ozawa, *Immunohistochemical localization of heparan sulfate proteoglycan in rat tibiae.* J Bone Miner Res, 1994. 9(8): p. 1289–99.

[107] Bernfield, M., et al., *Functions of cell surface heparan sulfate proteoglycans.* Annu Rev Biochem, 1999. 68: p. 729–77.

[108] Park, P.W., O. Reizes, and M. Bernfield, *Cell surface heparan sulfate proteoglycans: selective regulators of ligand-receptor encounters.* J Biol Chem, 2000. 275(39): p. 29923–6.

[109] Govindraj, P., et al., *Isolation and identification of the major heparan sulfate proteoglycans in the developing bovine rib growth plate.* J Biol Chem, 2002. 277(22): p. 19461–9.

[110] Iozzo, R.V., *Basement membrane proteoglycans: from cellar to ceiling.* Nat Rev Mol Cell Biol, 2005. 6(8): p. 646–56.

[111] Knox, S.M. and J.M. Whitelock, *Perlecan: how does one molecule do so many things?* Cell Mol Life Sci, 2006. 63(21): p. 2435–45.

[112] Kruegel, J. and N. Miosge, *Basement membrane components are key players in specialized extracellular matrices.* Cell Mol Life Sci, 2010. 67(17): p. 2879–95.

[113] Nicole, S., et al., *Perlecan, the major proteoglycan of basement membranes, is altered in patients with Schwartz-Jampel syndrome (chondrodystrophic myotonia).* Nat Genet, 2000. 26(4): p. 480–3.

[114] Arikawa-Hirasawa, E., et al., *Perlecan is essential for cartilage and cephalic development.* Nat Genet, 1999. 23(3): p. 354–8.

[115] Costell, M., et al., *Perlecan maintains the integrity of cartilage and some basement membranes.* J Cell Biol, 1999. 147(5): p. 1109–22.

[116] Chami, I., H. Boujnah, and S. Zmerli, *[Role of echography in the evaluation of renal masses. Apropos of 65 cases].* J Urol (Paris), 1982. 88(9): p. 577–85.

[117] Ishijima, M., et al., *Perlecan modulates VEGF signaling and is essential for vascularization in endochondral bone formation.* Matrix Biol, 2012. 31(4): p. 234–45.

[118] Solis, M.A., et al., *Hyaluronan regulates cell behavior: a potential niche matrix for stem cells.* Biochem Res Int, 2012. 2012: p. 346972.

[119] Astachov, L., et al., *Hyaluronan and mesenchymal stem cells: from germ layer to cartilage and bone.* Front Biosci (Landmark Ed), 2011. 16: p. 261–76.

[120] Schwartz, Z., et al., *Hyaluronic acid and chondrogenesis of murine bone marrow mesenchymal stem cells in chitosan sponges.* Am J Vet Res, 2011. 72(1): p. 42–50.

[121] Wu, S.C., et al., *Enhancement of chondrogenesis of human adipose derived stem cells in a hyaluronan-enriched microenvironment.* Biomaterials, 2010. 31(4): p. 631–40.

[122] Maioli, M., et al., *Hyaluronan esters drive Smad gene expression and signaling enhancing cardiogenesis in mouse embryonic and human mesenchymal stem cells.* PLoS One, 2010. 5(11): p. e15151.

[123] Termine, J.D., et al., *Osteonectin, a bone-specific protein linking mineral to collagen.* Cell, 1981. 26(1 Pt 1): p. 99–105.

[124] Brekken, R.A. and E.H. Sage, *SPARC, a matricellular protein: at the crossroads of cell-matrix.* Matrix Biol, 2000. 19(7): p. 569–80.

[125] Kupprion, C., K. Motamed, and E.H. Sage, *SPARC (BM-40, osteonectin) inhibits the mitogenic effect of vascular endothelial growth factor on microvascular endothelial cells.* J Biol Chem, 1998. 273(45): p. 29635–40.

[126] Yang, E., et al., *Frequent inactivation of SPARC by promoter hypermethylation in colon cancers.* Int J Cancer, 2007. 121(3): p. 567–75.

[127] Funk, S.E. and E.H. Sage, *The Ca2(+)-binding glycoprotein SPARC modulates cell cycle progression in bovine aortic endothelial cells.* Proc Natl Acad Sci U S A, 1991. 88(7): p. 2648–52.

[128] Tremble, P.M., et al., *SPARC, a secreted protein associated with morphogenesis and tissue remodeling, induces expression of metalloproteinases in fibroblasts through a novel extracellular matrix-dependent pathway.* J Cell Biol, 1993. 121(6): p. 1433–44.

[129] Nie, J. and E.H. Sage, *SPARC inhibits adipogenesis by its enhancement of beta-catenin signaling.* J Biol Chem, 2009. 284(2): p. 1279–90.

[130] Gruber, H.E., et al., *Cellular, but not matrix, immunolocalization of SPARC in the human intervertebral disc: decreasing localization with aging and disc degeneration.* Spine (Phila Pa 1976), 2004. 29(20): p. 2223–8.

[131] Gruber, H.E., et al., *Targeted deletion of the SPARC gene accelerates disc degeneration in the aging mouse.* J Histochem Cytochem, 2005. 53(9): p. 1131–8.

[132] Millecamps, M., et al., *Lumbar intervertebral disc degeneration associated with axial and radiating low back pain in ageing SPARC-null mice.* Pain, 2012. 153(6): p. 1167–79.

[133] Goyal, N., et al., *Immunohistochemical analysis of ageing and osteoarthritic articular cartilage.* J Mol Histol, 2010. 41(4-5): p. 193–7.

[134] Hecht, J.T. and E.H. Sage, *Retention of the matricellular protein SPARC in the endoplasmic reticulum of chondrocytes from patients with pseudoachondroplasia.* J Histochem Cytochem, 2006. 54(3): p. 269–74.

[135] Hynes, R.O., *Fibronectins.* Springer series in molecular biology. 1990, New York: Springer-Verlag. xv, 546 p.

[136] Matsuura, H., et al., *The oncofetal structure of human fibronectin defined by monoclonal antibody FDC-6. Unique structural requirement for the antigenic specificity provided by a glycosylhexapeptide.* J Biol Chem, 1988. 263(7): p. 3314–22.

[137] Pankov, R. and K.M. Yamada, *Fibronectin at a glance.* J Cell Sci, 2002. 115(Pt 20): p. 3861–3.

[138] Bentmann, A., et al., *Circulating fibronectin affects bone matrix, whereas osteoblast fibronectin modulates osteoblast function.* J Bone Miner Res, 2010. 25(4): p. 706–15.

[139] Stein, G.S. and J.B. Lian, *Molecular mechanisms mediating proliferation/differentiation interrelationships during progressive development of the osteoblast phenotype.* Endocr Rev, 1993. 14(4): p. 424–42.

[140] Sottile, J. and D.C. Hocking, *Fibronectin polymerization regulates the composition and stability of extracellular matrix fibrils and cell-matrix adhesions.* Mol Biol Cell, 2002. 13(10): p. 3546–59.

[141] George, E.L., et al., *Defects in mesoderm, neural tube and vascular development in mouse embryos lacking fibronectin.* Development, 1993. 119(4): p. 1079–91.

[142] Adams, J.C. and J. Lawler, *The thrombospondins.* Int J Biochem Cell Biol, 2004. 36(6): p. 961–8.

[143] Bornstein, P., *Matricellular proteins: an overview.* J Cell Commun Signal, 2009. 3(3-4): p. 163–5.

[144] Chen, H., M.E. Herndon, and J. Lawler, *The cell biology of thrombospondin-1.* Matrix Biol, 2000. 19(7): p. 597–614.

[145] Lawler, J., *The functions of thrombospondin-1 and-2.* Curr Opin Cell Biol, 2000. 12(5): p. 634–40.

[146] Unger, S. and J.T. Hecht, *Pseudoachondroplasia and multiple epiphyseal dysplasia: new etiologic developments.* Am J Med Genet, 2001. 106(4): p. 244–50.

[147] Hankenson, K.D., et al., *Thrombospondins and novel TSR-containing proteins, R-spondins, regulate bone formation and remodeling.* Curr Osteoporos Rep, 2010. 8(2): p. 68–76.

[148] Crawford, S.E., et al., *Thrombospondin-1 is a major activator of TGF-beta1 in vivo.* Cell, 1998. 93(7): p. 1159–70.

[149] Posey, K.L., et al., *Skeletal abnormalities in mice lacking extracellular matrix proteins, thrombospondin-1, thrombospondin-3, thrombospondin-5, and type IX collagen.* Am J Pathol, 2008. 172(6): p. 1664–74.

[150] Smadja, D.M., et al., *Thrombospondin-1 is a plasmatic marker of peripheral arterial disease that modulates endothelial progenitor cell angiogenic properties.* Arterioscler Thromb Vasc Biol, 2011. 31(3): p. 551–9.

[151] Amend, S.R., et al., *Thrombospondin-1 regulates bone homeostasis through effects on bone matrix integrity and nitric oxide signaling in osteoclasts.* J Bone Miner Res, 2015. 30(1): p. 106–15.

[152] Hankenson, K.D., et al., *Increased marrow-derived osteoprogenitor cells and endosteal bone formation in mice lacking thrombospondin 2.* J Bone Miner Res, 2000. 15(5): p. 851–62.

[153] Hankenson, K.D. and P. Bornstein, *The secreted protein thrombospondin 2 is an autocrine inhibitor of marrow stromal cell proliferation.* J Bone Miner Res, 2002. 17(3): p. 415–25.

[154] Briggs, M.D., et al., *Pseudoachondroplasia and multiple epiphyseal dysplasia due to mutations in the cartilage oligomeric matrix protein gene.* Nat Genet, 1995. 10(3): p. 330–6.

[155] Hecht, J.T., et al., *Mutations in exon 17B of cartilage oligomeric matrix protein (COMP) cause pseudoachondroplasia.* Nat Genet, 1995. 10(3): p. 325–9.

[156] Svensson, L., et al., *Cartilage oligomeric matrix protein-deficient mice have normal skeletal development.* Mol Cell Biol, 2002. 22(12): p. 4366–71.

[157] DiPietro, L.A., et al., *Thrombospondin 1 synthesis and function in wound repair.* Am J Pathol, 1996. 148(6): p. 1851–60.

[158] Agah, A., et al., *The lack of thrombospondin-1 (TSP1) dictates the course of wound healing in double-TSP1/TSP2-null mice.* Am J Pathol, 2002. 161(3): p. 831–9.

[159] Kyriakides, T.R., et al., *Mice that lack thrombospondin 2 display connective tissue abnormalities that are associated with disordered collagen fibrillogenesis, an increased vascular density, and a bleeding diathesis.* J Cell Biol, 1998. 140(2): p. 419–30.

[160] Gruber, H.E., et al., *Disruption of the thrombospondin-2 gene alters the lamellar morphology but does not permit vascularization of the adult mouse lumbar disc.* Arthritis Res Ther, 2008. 10(4): p. R96.

[161] Hsieh, J.L., et al., *Intraarticular gene transfer of thrombospondin-1 suppresses the disease progression of experimental osteoarthritis.* J Orthop Res, 2010. 28(10): p. 1300–6.

[162] Kudo, Y., et al., *Periostin: novel diagnostic and therapeutic target for cancer.* Histol Histopathol, 2007. 22(10): p. 1167–74.

[163] Litvin, J., et al., *Expression and function of periostin-isoforms in bone.* J Cell Biochem, 2004. 92(5): p. 1044–61.

[164] Tsai, T.T., et al., *Increased periostin gene expression in degenerative intervertebral disc cells.* Spine J, 2013. 13(3): p. 289–98.

[165] Bedore, J., A. Leask, and C.A. Seguin, *Targeting the extracellular matrix: matricellular proteins regulate cell-extracellular matrix communication within distinct niches of the intervertebral disc.* Matrix Biol, 2014. 37: p. 124–30.

[166] Takayama, G., et al., *Periostin: a novel component of subepithelial fibrosis of bronchial asthma downstream of IL-4 and IL-13 signals.* J Allergy Clin Immunol, 2006. 118(1): p. 98–104.

[167] Oku, E., et al., *Periostin and bone marrow fibrosis.* Int J Hematol, 2008. 88(1): p. 57–63.

[168] Rios, H., et al., *periostin null mice exhibit dwarfism, incisor enamel defects, and an early-onset periodontal disease-like phenotype.* Mol Cell Biol, 2005. 25(24): p. 11131–44.

[169] Orimo, H., *The mechanism of mineralization and the role of alkaline phosphatase in health and disease.* J Nippon Med Sch, 2010. 77(1): p. 4–12.

[170] Millan, J.L., *The role of phosphatases in the initiation of skeletal mineralization.* Calcif Tissue Int, 2013. 93(4): p. 299–306.

[171] Anderson, H.C., *Vesicles associated with calcification in the matrix of epiphyseal cartilage.* J Cell Biol, 1969. 41(1): p. 59–72.

[172] Ali, S.Y., S.W. Sajdera, and H.C. Anderson, *Isolation and characterization of calcifying matrix vesicles from epiphyseal cartilage.* Proc Natl Acad Sci U S A, 1970. 67(3): p. 1513–20.

[173] Henthorn, P.S. and M.P. Whyte, *Missense mutations of the tissue-nonspecific alkaline phosphatase gene in hypophosphatasia.* Clin Chem, 1992. 38(12): p. 2501–5.

[174] Hessle, L., et al., *Tissue-nonspecific alkaline phosphatase and plasma cell membrane glycoprotein-1 are central antagonistic regulators of bone mineralization.* Proc Natl Acad Sci U S A, 2002. 99(14): p. 9445–9.

[175] Scriver, C.R., *The metabolic and molecular bases of inherited disease.* 7th ed. 1995, New York: McGraw-Hill, Health Professions Division.

[176] Fedde, K.N., et al., *Alkaline phosphatase knock-out mice recapitulate the metabolic and skeletal defects of infantile hypophosphatasia.* J Bone Miner Res, 1999. 14(12): p. 2015–26.

[177] Ogbureke, K.U. and L.W. Fisher, *Expression of SIBLINGs and their partner MMPs in salivary glands.* J Dent Res, 2004. 83(9): p. 664–70.

[178] Ogbureke, K.U. and L.W. Fisher, *Renal expression of SIBLING proteins and their partner matrix metalloproteinases (MMPs).* Kidney Int, 2005. 68(1): p. 155–66.

[179] Sodek, J., B. Ganss, and M.D. McKee, *Osteopontin.* Crit Rev Oral Biol Med, 2000. 11(3): p. 279–303.

[180] Mi, Z., H. Guo, and P.C. Kuo, *Identification of osteopontin-dependent signaling pathways in a mouse model of human breast cancer.* BMC Res Notes, 2009. 2: p. 119.

[181] Choi, S.T., et al., *Osteopontin might be involved in bone remodelling rather than in inflammation in ankylosing spondylitis.* Rheumatology (Oxford), 2008. 47(12): p. 1775–9.

[182] Wang, K.X. and D.T. Denhardt, *Osteopontin: role in immune regulation and stress responses.* Cytokine Growth Factor Rev, 2008. 19(5-6): p. 333–45.

[183] Pardo, A., et al., *Up-regulation and profibrotic role of osteopontin in human idiopathic pulmonary fibrosis.* PLoS Med, 2005. 2(9): p. e251.

[184] Wilson, M.J., L. Liaw, and P. Koopman, *Osteopontin and related SIBLING glycoprotein genes are expressed by Sertoli cells during mouse testis development.* Dev Dyn, 2005. 233(4): p. 1488–95.

[185] Giachelli, C.M., et al., *Evidence for a role of osteopontin in macrophage infiltration in response to pathological stimuli in vivo.* Am J Pathol, 1998. 152(2): p. 353–8.

[186] O'Regan, A.W., et al., *Osteopontin is associated with T cells in sarcoid granulomas and has T cell adhesive and cytokine-like properties in vitro.* J Immunol, 1999. 162(2): p. 1024–31.

[187] Standal, T., et al., *Osteopontin is an adhesive factor for myeloma cells and is found in increased levels in plasma from patients with multiple myeloma.* Haematologica, 2004. 89(2): p. 174–82.

[188] McKee, M.D. and A. Nanci, *Osteopontin: an interfacial extracellular matrix protein in mineralized tissues.* Connect Tissue Res, 1996. 35(1–4): p. 197–205.

[189] Hunter, G.K., et al., *Nucleation and inhibition of hydroxyapatite formation by mineralized tissue proteins.* Biochem J, 1996. 317(Pt 1): p. 59–64.

[190] Qiu, S.R., et al., *Molecular modulation of calcium oxalate crystallization by osteopontin and citrate.* Proc Natl Acad Sci U S A, 2004. 101(7): p. 1811–5.

[191] Giachelli, C.M. and S. Steitz, *Osteopontin: a versatile regulator of inflammation and biomineralization.* Matrix Biol, 2000. 19(7): p. 615–22.

[192] Moreno-Layseca, P. and C.H. Streuli, *Signalling pathways linking integrins with cell cycle progression.* Matrix Biol, 2014. 34: p. 144–53.

[193] Rittling, S.R., et al., *Mice lacking osteopontin show normal development and bone structure but display altered osteoclast formation in vitro.* J Bone Miner Res, 1998. 13(7): p. 1101–11.

[194] Koyama, Y., et al., *Osteopontin deficiency suppresses high phosphate load-induced bone loss via specific modulation of osteoclasts.* Endocrinology, 2006. 147(6): p. 3040–9.

[195] Ishijima, M., et al., *Osteopontin is associated with nuclear factor kappaB gene expression during tail-suspension-induced bone loss.* Exp Cell Res, 2006. 312(16): p. 3075–83.

[196] Boskey, A.L., et al., *Osteopontin deficiency increases mineral content and mineral crystallinity in mouse bone.* Calcif Tissue Int, 2002. 71(2): p. 145–54.

[197] Chellaiah, M.A., et al., *Osteopontin deficiency produces osteoclast dysfunction due to reduced CD44 surface expression.* Mol Biol Cell, 2003. 14(1): p. 173–89.

[198] Chellaiah, M.A., et al., *Rho-dependent Rho kinase activation increases CD44 surface expression and bone resorption in osteoclasts.* J Biol Chem, 2003. 278(31): p. 29086–97.

[199] Franzen, A., et al., *Altered osteoclast development and function in osteopontin deficient mice.* J Orthop Res, 2008. 26(5): p. 721–8.

[200] Bianco, P., et al., *Expression of bone sialoprotein (BSP) in developing human tissues.* Calcif Tissue Int, 1991. 49(6): p. 421–6.

[201] Gorski, J.P., et al., *Extracellular bone acidic glycoprotein-75 defines condensed mesenchyme regions to be mineralized and localizes with bone sialoprotein during intramembranous bone formation.* J Biol Chem, 2004. 279(24): p. 25455–63.

[202] Chen, J., H.S. Shapiro, and J. Sodek, *Development expression of bone sialoprotein mRNA in rat mineralized connective tissues.* J Bone Miner Res, 1992. 7(8): p. 987–97.

[203] Malaval, L., et al., *Bone sialoprotein plays a functional role in bone formation and osteoclastogenesis.* J Exp Med, 2008. 205(5): p. 1145–53.

[204] Boudiffa, M., et al., *Bone sialoprotein deficiency impairs osteoclastogenesis and mineral resorption in vitro.* J Bone Miner Res, 2010. 25(12): p. 2669–79.

[205] Nielsen-Marsh, C.M., et al., *Osteocalcin protein sequences of Neanderthals and modern primates.* Proc Natl Acad Sci U S A, 2005. 102(12): p. 4409–13.

[206] Villafan-Bernal, J.R., S. Sanchez-Enriquez, and J.F. Munoz-Valle, *Molecular modulation of osteocalcin and its relevance in diabetes (Review).* Int J Mol Med, 2011. 28(3): p. 283–93.

[207] Hauschka, P.V., et al., *Osteocalcin and matrix Gla protein: vitamin K-dependent proteins in bone.* Physiol Rev, 1989. 69(3): p. 990–1047.

[208] Glowacki, J. and J.B. Lian, *Impaired recruitment and differentiation of osteoclast progenitors by osteocalcin-deplete bone implants.* Cell Differ, 1987. 21(4): p. 247–54.

[209] Ducy, P., et al., *Increased bone formation in osteocalcin-deficient mice.* Nature, 1996. 382(6590): p. 448–52.

[210] Ducy, P., *The role of osteocalcin in the endocrine cross-talk between bone remodelling and energy metabolism.* Diabetologia, 2011. 54(6): p. 1291–7.

[211] Karsenty, G. and M. Ferron, *The contribution of bone to whole-organism physiology.* Nature, 2012. 481(7381): p. 314–20.

[212] Bork, P., *The modular architecture of a new family of growth regulators related to connective tissue growth factor.* FEBS Lett, 1993. 327(2): p. 125–30.

[213] Voorberg, J., et al., *Assembly and routing of von Willebrand factor variants: the requirements for disulfide-linked dimerization reside within the carboxy-terminal 151 amino acids.* J Cell Biol, 1991. 113(1): p. 195–205.

[214] Holt, G.D., M.K. Pangburn, and V. Ginsburg, *Properdin binds to sulfatide [Gal(3-SO4)beta 1-1 Cer] and has a sequence homology with other proteins that bind sulfated glycoconjugates.* J Biol Chem, 1990. 265(5): p. 2852–5.

[215] McDonald, N.Q. and W.A. Hendrickson, *A structural superfamily of growth factors containing a cystine knot motif.* Cell, 1993. 73(3): p. 421–4.

[216] Desnoyers, L., *Structural basis and therapeutic implication of the interaction of CCN proteins with glycoconjugates.* Curr Pharm Des, 2004. 10(31): p. 3913–28.

[217] Perbal, B., *The CCN3 (NOV) cell growth regulator: a new tool for molecular medicine.* Expert Rev Mol Diagn, 2003. 3(5): p. 597–604.

[218] Kubota, S. and M. Takigawa, *CCN family proteins and angiogenesis: from embryo to adulthood.* Angiogenesis, 2007. 10(1): p. 1–11.

[219] Juric, V., C.C. Chen, and L.F. Lau, *TNFalpha-induced apoptosis enabled by CCN1/CYR61: pathways of reactive oxygen species generation and cytochrome c release.* PLoS One, 2012. 7(2): p. e31303.

[220] Kawata, K., et al., *Role of LRP1 in transport of CCN2 protein in chondrocytes.* J Cell Sci, 2012. 125(Pt 12): p. 2965–72.

[221] Shakibaei, M., C. Csaki, and A. Mobasheri, *Diverse roles of integrin receptors in articular cartilage.* Adv Anat Embryol Cell Biol, 2008. 197: p. 1–60.

[222] Aszodi, A., et al., *Beta1 integrins regulate chondrocyte rotation, G1 progression, and cytokinesis.* Genes Dev, 2003. 17(19): p. 2465–79.

[223] Benassi, M.S., et al., *NG2 expression predicts the metastasis formation in soft-tissue sarcoma patients.* J Orthop Res, 2009. 27(1): p. 135–40.

[224] Pfaffle, M., et al., *Anchorin CII, a collagen-binding chondrocyte surface protein of the calpactin family.* Prog Clin Biol Res, 1990. 349: p. 147–57.

[225] Knudson, W., et al., *Hyaluronan oligosaccharides perturb cartilage matrix homeostasis and induce chondrocytic chondrolysis.* Arthritis Rheum, 2000. 43(5): p. 1165–74.

[226] Nishida, T., et al., *CCN2 (Connective Tissue Growth Factor) is essential for extracellular matrix production and integrin signaling in chondrocytes.* J Cell Commun Signal, 2007. 1(1): p. 45–58.

[227] Goessler, U.R., et al., *Differential modulation of integrin expression in chondrocytes during expansion for tissue engineering.* In Vivo, 2005. 19(3): p. 501–7.

[228] Garciadiego-Cazares, D., et al., *Coordination of chondrocyte differentiation and joint formation by alpha5beta1 integrin in the developing appendicular skeleton.* Development, 2004. 131(19): p. 4735–42.

[229] Iannone, F., et al., *[Phenotyping of chondrocytes from human osteoarthritic cartilage: chondrocyte expression of beta integrins and correlation with anatomic injury].* Reumatismo, 2001. 53(2): p. 122–130.

[230] Shimo, T., et al., *Retinoid signaling regulates CTGF expression in hypertrophic chondrocytes with differential involvement of MAP kinases.* J Bone Miner Res, 2005. 20(5): p. 867–77.

[231] Kawaki, H., et al., *Cooperative regulation of chondrocyte differentiation by CCN2 and CCN3 shown by a comprehensive analysis of the CCN family proteins in cartilage.* J Bone Miner Res, 2008. 23(11): p. 1751–64.

[232] Lau, L.F., *CCN1/CYR61: the very model of a modern matricellular protein.* Cell Mol Life Sci, 2011. 68(19): p. 3149–63.

[233] Kubota, S. and M. Takigawa, *The role of CCN2 in cartilage and bone development.* J Cell Commun Signal, 2011. 5(3): p. 209–17.

[234] Abreu, J.G., et al., *Connective-tissue growth factor (CTGF) modulates cell signalling by BMP and TGF-beta.* Nat Cell Biol, 2002. 4(8): p. 599–604.

[235] Song, J.J., et al., *Connective tissue growth factor (CTGF) acts as a downstream mediator of TGF-beta1 to induce mesenchymal cell condensation.* J Cell Physiol, 2007. 210(2): p. 398–410.

[236] Pala, D., et al., *Focal adhesion kinase/Src suppresses early chondrogenesis: central role of CCN2.* J Biol Chem, 2008. 283(14): p. 9239–47.

[237] Tomita, N., et al., *Cartilage-specific over-expression of CCN family member 2/connective tissue growth factor (CCN2/CTGF) stimulates insulin-like growth factor expression and bone growth.* PLoS One, 2013. 8(3): p. e59226.

[238] Wan, M. and X. Cao, *BMP signaling in skeletal development.* Biochem Biophys Res Commun, 2005. 328(3): p. 651–7.

[239] Ivkovic, S., et al., *Connective tissue growth factor coordinates chondrogenesis and angiogenesis during skeletal development.* Development, 2003. 130(12): p. 2779–91.

[240] Scarfi, S., *Use of bone morphogenetic proteins in mesenchymal stem cell stimulation of cartilage and bone repair.* World J Stem Cells, 2016. 8(1): p. 1–12.

[241] Shu, B., et al., *BMP2, but not BMP4, is crucial for chondrocyte proliferation and maturation during endochondral bone development.* J Cell Sci, 2011. 124(Pt 20): p. 3428–40.

[242] Minamizato, T., et al., *CCN3/NOV inhibits BMP-2-induced osteoblast differentiation by interacting with BMP and Notch signaling pathways.* Biochem Biophys Res Commun, 2007. 354(2): p. 567–73.

[243] Maeda, A., et al., *CCN family 2/connective tissue growth factor modulates BMP signalling as a signal conductor, which action regulates the proliferation and differentiation of chondrocytes.* J Biochem, 2009. 145(2): p. 207–16.

[244] Ono, M., et al., *WISP-1/CCN4 regulates osteogenesis by enhancing BMP-2 activity.* J Bone Miner Res, 2011. 26(1): p. 193–208.

[245] Mercurio, S., et al., *Connective-tissue growth factor modulates WNT signalling and interacts with the WNT receptor complex.* Development, 2004. 131(9): p. 2137–47.

[246] French, D.M., et al., *WISP-1 is an osteoblastic regulator expressed during skeletal development and fracture repair.* Am J Pathol, 2004. 165(3): p. 855–67.

[247] Yanagita, T., et al., *Expression and physiological role of CCN4/Wnt-induced secreted protein 1 mRNA splicing variants in chondrocytes.* FEBS J, 2007. 274(7): p. 1655–65.

[248] Safadi, F.F., et al., *Expression of connective tissue growth factor in bone: its role in osteoblast proliferation and differentiation in vitro and bone formation in vivo.* J Cell Physiol, 2003. 196(1): p. 51–62.

[249] Tran, C.M., I.M. Shapiro, and M.V. Risbud, *Molecular regulation of CCN2 in the intervertebral disc: lessons learned from other connective tissues.* Matrix Biol, 2013. 32(6): p. 298–306.

[250] Peng, B., et al., *Expression and role of connective tissue growth factor in painful disc fibrosis and degeneration.* Spine (Philadelphia, PA 1976), 2009. 34(5): p. E178–82.

[251] Lorenz, H. and W. Richter, *Osteoarthritis: cellular and molecular changes in degenerating cartilage.* Prog Histochem Cytochem, 2006. 40(3): p. 135–63.

[252] Speziali, A., et al., *Chondropenia: current concept review.* Musculoskelet Surg, 2015. 99(3): p. 189–200.

[253] Olivotti, L., et al., *Maximal endothelial tissue plasminogen activator release is not impaired in patients with acute coronary syndromes before heparin treatment.* Blood Coagul Fibrinolysis, 2001. 12(4): p. 261–7.

[254] Mariani, E., L. Pulsatelli, and A. Facchini, *Signaling pathways in cartilage repair.* Int J Mol Sci, 2014. 15(5): p. 8667–98.

[255] Umlauf, D., et al., *Cartilage biology, pathology, and repair.* Cell Mol Life Sci, 2010. 67(24): p. 4197–211.

[256] Miosge, N., et al., *Expression of collagen type I and type II in consecutive stages of human osteoarthritis.* Histochem Cell Biol, 2004. 122(3): p. 229–36.

[257] Cs-Szabo, G., et al., *Changes in messenger RNA and protein levels of proteoglycans and link protein in human osteoarthritic cartilage samples.* Arthritis Rheum, 1997. 40(6): p. 1037–45.

[258] Cawston, T.E. and A.J. Wilson, *Understanding the role of tissue degrading enzymes and their inhibitors in development and disease.* Best Pract Res Clin Rheumatol, 2006. 20(5): p. 983–1002.

[259] Wang, M., et al., *Recent progress in understanding molecular mechanisms of cartilage degeneration during osteoarthritis.* Ann N Y Acad Sci, 2011. 1240: p. 61–9.

[260] Kumar, S., et al., *Identification and initial characterization of 5000 expressed sequenced tags (ESTs) each from adult human normal and osteoarthritic cartilage cDNA libraries.* Osteoarthr Cartilage, 2001. 9(7): p. 641–53.

[261] Zhang, H., C.C. Liew, and K.W. Marshall, *Microarray analysis reveals the involvement of beta-2 microglobulin (B2M) in human osteoarthritis.* Osteoarthr Cartilage, 2002. 10(12): p. 950–60.

[262] Omoto, S., et al., *Expression and localization of connective tissue growth factor (CTGF/Hcs24/CCN2) in osteoarthritic cartilage.* Osteoarthr Cartilage, 2004. 12(10): p. 771–8.

[263] Nishida, T., et al., *Regeneration of defects in articular cartilage in rat knee joints by CCN2 (connective tissue growth factor).* J Bone Miner Res, 2004. 19(8): p. 1308–19.

[264] Meng, J., et al., *Microarray analysis of differential gene expression in temporomandibular joint condylar cartilage after experimentally induced osteoarthritis.* Osteoarthr Cartilage, 2005. 13(12): p. 1115–25.

[265] Komatsu, M., et al., *Expression profiles of human CCN genes in patients with osteoarthritis or rheumatoid arthritis.* J Orthop Sci, 2015. 20(4): p. 708–16.

[266] Roddy, K.A. and C.A. Boulter, *Targeted mutation of NOV/CCN3 in mice disrupts joint homeostasis and causes osteoarthritis-like disease.* Osteoarthr Cartilage, 2015. 23(4): p. 607–15.

[267] Yu, C., et al., *NOV (CCN3) regulation in the growth plate and CCN family member expression in cartilage neoplasia.* J Pathol, 2003. 201(4): p. 609–15.

[268] Damron, T.A., W.G. Ward, and A. Stewart, *Osteosarcoma, chondrosarcoma, and Ewing's sarcoma: National Cancer Data Base Report.* Clin Orthop Relat Res, 2007. 459: p. 40–7.

[269] Dorfman, H.D. and B. Czerniak, *Bone cancers.* Cancer, 1995. 75(1 Suppl): p. 203–10.

[270] Sabile, A.A., et al., *Cyr61 expression in osteosarcoma indicates poor prognosis and promotes intratibial growth and lung metastasis in mice.* J Bone Miner Res, 2012. 27(1): p. 58–67.

[271] Fromigue, O., et al., *CYR61 downregulation reduces osteosarcoma cell invasion, migration, and metastasis.* J Bone Miner Res, 2011. 26(7): p. 1533–42.

[272] Manara, M.C., et al., *The expression of ccn3(nov) gene in musculoskeletal tumors.* Am J Pathol, 2002. 160(3): p. 849–59.

[273] Wu, C.L., et al., *Ras activation mediates WISP-1-induced increases in cell motility and matrix metalloproteinase expression in human osteosarcoma.* Cell Signal, 2013. 25(12): p. 2812–22.

[274] Perbal, B., et al., *Prognostic relevance of CCN3 in Ewing sarcoma.* Hum Pathol, 2009. 40(10): p. 1479–86.

[275] Li, Y., et al., *A fibrillar collagen gene, Col11a1, is essential for skeletal morphogenesis.* Cell, 1995. 80(3): p. 423–30.

[276] Fassler, R., et al., *Mice lacking alpha 1 (IX) collagen develop noninflammatory degenerative joint disease.* Proc Natl Acad Sci U S A, 1994. 91(11): p. 5070–4.

[277] Rosati, R., et al., *Normal long bone growth and development in type X collagen-null mice.* Nat Genet, 1994. 8(2): p. 129–35.

[278] Wong, M., et al., *Cyr61, product of a growth factor-inducible immediate-early gene, regulates chondrogenesis in mouse limb bud mesenchymal cells.* Dev Biol, 1997. 192(2): p. 492–508.

[279] Zhang, Y., et al., *CCN1 Regulates Chondrocyte Maturation and Cartilage Development.* J Bone Miner Res, 2015.

[280] Schutze, N., et al., *Differential expression of CCN-family members in primary human bone marrow-derived mesenchymal stem cells during osteogenic, chondrogenic and adipogenic differentiation.* Cell Commun Signal, 2005. 3(1): p. 5.

[281] Sonnylal, S., et al., *Selective expression of connective tissue growth factor in fibroblasts in vivo promotes systemic tissue fibrosis.* Arthritis Rheum, 2010. 62(5): p. 1523–32.

[282] Fujisawa, T., et al., *CCN family 2/connective tissue growth factor (CCN2/CTGF) stimulates proliferation and differentiation of auricular chondrocytes.* Osteoarthr Cartilage, 2008. 16(7): p. 787–95.

[283] Takigawa, M., et al., *Role of CTGF/HCS24/ecogenin in skeletal growth control.* J Cell Physiol, 2003. 194(3): p. 256–66.

[284] Shimo, T., et al., *Connective tissue growth factor induces the proliferation, migration, and tube formation of vascular endothelial cells in vitro, and angiogenesis in vivo.* J Biochem, 1999. 126(1): p. 137–45.

[285] Li, N.G., et al., *New hope for the treatment of osteoarthritis through selective inhibition of MMP-13.* Curr Med Chem, 2011. 18(7): p. 977–1001.

[286] Lohmander, L.S., P.J. Neame, and J.D. Sandy, *The structure of aggrecan fragments in human synovial fluid. Evidence that aggrecanase mediates cartilage degradation in inflammatory joint disease, joint injury, and osteoarthritis.* Arthritis Rheum, 1993. 36(9): p. 1214–22.

[287] Krzeski, P., et al., *Development of musculoskeletal toxicity without clear benefit after administration of PG-116800, a matrix metalloproteinase inhibitor, to patients with knee osteoarthritis: a randomized, 12-month, double-blind, placebo-controlled study.* Arthritis Res Ther, 2007. 9(5): p. R109.

[288] Itoh, S., et al., *CCN family member 2/connective tissue growth factor (CCN2/CTGF) has anti-aging effects that protect articular cartilage from age-related degenerative changes.* PLoS One, 2013. 8(8): p. e71156.

[289] Aoyama, E., et al., *CCN2 enhances RANKL-induced osteoclast differentiation via direct binding to RANK and OPG.* Bone, 2015. 73: p. 242–8.

[290] Bleau, A.M., et al., *Antiproliferative activity of CCN3: involvement of the C-terminal module and post-translational regulation.* J Cell Biochem, 2007. 101(6): p. 1475–91.

[291] Ding, L., et al., *Effects of CCN3 on rat cartilage endplate chondrocytes cultured under serum deprivation in vitro.* Mol Med Rep, 2016.

[292] Matsushita, Y., et al., *CCN3 protein participates in bone regeneration as an inhibitory factor.* J Biol Chem, 2013. 288(27): p. 19973–85.

[293] Yoshioka, Y., et al., *CCN4/WISP-1 positively regulates chondrogenesis by controlling TGF-beta3 function.* Bone, 2016. 83: p. 162–70.

[294] Hurvitz, J.R., et al., *Mutations in the CCN gene family member WISP3 cause progressive pseudorheumatoid dysplasia.* Nat Genet, 1999. 23(1): p. 94–8.

[295] Baker, N., et al., *Dual regulation of metalloproteinase expression in chondrocytes by Wnt-1-inducible signaling pathway protein 3/CCN6.* Arthritis Rheum, 2012. 64(7): p. 2289–99.

[296] Hadjiargyrou, M., W. Ahrens, and C.T. Rubin, *Temporal expression of the chondrogenic and angiogenic growth factor CYR61 during fracture repair.* J Bone Miner Res, 2000. 15(6): p. 1014–23.

[297] Nakata, E., et al., *Expression of connective tissue growth factor/hypertrophic chondrocyte-specific gene product 24 (CTGF/Hcs24) during fracture healing.* Bone, 2002. 31(4): p. 441–7.

[298] Athanasopoulos, A.N., et al., *Vascular endothelial growth factor (VEGF)-induced up-regulation of CCN1 in osteoblasts mediates proangiogenic activities in endothelial cells and promotes fracture healing.* J Biol Chem, 2007. 282(37): p. 26746–53.

Chapter 9

Biophysical Properties of the Basal Lamina: A Highly Selective Extracellular Matrix

Fabienna Arends and Oliver Lieleg

Additional information is available at the end of the chapter

http://dx.doi.org/10.5772/62519

Abstract

In this chapter, we discuss a specialized version of the extracellular matrix, the basal lamina. We focus on biophysical approaches which helped in identifying the mechanistic principles that allow the basal lamina to act as a selective permeability barrier. We discuss the physicochemical interactions that entail binding of molecules or nanoparticles to the basal lamina matrix and outline physiological scenarios where altered selective permeability properties of the basal lamina might contribute to physiological (mal) function.

Keywords: laminin, collagen, entactin, microstructure, viscoelastic properties, permeability

1. Molecular composition of the basal lamina

The basal lamina constitutes a thin extracellular matrix, which is located between the connective tissue and the basolateral side of a cell layer. This cellular layer can consist of either endothelial or epithelial cells, and those cell types secrete the different molecular components of the basal lamina. The main components of the basal lamina are laminin, collagen IV, the perlecan complex, and entactin, which are also known as nidogen [1, 2]. Together, those macromolecules form a complex network as illustrated in **Figure 1**. In addition, the basal lamina may contain several proteases such as matrix metalloproteinase-2 (MMP-2), MMP-9, and growth factors such as transforming growth factor beta (TGF-β), insulin-like growth factor (IGF) and fibroblast growth factor (FGF) [3].

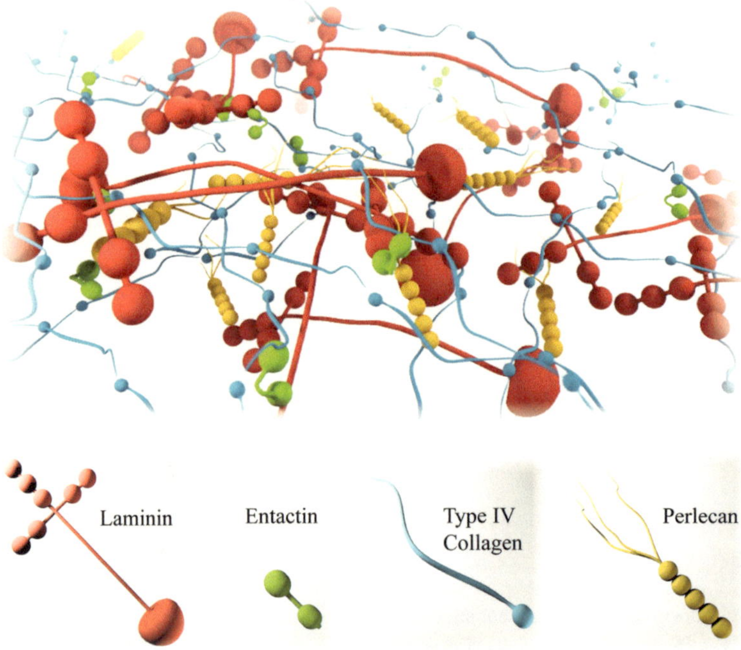

Figure 1. Schematic illustration of the basal lamina network. Both laminin and collagen IV assemble into a sheet-like network. Those two networks are cross-linked by entactin as well as the perlecan complex.

Laminin is a glycoprotein mainly found in basement membranes such as the basal lamina and is composed of three polypeptide chains: α-chain, β-chain, and γ-chain [4]. These three chains assemble into a cross-like structure, where the short arms of the cross are formed by the N-termini of the three subunits. The long arm of this cross-like structure is formed by all three subunits which assemble into an α-helical coiled-coil structure with a globular end [4, 5]. Laminin self-assembles into a sheet-like structure by binding the short arms of different laminins to each other [6]. The polymerized laminin network is anchored to the underlying cell layer via integrin interactions mediated by the globular end of the cross-like structure [4, 7]. Collagen IV is a collagen variant mostly found in the basal lamina and forms a helix similar to collagen I [8]. Type IV collagen self-assembles via covalent bonding, disulfide cross-linking, and non-covalent side-by-side interactions into a sheet-like structure [2]. Both sheet-like structures, the laminin and collagen IV network, do not interact with each other; however, both laminin and collagen IV can bind to perlecan as well as entactin. As a consequence, the latter two molecules act as cross-linkers between the two sheet-like structures, thus maintaining the complex architecture of the basal lamina [9]. The perlecan complex is a basal lamina-specific proteoglycan. In general, proteoglycans consist of a protein core with glycosaminoglycans covalently attached to the protein. Thus, the perlecan complex consists of perlecan as core protein and 2–15 heparan sulfate (HS) side chains [10]. Also entactin is a glycoprotein and consists of three globular units connected by rod-like structures [11]. Two of the three globular units (G1 and G2) are situated at the N-terminus of entactin. The third globular unit (G3) is found at the C-terminus of the protein. G3 strongly binds to the γ-laminin short arm but can also bind to collagen IV. In contrast, G2 only binds to collagen type IV [11], thus connecting the networks built by laminin and collagen type IV.

In addition to building a complex network and serving as an anchoring matrix for a neighboring cell layer, all of these basal lamina components can directly influence the cell fate: laminin, in combination with collagen IV supports cell attachment, differentiation, migration, and growth [12]. It was suggested that in addition to fibronectin, type IV collagen and laminin are involved in the formation of tight junctions [13]. Laminin and collagen IV are also key players in establishing the mechanical stability of the basal lamina [10]. As mentioned above, the proteoglycan perlecan consists of a core protein to which HS, a heavily charged glycosaminoglycan, is attached. In addition to acting as a cross-linker between laminin and collagen IV, perlecan and, in particular, the highly charged HS chains are responsible for the hydration of the matrix and contribute to the selective filtering properties of the basal lamina [14–16].

Although this highly specific structure–function relationship suggests that the microarchitecture of the basal lamina might be rather static, proteolysis of extracellular matrix (ECM) components and thus matrix remodeling is a process which continuously takes place *in vivo*. Remodeling of the ECM is, for example, a crucial part of wound healing and cell differentiation [17]. In addition, the degradation of ECM components can be responsible for cell apoptosis but, depending on the ECM component degraded, can also enhance cell viability [18]. In particular, the degradation of laminin is thought to be harmful for cells: In a study conducted in mice, it was suggested that the breakdown of laminin by the MMP-9 induces neuronal apoptosis but can be prevented by the addition of MMP-9 inhibitors [19].

Moreover, the degradation of laminin does not only result in cell apoptosis but also impacts the stability of the basal lamina [20]. Since laminin interacts with the integrins on the cell surface and anchors the cells onto the basal lamina, a breakdown of laminin results in a separation of the basal lamina from the endothelial/epithelial cell layer which in turn induces a loss of cell–matrix communication [21, 22]. It was shown in an *in vivo* study in a mouse model that when the second structural main component of the basal lamina, collagen IV, is knocked out, embryos develop normal during the first few days, but after 10 days of development lethality occurs [23]. It was suggested that collagen IV is essential for the function and integrity of the basal lamina when mechanical stress increases. However, collagen IV seems to be unimportant in the assembly of the basal lamina at early embryonic states [23]. Similar results were obtained when an enzyme, which catalyzes the assembly of collagen IV, was modified and thus nonfunctional. In these mice, collagen IV was present but did not assemble properly and the mouse embryos died after 10 days [24].

In contrast to those structural main components, loss of the small cross-linking molecule entactin seems to have a weaker influence on basal lamina structure and function. In mice, the inactivation or mutation of the gene encoding entactin results in a normal basal lamina phenotype, and the viability of the mutant mice seems not be strongly impaired by a loss of entactin [25–27]. Exceptions are the lung and the kidney, organs which fulfill important filtering tasks and thus contain a huge amount of basal lamina. Here, a loss of entactin cross-linking function entailed strong alteration of those tissues during embryonal development and ultimately led to death immediately after birth [28]. Of course, alterations in basal lamina properties can also have less severe consequences. For instance, long-term diabetes patients not only often suffer from retinopathies but also show an increased thickness and stiffness in the ocular basal lamina. Here, however, the higher amount of basal lamina proteins is due to the expression of diabetes-specific proteins whereas the production of the normal basal lamina components is not increased [29].

2. Selective permeability of the basal lamina *in vivo*

In the human body, the basal lamina always supports a cell layer of either endothelial or epithelial cells (**Figure 2**). Together, these two layers form a complex barrier which selectively regulates the entrance and distribution of molecules from or into the connective tissue. Molecules which are selectively transported across the basal lamina include growth factors, nutrients, and hormones. Examples for such basal lamina/cell barriers are found in the skin, the kidney, the blood–brain barrier, and the vascular system [10, 30–34].

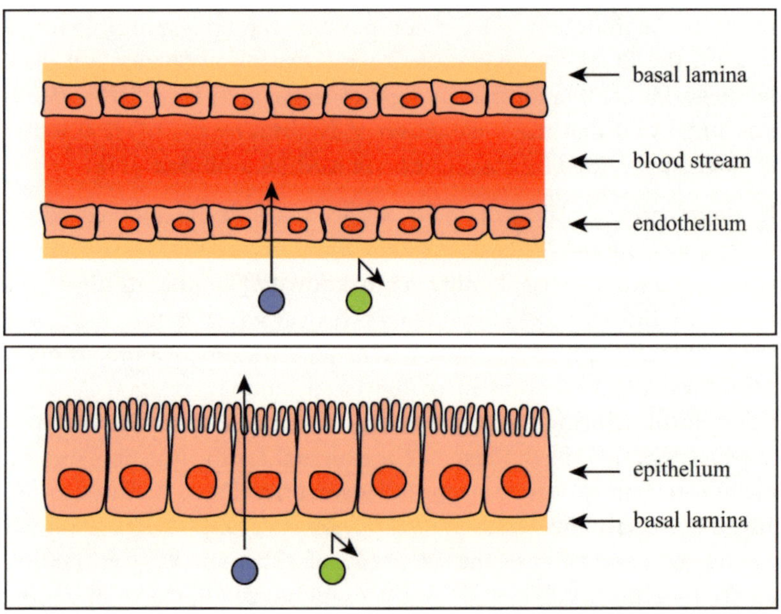

Figure 2. Illustration of complex barriers consisting of a cell layer and an adjacent basal lamina layer. The inner layer of blood vessels is constituted by endothelial cells with a basal lamina layer located on the outer side of the blood vessel. Also epithelial cells are supported with a thin layer of basal lamina. In both examples, selective permeability of the complex cell/biopolymer barrier toward molecules is observed, that is, some molecules can penetrate the barrier whereas others are rejected.

The skin poses one of the largest and, in most cases, the first barrier for foreign compounds. In addition to this protective function, the skin also regulates the uptake of oxygen and prevents the loss of water from the underlying tissue [32]. In kidney tissue, the basal lamina is, in combination with the epithelial cells, responsible for filtering [10], and defects in the basal lamina can result in kidney malfunction [34]. The blood–brain barrier [30] protects the brain tissue from pathogens and neurotoxic molecules, whereas it allows the passage of regulatory molecules such as hormones from the blood stream into the cerebrospinal fluid [35]. A similar structure is present in the vascular system. Here, the first barrier is established by endothelial cells which rest on a thin layer of basal lamina on their basolateral side [10]. Nutrients, growth factors, proteins, hormones, and polysaccharides are prevented from leaking from the blood stream into the connective tissue by tight junctions between the endothelial cells [31, 35–37]. If the integrity of these tight junctions is impaired, the basal lamina becomes directly accessible for blood compounds. Moreover, if the basal lamina layer is damaged, the translocation of solutes from the blood stream into the connective tissue is increased, even if the tight junctions are intact [20]. Of course, for molecules which need to traverse from the connective tissue into the blood stream, the basal lamina is encountered first before the endothelial cells are reached. In this scenario, the basal lamina layer constitutes the primary barrier.

A detailed knowledge of the molecular interactions which determine the selective filtering properties of the basal lamina is especially interesting for the design of new drug carrier vehicles for targeted drug delivery applications. One example for such an application is the specific targeting of tumors. In tumor tissue, the influence of the basal lamina barrier becomes even more important since tumors usually show an increased production of ECM [38]. Drug carriers are often injected intravenously; thus, the vascular system poses the critical barrier which the drug carriers have to pass. Here, the passage of drugs/drug carrier vehicles from the blood stream into the adjoining tissue is primarily regulated by the endothelial cells. However, in most cases, the endothelium around tumors is leaky. This is also known as an "enhanced permeability and retention effect" (EPR). Since the barrier function of the endothelium is impaired by the tumor, the basal lamina becomes directly accessible for compounds from the blood stream. In such a situation, the passage of drug carrier systems and their incorporated drugs is mainly regulated by the basal lamina.

In all of these examples, the selective barrier properties of the basal lamina are key for regulating complex biological processes. To possess such a high selectivity toward molecules or drug carrier particles, that is, deciding which of them are allowed to pass and which are rejected, an advanced molecular filter system based on various interactions is needed. Understanding the physical interactions between drug carriers and the complex multicomponent, basal lamina is crucial to efficiently adjust the surface parameters of drug carriers in such a way either that they are able to easily penetrate the basal lamina barrier or that they accumulate at the basal lamina interface. Studying the mechanistic principles which govern the selective permeability properties of the basal lamina layer *in vivo* is, however, very difficult: On the one hand, the basal lamina has a thickness of only a few hundred nanometers which would require optical experiments with a supreme spatial resolution such as PALM/STORM or STED microscopy [39]. On the other hand, the presence of a plethora of molecules, dynamic alterations in the basal lamina composition by enzymatic processes, or generation of new basal lamina components by the adjacent cell layer further complicates the interpretation of *in vivo* permeability studies and the correlation of the experimental results with physicochemical principles. Thus, a detailed investigation of the selective permeability properties of a complex biopolymer barrier such as the basal lamina requires a reliable *in vitro* model system, which is available in quantities large enough to conduct systematic tests while reproducing the behavior of the *in vivo* basal lamina layer.

3. Basal lamina model systems

A suitable source for the purification of an extracellular matrix that mimics the basal lamina is the Engelbreth–Holm–Swarm sarcoma of mice. This tumor produces, in contrast to healthy tissue, large amounts of ECM with laminin and collagen IV being the main components [3]. Depending on the question asked, individual macromolecular components of the basal lamina may be sufficient to take over the role of the complex biopolymer mixture. For instance, adhesion of cells to solid substrates is promoted similarly well by laminin coatings as by coatings with the multicomponent ECM [40, 41]. For other basal lamina properties such as viscoelasticity and selective permeability, it is crucial that the biological complexity of the system is maintained so that the full spectrum of basal lamina function is obtained.

The high abundance of ECM in Engelbreth–Holm–Swarm tumor tissue makes it possible to purify reasonable amounts of this multicomponent matrix as required for systematic *in vitro* experiments. A first purification protocol for this ECM was established by Kleinman et al. [42, 43] in the 1980s. The extract is liquid at temperatures between 4°C and approx. 15°C and forms a gel at higher temperatures. In its gel form, the matrix was tested for its biological activity, and it was shown in several studies that the purified ECM successfully promotes the differentiation of various cell types [44–47]. Cells can be either plated on top of the gel, thus simulating a two-dimensional (2D) environment, or they can be embedded into a 3D ECM matrix. Which configuration is chosen depends on the detailed experimental setup, the cell type used and the biological question. For instance, cell migration experiments can be conducted both on flat surfaces which have been coated by ECM components and in 3-dimensional basal lamina gels [48, 49].

The purification protocol of Kleinman et al. is used by several companies for the commercial production of ECM. Although these commercial ECM variants are extracted according to the same purification protocol, significant differences in the behavior of cells embedded into those gels have recently been described [50]: The migration behavior of leukocyte-like dHL-60 cells in four different commercially available ECM gel variants differed strongly even though the gels were prepared at matching total protein concentrations. Moreover, in one of the ECM gels, life–dead stains demonstrated a significantly increased percentage of nonviable cells. At the same time, for this gel variant, there was an additional band visible when the gel was analyzed by SDS-PAGE. Mass spectrometry showed that this additional band contained laminin fragments which indeed are suspected to be harmful for cells. This result demonstrates the dilemma a researcher is exposed to when working with commercial model systems: On the one hand, the relatively easy availability of the material in reasonable quantities allows for conducting *in vitro* experiments which otherwise would not be possible. On the other hand,

http://dx.doi.org/10.5772/62519

comparability to results from other researchers is often difficult if different vendor sources for the biopolymer mixture are used: The data obtained need to be interpreted with great care and ideally should be double-checked with a second, independent ECM preparation. SDS-PAGE analysis also suggested that the commercial ECM variants differed in terms of the relative concentration of basal lamina components. Whereas, in all ECM preparations, the bands corresponding to collagen IV and laminin were clearly most pronounced, the strongest variability occurred in a band around 50 kDa which matches the molecular weight of entactin. Since this molecule acts as a cross-linker between laminin and collagen IV, it is reasonable to assume that variations in its relative concentration will also affect the structure and permeability properties of the ECM gels and, ultimately, the migration behavior of cells in those gels. However, biochemical techniques are not able to predict those gel parameters, which is why physical methods are required to further characterize the different basal lamina model systems.

4. Physical properties of basal lamina gels *in vitro*

In addition to the biochemical structure of its constituents, the following three physical parameters dictate the behavior of molecules, nanoparticles, or cells within the basal lamina: the microstructure, the mechanical properties, and the permeability of the hydrogel. In biopolymer networks, the microstructure of the system has direct implications on both the viscoelastic properties of the network [51] and its permeability properties [31]. Thus, imaging methods for visualizing biopolymer networks such as the basal lamina are discussed first.

4.1. Microstructure of ECM gels

There are various methods to evaluate the structure of a material, and those methods can be subdivided into the following categories: surface imaging techniques, near-field/contact-based techniques and far-field imaging. Which method is used to resolve the structure depends on the material and on the experimental question: Do I require information on the surface topology or on the inner structure of the material? For biological samples, a fixation is needed for most of the imaging techniques so that the structure does not change over time or during the sample preparation process. One technique which is often used to image biological samples is fluorescence confocal microscopy as this method can visualize the 3D structure of a biopolymer network. However, most biological samples are not fluorescent by themselves and thus a fluorescent dye has to be used to stain the structure of interest. For many target proteins, commercial antibodies are available to which a fluorescent dye is attached. Before such a staining with antibodies is performed, the samples are typically fixed to ensure that the structure of the biopolymer network is not altered by the antibody application and the following washing step.

A suitable technique for imaging the surface of a biopolymer material is scanning electron microscopy (SEM). For this technique, the sample surface needs to be electrically conductive. Since this is typically not the case for biological samples, the application of a thin conductive layer, for example gold, is necessary. Depending on the type of SEM used for imaging, the samples are also exposed to a vacuum for imaging; this requires sample fixation and subsequent dehydration as typical additional preparation steps for this imaging method. It is clear that sample preparation steps necessary for both imaging techniques may introduce artifacts, that is, alterations of the microstructure of the biopolymer network. However, as the preparation steps for both techniques are different, the obtained pictures are reliable if both imaging methods return comparable structures.

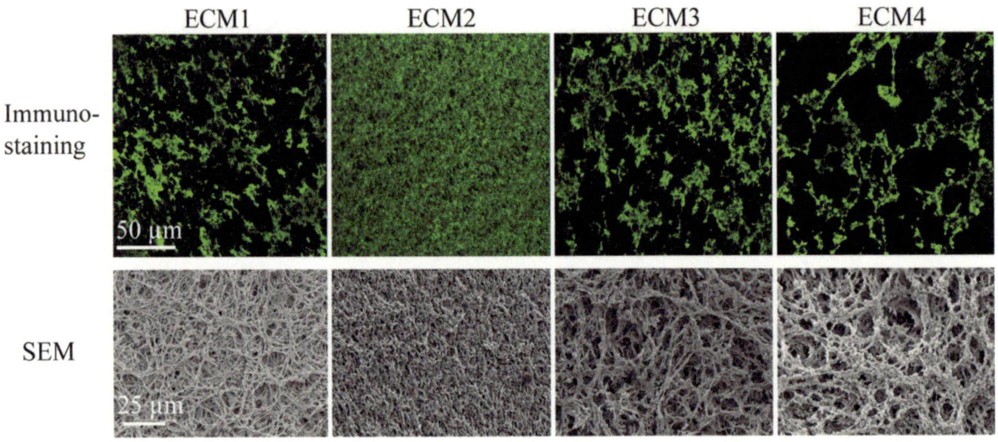

Figure 3. Microstructure of four ECM gel variants. Images obtained with immunostaining/confocal microscopy and with a SEM are compared. Both methods show a denser network for the gel variant 2 (ECM2) than for the other three variants. Edited figure with permission from Arends et al. [50]. © 2015 Arends et al. Published under CC BY license.

Example images of basal lamina model systems are shown in **Figure 3**, where the microstructure of four different ECM gel variants is compared. For the fluorescent confocal image, the ECM component laminin was stained with a fluorescent antibody and an optical slice with a thickness of 0.9 μm was acquired inside the 3-dimensional gel. Here, the ECM variant in which leukocyte migration was slowed down most (ECM2) shows the lowest porosity. The same difference in the microarchitecture of the ECM gels is obtained when SEM is used for imaging: ECM2 shows the highest density, whereas the other gel variants exhibit a comparable network structure. As the four gel variants have all been reconstituted at identical total protein concentrations, the observed structural difference is most likely due to the higher content of the cross-linking molecule entactin in this ECM variant as detected by SDS-PAGE.

4.2. Viscoelastic properties of ECM gels

Especially for cell differentiation, the mechanical properties of the ECM play an important role. Using artificial hydrogels such as cross-linked polyacrylamide gels [52], it was shown that cell differentiation can be directed by the stiffness of the substrate. The ECM is a viscoelastic material, that is, its mechanical behavior combines both viscous as well as elastic properties. Those viscoelastic properties can be probed macroscopically with a shear rheometer as

http://dx.doi.org/10.5772/62519

illustrated in **Figure 4** as well as microscopically with single-particle tracking microrheology or AFM indentation.

In macroscopic shear rheology, the sample is placed between two plates, a stationary bottom plate and a rotating top plate for shear stress application. The bottom plate can be heated or cooled depending on the desired temperature conditions. The top plates are available in various diameters and shapes and are chosen according to the sample properties and the quantity to be measured. For determining viscoelastic properties, the top plate is typically oscillated at different frequencies, either at a fixed strain or at a fixed torque. Small torques during such a measurement ensure that the material response is quantified in the linear response regime, where Hooke's law holds, that is, where the ratio of stress and strain is independent from the amplitude of the applied force.

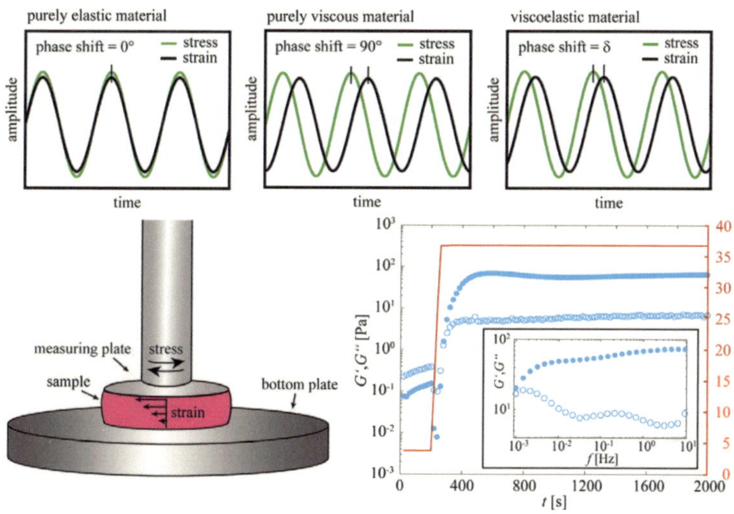

Figure 4. The viscoelastic behavior of ECM gels can be quantified by shear rheology. The sample is placed between a stationary bottom plate and a measuring plate, and then, an oscillating shear stress is induced. The temporal delay (phase shift) of the material response is measured. For a purely elastic material, the phase shift is 0°; in contrast, a phase shift of 90° is obtained for a purely viscous substance. For a viscoelastic material, the phase shift can assume any value between 0° and 90° (adapted from an illustration by Stefan Grumbein). A typical gelation curve for a basal lamina gel at a concentration of 8.3 mg/mL is shown at the bottom right and was acquired using a plate–plate geometry oscillating at a frequency of 1 Hz. For the first 200 s, the measurement was performed at 4°C. At this temperature, the ECM is in its liquid state, thus the loss modulus (open circles) dominates over the storage modulus (full circles). When the temperature is increased to 37°C, gelation is initiated and the storage modulus dominates. After a few minutes, a plateau value is reached. A typical frequency spectrum after gelation is shown as inset. The storage modulus dominates over the loss modulus over four decades of frequencies.

The viscoelastic properties of the ECM can then be described by the storage modulus G' and the loss modulus G''. Here, G' is a measure for the elastic properties and G'' for the viscous properties of the ECM gel. At low temperatures around 4°C, the ECM is in a liquid state. Here, its viscous properties dominate and the loss modulus is larger than the storage modulus. When the ECM gel is heated to room temperature or above, a gelation process is initiated which results in an increased storage modulus: Within a few minutes after the temperature increase is applied, the storage modulus starts to dominate over the loss modulus and then increases further until it reaches a plateau value (see **Figure 4**). In general, there are several parameters determining the absolute value of the plateau elasticity: The higher the concentration of proteins/polymers the higher is typically the storage modulus [53]. In addition to the concentration of protein, also the type of polymer/polymer interaction plays a role. The storage modulus of an entangled solution is usually lower than for a cross-linked network. In the study conducted in [50], it was shown that the amount of the cross-linking molecule entactin influences the network stiffness: The higher the concentration of entactin the higher the storage modulus and thus the elasticity of the formed matrix. Typical values for the elastic modulus obtained for ECM gels at a total protein concentration of 3.5 mg/mL are in the range of 1-10 Pa which is very soft and lies in the range of moduli reported to induce neuron-like differentiation of stem cells [52].

In general, the viscoelastic properties of a biopolymer network may depend strongly on the probing frequency [51], especially if the network constituents are only entangled. For cross-linked systems, however, a pronounced plateau in the frequency-dependent shear moduli is expected, and exactly such behavior is also observed for ECM gels (**Figure 4**).

The absolute values of the viscoelastic parameters obtained with macrorheology may not necessarily reflect the local stiffness of a biopolymer network. Thus, microrheological techniques such as bead microrheology [54] or AFM nanoindentation [55] have been introduced and already applied to other biopolymer systems such as cytoskeletal networks [56–58] or cartilage [59]. With those nano-/microscopic techniques, it is also possible to spatially map the mechanical properties of native basement membranes [60, 61], which might give insights important for cellular processes such as differentiation or migration.

4.3. Permeability of ECM gels

One of the major tasks of the basal lamina is to act as a molecular filter. Here, the exclusion of particles or molecules according to their size is one of the simplest mechanisms for establishing permeability: A mesh size smaller or in the order of the particle diameter will prevent the entrance of particles into the network; conversely, if particles have already entered the network, they will be efficiently trapped within the biopolymer matrix. However, this filter mechanism is not very sophisticated as it cannot differentiate between objects of the same size. Thus, a second filter mechanism based on binding interactions between diffusing particles/molecules and the basal lamina constituents has been put forward to contribute to the selective permeability properties of biopolymer hydrogels such as the basal lamina [31]. With the ECM model system discussed above, the physicochemical principles governing the high selectivity of basal lamina gels can be studied systematically.

To probe the interactions between particles and the ECM, single-particle tracking (SPT) can be employed. In contrast to SPT used for microrheology [62], the diameter of the particles embedded into the ECM should be small compared to the mesh size of the gel. Only then one can be sure that the particle motion is not geometrically restricted by the network microarchitecture—which demonstrates the importance of obtaining structural information on the system prior to commencing SPT experiments. In SPT measurements, the diffusive movement of particles within the gel is recorded via light microscopy and every single particle is evaluated separately. The trajectory of motion of each particle, in particular the x- and y-position, is extracted from recorded movies for every frame of the movie—typically over a time course of several seconds up to a minute (depending on the temporal resolution of the image acquisition process, **Figure 5**). These data are then used to calculate the mean squared displacement (*MSD*) of every particle according to the following

$$MSD(\tau) = \frac{1}{N}\sum_{i=1}^{N}[\vec{r}\left(i\Delta t+\tau\right)-\vec{r}\left(i\Delta t\right)]^2 \tag{1}$$

Here, N denotes the total number of recorded frames, $\dot{r}(t)$ is the position of the particle at time t, and τ denotes the time interval between two particle positions within a given trajectory. For diffusive processes, the *MSD* typically grows with time as a power law τ^{α}, with the exponent α characterizing the type of diffusive motion: One can distinguish sub-diffusive (α<1), normal diffusive (α=1), or superdiffusive behavior (α>1), the latter of which is typically linked to active transport phenomena or liquid flow.

Such SPT experiments revealed that both positively and negatively charged microparticles were efficiently immobilized in the ECM gel, whereas PEGylated (and thus only weakly charged) polystyrene particles of identical size were able to diffuse almost freely within the gel [63]. Equivalent results were obtained with liposome particles and suggested that free diffusion within the ECM matrix is only possible as long as the particle surface charge (as quantified by the zeta potential) lies within a window ranging from intermediate negative charge to low positive charge. Enzymatic digestion of the ECM component HS entailed a mobilization of positively charged particles. This finding suggested that the polyanionic HS chains present in the perlecan complex critically contribute to the selective properties of the ECM gel—likely through trapping of positively charged objects by means of electrostatic binding.

The notion that electrostatic binding interactions contribute to particle trapping in ECM gels was confirmed by experiments conducted at elevated ionic strength of the hydrogel buffer. Increased salt concentrations lead to charge screening effects by the formation of a layer of counter ions around the surface of charged objects such as particles or hydrogel polymers. As a consequence, the strength of electrostatic interactions at a given separation distance between two objects is reduced—a process which is described by the Debye–Hückel theory [64]. At physiological concentrations of KCl, both positively and negatively charged polystyrene microparticles are immobilized in ECM gels. However, when the KCl concentration is increased, a fraction of the particles becomes mobile [63, 65]. This mobilization does not have to be permanent as individual particles can dynamically switch between a freely diffusing and bound state over time, and—while in the bound state—also between a weakly and strongly bound configuration. As shown in **Figure 5**, the degree of particle mobilization depends both on the ion concentration and valency which is consistent with the Debye–Hückel theory. However, particle mobilization efficiency seems also to depend on the particular ion species as identical concentrations of the divalent ions Mg^{2+} and Ca^{2+} lead to different experimental outcome [65]. This ion-specific effect suggests that, in addition to electrostatic forces, also hydrophobic interactions are likely to contribute to the selective filtering properties of the basal lamina.

Systematic permeability studies with artificial particles were very helpful to unravel the physical mechanisms which are responsible for the trapping of solutes in the basal lamina. However, most compounds which encounter the basal lamina layer under physiological conditions are small molecules rather than microparticles. To investigate the selective properties of the basal lamina toward small molecules, a microfluidic setup (**Figure 6**) was recently introduced [66]. Here, customized peptides with tailored amino acid sequences and thus different net charges were used as diffusion probes. To ensure optimal comparability, the molecular weight of those oligopeptides was kept constant. The penetration behavior of those peptides into an ECM gel was visualized by fluorescent microscopy, and similar to the SPT experiments discussed above also the behavior of those molecules critically depended on their charge. Positively charged peptides accumulated at the gel/buffer interface, whereas nega-

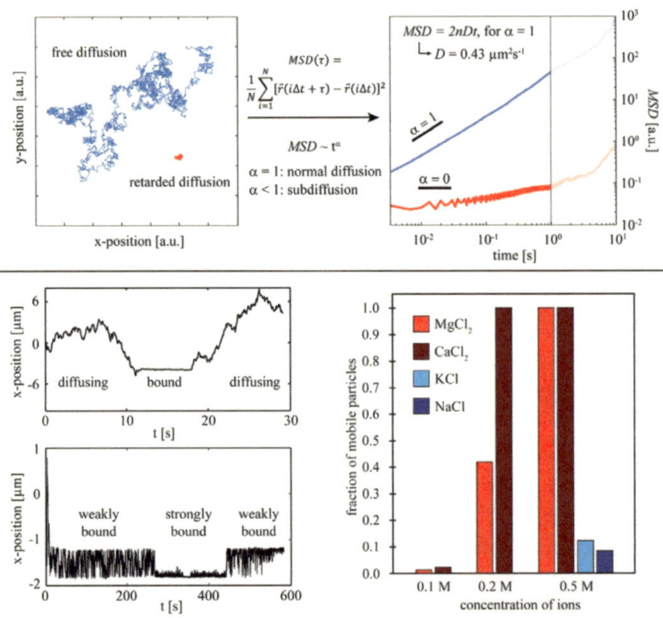

Figure 5. Single-particle tracking experiments can locally map the permeability of ECM gels. *Upper panel*: Exemplary trajectories and the corresponding MSD curves of a particle showing free diffusion and a particle showing retarded diffusion. For freely diffusing particles, the dependence of the MSD on time is linear and a diffusion coefficient D can be calculated according to the formula shown in the graph. For the calculation of D, only the first 10% of the MSD data is used to avoid errors arising from statistical uncertainties. *Lower panel*: Trajectories of a particle which transiently switches between a diffusing and a bound state and of a particle which alternates between a strongly and a weakly bound configuration. The states of motion can be distinguished based on the fluctuation amplitude of the particle. Both trajectories were obtained at a salt concentration of 1 M KCl. The histogram shows that in ECM gels, the fraction of mobile particles depends on the concentration, valency, and detailed species of the ions used. Adapted with permission from Arends et al. [65]. Copyright ©2013 American Chemical Society.

tively charged peptides did not. Moreover, when the net charge of the positively charged peptides was increased, the accumulation propensity of the molecules at the gel interface was increased as well. Of course, such an artificial microfluidic setup does not reproduce the complex situation of the basal lamina interface found *in vivo*. However, peptide injection tests in the connective tissue of living mice demonstrated a similar charge-selective accumulation behavior at the basal lamina layer of blood vessels as observed on-chip with the simplified ECM/buffer interface. This underscores the great potential basal lamina model systems and biophysical characterization methods hold for gaining a better insight into the mechanistic principles that establish the complex properties of the basal lamina.

5. Outlook

Here, we have summarized selected aspects of our current understanding how the biochemical composition of the basal lamina is mirrored in the complex microarchitecture as well as the multi-facetted material properties of the biopolymer network. Deciphering the physicochemical principles which dictate the microstructure, viscoelastic properties, and selective permeability of the basal lamina layer are not only interesting for cell biology studies [67, 68], drug

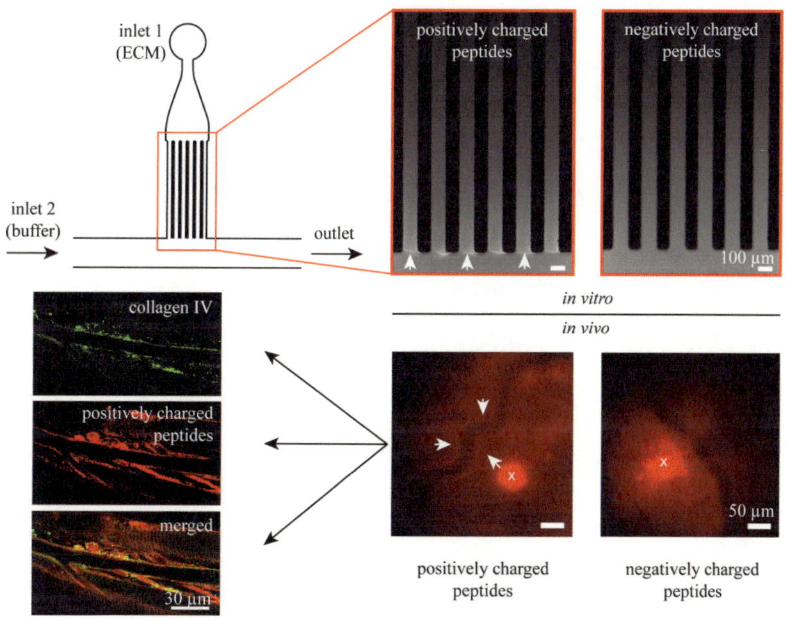

Figure 6. Illustration of a microfluidic setup used to probe the diffusive penetration of peptides from a buffer compartment into a basal lamina gel. Typical intensity profile images for positively and negatively charged peptides are shown next to the microfluidic channel. For positively charged peptides, an intensity peak is observed at the buffer/gel interface, whereas such an accumulation does not occur for negatively charged peptides. Similar results are obtained when those peptides are injected into the connective tissue of mice. Tissue immunostaining confirmed that the positively charged peptides colocalize with collagen IV, a main component of the basal lamina. Adapted with permission from Arends et al. [66]. © 2015 Arends et al. Published under CC BY license.

delivery questions and tumor treatment [69–72], but might also have strong implications for tattoo removal applications: Here, ink nanoparticles trapped in the skin tissue have to be mobilized, for example, by soaking the tissue in salt solutions, so that they can be washed out from the skin rather than removed by painful and scar-inducing laser treatment. The lessons learned from systematically unraveling the physical and chemical mechanisms, which give rise to the complex properties of the basal lamina, may also help in the rational design of artificial hydrogel systems for tissue engineering approaches: The synthesis of complex macromolecules with well-defined chemical properties may allow for constructing hydrogels with both tailored mechanical properties and selective permeability behavior.

Acknowledgements

We thank Iris König-Decker for providing graphics. We thank Kathrin Boettcher, Benjamin Käsdorf and Corinna Lieleg for critical reading of the manuscript. Financial support from the Deutsche Forschungsgemeinschaft (DFG) through project B7 within the framework of the SFB 1032 is gratefully acknowledged.

Author details

Fabienna Arends[1,2] and Oliver Lieleg[1,2*]

*Address all correspondence to: oliver.lieleg@tum.de

1 Institute of Medical Engineering IMETUM, Technical University of Munich, Boltzmannstrasse 11, Garching, Germany

2 Department of Mechanical Engineering, Technical University of Munich, Boltzmannstrasse 15, Garching, Germany

References

[1] Timpl R. Macromolecular organization of basement membranes. Current Opinion in Cell Biology. 1996;8(5):618–24.

[2] Yurchenco PD, Orear JJ. Basal lamina assembly. Current Opinion in Cell Biology. 1994;6(5):674–81.

[3] Kleinman HK, Martin GR. Matrigel. Basement membrane matrix with biological activity. Seminars in Cancer Biology. 2005;15(5):378–86.

[4] Aumailley M. The laminin family. Cell Adhesion & Migration. 2013;7(1):48–55.

[5] Paulsson M, Deutzmann R, Timpl R, Dalzoppo D, Odermatt E, Engel J. Evidence for coiled-coil alpha-helical regions in the long arm of laminin. The Embo Journal. 1985;4(2):309–16.

[6] Cheng YS, Champliaud MF, Burgeson RE, Marinkovich MP, Yurchenco PD. Self-assembly of laminin isoforms. Journal of Biological Chemistry. 1997;272(50):31525–32.

[7] Aumailley M, Pesch M, Tunggal L, Gaill F, Fassler R. Altered synthesis of laminin 1 and absence of basement membrane component deposition in beta 1 integrin-deficient embryoid bodies. Journal of Cell Science. 2000;113(2):259–68.

[8] Khoshnoodi J, Pedchenko V, Hudson BG. Mammalian collagen IV. Microscopy Research and Technique. 2008;71(5):357–70.

[9] Mouw JK, Ou G, Weaver VM. Extracellular matrix assembly: a multiscale deconstruction. Nature Reviews Molecular Cell Biology. 2014;15(12):771–85.

[10] Alberts B, et al. Molecular biology of the cell. 4th ed. Garland Science, New York; 2002.

[11] Yurchenco PD, Patton BL. Developmental and pathogenic mechanisms of basement membrane assembly. Current Pharmaceutical Design. 2009;15(12):1277–94.

[12] Kalluri R. Basement membranes: structure, assembly and role in tumour angiogenesis. Nature Reviews Cancer. 2003;3(6):422–33.

[13] Tilling T, Korte D, Hoheisel D, Galla HJ. Basement membrane proteins influence brain capillary endothelial barrier function in vitro. Journal of Neurochemistry. 1998;71(3): 1151–7.

[14] Iozzo RV. Matrix proteoglycans: from molecular design to cellular function. Annual Review of Biochemistry. 1998;67:609–52.

[15] Kresse H, Schonherr E. Proteoglycans of the extracellular matrix and growth control. Journal of Cell Physiology. 2001;189(3):266–74.

[16] Yanagishita M. Function of proteoglycans in the extracellular-matrix. Acta Pathology Japan. 1993;43(6):283–93.

[17] Streuli C. Extracellular matrix remodelling and cellular differentiation. Current Opinion in Cell Biology. 1999;11(5):634–40.

[18] Werb Z. ECM and cell surface proteolysis: regulating cellular ecology. Cell. 1997;91(4): 439–42.

[19] Gu ZZ, Cui J, Brown S, Fridman R, Mobashery S, Strongin AY, et al. A highly specific inhibitor of matrix metalloproteinase-9 rescues laminin from proteolysis and neurons from apoptosis in transient focal cerebral ischemia. Journal of Neuroscience. 2005;25(27):6401–8.

[20] Wang CX, Shuaib A. Critical role of microvasculature basal lamina in ischemic brain injury. Progress in Neurobiology. 2007;83(3):140–8.

[21] Kitajewski J. Endothelial laminins underlie the tip cell microenvironment. Embo Reports. 2011;12(11):1087–8.

[22] Timpl R, Brown JC. The laminins. Matrix Biology. 1994;14(4):275–81.

[23] Poschl E, Schlotzer-Schrehardt U, Brachvogel B, Saito K, Ninomiya Y, Mayer U. Collagen IV is essential for basement membrane stability but dispensable for initiation of its assembly during early development. Development. 2004;131(7):1619–28.

[24] Holster T, Pakkanen O, Soininen R, Sormunen R, Nokelainen M, Kivirikko KI, et al. Loss of assembly of the main basement membrane collagen, type IV, but not fibril-forming collagens and embryonic death in collagen prolyl 4-hydroxylase I null mice. The Journal of Biological Chemistry. 2007;282(4):2512–9.

[25] Kang SH, Kramer JM. Nidogen is nonessential and not required for normal type IV collagen localization in *Caenorhabditis elegans*. Molecular Biology of the Cell. 2000;11(11):3911–23.

[26] Murshed M, Smyth N, Miosge N, Karolat J, Krieg T, Paulsson M, et al. The absence of nidogen 1 does not affect murine basement membrane formation. Molecular and Cellular Biology. 2000;20(18):7007–12.

[27] Schymeinsky J, Nedbal S, Miosge N, Poschl E, Rao C, Beier DR, et al. Gene structure and functional analysis of the mouse nidogen-2 gene: nidogen-2 is not essential for basement membrane formation in mice. Molecular and Cellular Biology. 2002;22(19): 6820–30.

[28] Willem M, Miosge N, Halfter W, Smyth N, Jannetti I, Burghart E, et al. Specific ablation of the nidogen-binding site in the laminin gamma1 chain interferes with kidney and lung development. Development. 2002;129(11):2711–22.

[29] To M, Goz A, Camenzind L, Oertle P, Candiello J, Sullivan M, et al. Diabetes-induced morphological, biomechanical, and compositional changes in ocular basement membranes. Experimental Eye Research. 2013;116:298–307.

[30] Cecchelli R, Berezowski V, Lundquist S, Culot M, Renftel M, Dehouck MP, et al. Modelling of the blood–brain barrier in drug discovery and development. Nature Reviews Drug Discovery. 2007;6(8):650–61.

[31] Lieleg O, Ribbeck K. Biological hydrogels as selective diffusion barriers. Trends in Cell Biology. 2011;21(9):543–51.

[32] Proksch E, Brandner JM, Jensen JM. The skin: an indispensable barrier. Experimental Dermatology. 2008;17(12):1063–72.

[33] Sebinger DDR, Ofenbauer A, Gruber P, Malik S, Werner C. ECM modulated early kidney development in embryonic organ culture. Biomaterials. 2013;34(28):6670–82.

[34] Miner JH. Renal basement membrane components. Kidney International. 1999;56(6): 2016–24.

[35] Ballabh P, Braun A, Nedergaard M. The blood–brain barrier: an overview—structure, regulation, and clinical implications. Neurobiology of Disease. 2004;16(1):1–13.

[36] Gabe SM. Gut barrier function and bacterial translocation in humans. Clinical Nutrition. 2001;20:107–12.

[37] Tsang KY, Cheung MC, Chan D, Cheah KS. The developmental roles of the extracellular matrix: beyond structure to regulation. Cell Tissue Research. 2010;339(1):93–110.

[38] Benton G, Kleinman HK, George J, Arnaoutova I. Multiple uses of basement membrane-like matrix BME/Matrigel) in vitro and in vivo with cancer cells. International Journal of Cancer. 2011;128(8):1751–7.

[39] Godin AG, Lounis B, Cognet L. Super-resolution microscopy approaches for live cell imaging. Biophysical Journal. 2014;107(8):1777–84.

[40] Koh HS, Yong T, Chan CK, Ramakrishna S. Enhancement of neurite outgrowth using nano-structured scaffolds coupled with laminin. Biomaterials. 2008;29(26):3574–82.

[41] Hidalgo-Bastida LA, Barry JJ, Everitt NM, Rose FR, Buttery LD, Hall IP, et al. Cell adhesion and mechanical properties of a flexible scaffold for cardiac tissue engineering. Acta Biomaterialia. 2007;3(4):457–62.

[42] Kleinman HK, Mcgarvey ML, Hassell JR, Star VL, Cannon FB, Laurie GW, et al. Basement-membrane complexes with biological-activity. Biochemistry. 1986;25(2): 312–8.

[43] Kleinman HK, Mcgarvey ML, Liotta LA, Robey PG, Tryggvason K, Martin GR. isolation and characterization of type-iv procollagen, laminin, and heparan-sulfate proteoglycan from the EHS sarcoma. Biochemistry. 1982;21(24):6188–93.

[44] Vukicevic S, Luyten FP, Kleinman HK, Reddi AH. Differentiation of canalicular cell processes in bone-cells by basement-membrane matrix components—regulation by discrete domains of laminin. Cell. 1990;63(2):437–45.

[45] Li ML, Aggeler J, Farson DA, Hatier C, Hassell J, Bissell MJ. Influence of a reconstituted basement-membrane and its components on casein gene-expression and secretion in mouse mammary epithelial-cells. Proceedings of the National Academy of Sciences of the United States of America. 1987;84(1):136–40.

[46] Kibbey MC, Royce LS, Dym M, Baum BJ, Kleinman HK. Glandular-like morphogenesis of the human submandibular tumor cell line A253 on basement membrane components. Experimental Cell Research. 1992;198(2):343–51.

[47] Hadley MA, Byers SW, Suarezquian CA, Kleinman HK, Dym M. Extracellular-matrix regulates Sertoli-cell differentiation, testicular cord formation, and germ-cell development in vitro. Journal of Cell Biology. 1985;101(4):1511–22.

[48] Zaman MH, Trapani LM, Siemeski A, MacKellar D, Gong H, Kamm RD, et al. Migration of tumor cells in 3D matrices is governed by matrix stiffness along with cell-matrix

adhesion and proteolysis (vol. 103, p. 10889, 2006). Proceedings of the National Academy of Sciences of the United States of America. 2006;103(37):13897.

[49] Valster A, Tran NL, Nakada M, Berens ME, Chan AY, Symons M. Cell migration and invasion assays. Methods. 2005;37(2):208–15.

[50] Arends F, Nowald C, Pflieger K, Boettcher K, Zahler S, Lieleg O (2015) The Biophysical Properties of Basal Lamina Gels Depend on the Biochemical Composition of the Gel. PloS ONE 10(2): e0118090. doi:10.1371/journal.pone.0118090

[51] Lieleg O, Claessens MMAE, Bausch AR. Structure and dynamics of cross-linked actin networks. Soft Matter. 2010;6(2):218–25.

[52] Engler AJ, Sen S, Sweeney HL, Discher DE. Matrix elasticity directs stem cell lineage specification. Cell. 2006;126(4):677–89.

[53] Massensini AR, Ghuman H, Saldin LT, Medberry CJ, Keane TJ, Nicholls FJ, et al. Concentration-dependent rheological properties of ECM hydrogel for intracerebral delivery to a stroke cavity. Acta Biomaterialia. 2015;27:116–30.

[54] Chen DTN, Wen Q, Janmey PA, Crocker JC, Yodh AG. Rheology of soft materials. Annu. Rev. Condens. Matter Phys. 2010;1:301–22.

[55] Takeyasu K. Atomic force microscopy in nanobiology, Pan Stanford Publishing Pte. Ltd., Singapore, 2014

[56] Luan Y, Lieleg O, Wagner B, Bausch AR. Micro- and macrorheological properties of isotropically cross-linked actin networks. Biophysics Journal. 2008;94(2):688–93.

[57] Lu L, Oswald SJ, Ngu H, Yin FCP. Mechanical properties of actin stress fibers in living cells. Biophysical Journal. 2008;95(12):6060–71.

[58] Kirmizis D, Logothetidis S. Atomic force microscopy probing in the measurement of cell mechanics. International Journal of Nanomedicine. 2010;5:137–45.

[59] Nia HT, Bozchalooi IS, Li Y, Han L, Hung HH, Frank E, et al. High-bandwidth AFM-based rheology reveals that cartilage is most sensitive to high loading rates at early stages of impairment. Biophysical Journal. 2013;104(7):1529–37.

[60] Candiello J, Balasubramani M, Schreiber EM, Cole GJ, Mayer U, Halfter W, et al. Biomechanical properties of native basement membranes. The FEBS Journal. 2007;274(11):2897–908.

[61] Last JA, Liliensiek SJ, Nealey PF, Murphy CJ. Determining the mechanical properties of human corneal basement membranes with atomic force microscopy. Journal of Structural Biology. 2009;167(1):19–24.

[62] Valentine MT, Perlman ZE, Gardel ML, Shin JH, Matsudaira P, Mitchison TJ, et al. Colloid surface chemistry critically affects multiple particle tracking measurements of biomaterials. Biophysical Journal. 2004;86(6):4004–14.

[63] Lieleg O, Baumgärtel RM, Bausch AR. Selective filtering of particles by the extracellular matrix: an electrostatic bandpass. Biophysical Journal. 2009;97(6):1569–77.

[64] Hunter RJ. Zeta potential in colloid science: principles and applications. Academic Press, London; 1989.

[65] Arends F, Baumgartel R, Lieleg O. Ion-specific effects modulate the diffusive mobility of colloids in an extracellular matrix gel. Langmuir: the ACS Journal of Surfaces and Colloids. 2013;29(51):15965–73.

[66] Arends F, Sellner S, Seifert P, Gerland U, Rehberg M, Lieleg O. A microfluidics approach to study the accumulation of molecules at basal lamina interfaces. *Lab Chip*, 2015,, 3326–3334., 15

[67] Kaemmerer E, Melchels FP, Holzapfel BM, Meckel T, Hutmacher DW, Loessner D. Gelatine methacrylamide-based hydrogels: an alternative three-dimensional cancer cell culture system. Acta Biomaterialia. 2014;10(6):2551–62.

[68] Seto SP, Casas ME, Temenoff JS. Differentiation of mesenchymal stem cells in heparin-containing hydrogels via coculture with osteoblasts. Cell Tissue Research. 2012;347(3): 589–601.

[69] Chauhan VP, Stylianopoulos T, Boucher Y, Jain RK. Delivery of molecular and nanoscale medicine to tumors: transport barriers and strategies. Annual Review of Chemical and Biomolecular. 2011;2:281–98.

[70] Holback H, Yeo Y. Intratumoral drug delivery with nanoparticulate carriers. Pharmaceutical Research. 2011;28(8):1819–30.

[71] Ernsting MJ, Murakami M, Roy A, Li SD. Factors controlling the pharmacokinetics, biodistribution and intratumoral penetration of nanoparticles. Journal Controlled Release. 2013;172(3):782–94.

[72] Bertrand N, Wu J, Xu X, Kamaly N, Farokhzad OC. Cancer nanotechnology: the impact of passive and active targeting in the era of modern cancer biology. Advanced Drug Delivery Reviews. 2014;66:2–25.

Chapter 10

The Role of Extracellular Matrix Proteins in the Urinary Tract: A Literature Review

Cevdet Kaya and Bahadır Şahin

Additional information is available at the end of the chapter

http://dx.doi.org/10.5772/62807

Abstract

The extracellular matrix (ECM) is a noncellular component with a crucial role on tissue morphogenesis, differentiation and hemostasis within all tissues and organs. With advancement in the technology and increased data on ECM components, it was realized that many conditions in urinary tract have a close relation with the composition of ECM in the affected tissue. According to some basic research studies, ECM composition may give us important information about the prognosis and progression of disease in addition to the cause and pathophysiology of the diseases such as congenital ureterovesical and ureteropelvic junction obstruction. Afterwards, with better understanding of ECM one can develop new treatment and follow-up models. This chapter will summarize the evidence-based role of ECM in urinary tract conditions.

Keywords: urinary tract, extracellular matrix, immunohistochemistry, ureteropelvic junction, ureterovesical junction

1. Extracellular matrix

The extracellular matrix (ECM) is a noncellular component within all tissues and organs, and it is essential for the scaffolding of cellular constituents and also it plays a crucial role on tissue morphogenesis, differentiation and hemostasis [1]. It is an anchoring platform for epithelia, forms the basement membrane, and also surrounds capillaries and neural cells, and is part of the connective tissue [2].

In general, ECM molecules can be classified as fiber-forming and non–fiber-forming molecules [3, 4]. Collagens, elastins, laminins and fibronectins are the main fiber-forming ECM proteins. Proteoglycans which are main non–fiber-forming molecules fill the majority of the extracellular interstitial space and they have a wide variety of functions that reflect their unique buffering, hydration, binding and force–resistance properties.

2. Fiber-forming ECM elements

2.1. Collagen

Collagens are the most abundant proteins in ECM and the whole human body [5]. Collagens have many functions, which depend on the type and tissue they are in. Depending on the tissue, collagen fibers can provide tensile strength, can regulate cell adhesion, can support chemotaxis and migration and can direct tissue development [6]. Generally in normal physiologic states, different types of collagen fibers form a heterogeneous mixture but usually there is a dominant type of collagen in every given tissue.

The five most common collagen types are the following:

Type I: skin, tendon, vascular ligature, organs and bone (main component of the organic part of bone)

Type II: main collagenous component of cartilage

Type III: main component of reticular fibers

Type IV: forms basal lamina and basement membrane

Type V: placenta, cell surfaces and hair

A majority of collagen molecules are in the form of triple strands, which form supramolecular complexes like fibrils and networks depending on the type of collagen. Network collagens are incorporated into the basal membrane and fibrous collagens form a skeleton for the collagen fibril bundles in the interstitium [1].

2.2. Elastins

Elastins are the main ECM element, which gives tissues elasticity and allows a tissue to stretch and return to its original state if needed. Their tight association with collagen fibrils crucially limits their stretchability. Fibroblasts and smooth muscle cells secrete elastin in the form of its precursor, tropoelastin. Secreted tropoelastin molecules assemble into elastin fibers. Elastin fibers are covered by glycoprotein microfibrils. The most common glycoprotein covering elastin fibers is fibrillin. The presence of fibrillins is also essential for the integrity of elastin [1].

2.3. Laminins

Laminins are glycoproteins that form heterodimers containing 1 α, 1 β and 1 γ chain. They are synthesized in podocytes and endothelial cells. Laminin trimerization occurs inside the cell in the endoplasmic reticulum. Once trimerization completed they are secreted into the extracellular space and they polymerize to form a supramolecular network. Laminin polymerization initiates basement membrane formation and sends signal to adjacent cells [7].

2.4. Fibronectins

Fibronectins are glycoproteins that establish connection between cells and collagen fibers in the ECM. It is secreted as a dimer form, which are joined by two disulfide bonds. Fibronectin fibers have several binding sites through which they form a connection with other fibronectin dimers, collagen fibers, heparin and cell-surface integrin receptors [1].

Fibronectins have vital importance for mediating cell attachment and function and they have an important role in the organization process of the interstitial ECM. During tissue development, fibronectins are important for cell migration, and also studies showed that they have roles on cardiovascular disease and tumor metastasis [6].

3. Non–fiber-forming ECM molecules

3.1. Proteoglycans

Proteoglycans are heavily glycosylated proteins. They contain glycosaminoglycan (GAG) chain, which is linked with a protein core covalently. They are classified by their size (large and small) or the nature of their GAG chains. The main classes of proteoglycans are heparan sulfates, keratan sulfates and chondroitin sulfates. Their extreme hydrophilic nature and ability to adapt extended conformations make them easier to form hydrogel, and structures formed with these molecules can withstand high compressive forces.

Proteoglycans and their most common locations:

Heparan sulfate: basement membranes and components of cell surfaces.

Chondroitin sulfate: cartilage, heart valves and bone.

Keratan sulfate: cornea and bone.

3.2. Hyaluronic acid

Hyaluronic acid is a polysaccharide, which does not contain protein core and it consists of alternating residues of D-glucuronic acid and N-acetylglucosamine. Hyaluronic acid forms a coat around chondrocytes and they provide resilience of articular cartilage tissue. In extracellular space, hyaluronic acid also provides the ability to resist compression by absorbing significant amounts of water and providing a counteracting swelling force.

4. ECM remodeling and matrix metalloproteinases

Each tissue has an ECM with a unique composition, which is formed by biochemical and biophysical interaction between the various cellular components (e.g., fibroblast and endothelial elements) and the evolving cellular and protein microenvironment [1]. There is a constant turnover of ECM in the body and it is being regulated with either enzymatically or nonenzymatically which makes ECM a dynamic structure.

Matrix metalloproteinases (MMPs) are a large family of endopeptidases, which are calcium-dependent zinc-containing enzymes. They are responsible for the degradation of the ECM in which they assist the tissue remodeling and they play a central role in tissue homeostasis [8]. They are present in both pathologic and normal tissues performing proteolytic action [9]. The main cell types excreting MMPs are macrophages, fibroblasts, osteoblasts, endothelial cells, neutrophils and lymphocytes [8].

MMPs have been studied in many conditions. It has been found that they have an impact on tumor cell behavior as a result of their ability to make alterations on cell surface receptors, growth factors, cell adhesion molecules and cytokines. Furthermore, MMPs are able to produce apoptosis-resistant cells, which leads to generation of an aggressive phenotype. MMPs may also regulate angiogenesis positively or negatively in cancer depending on activation of proangiogenic factors or generation of angiogenesis inhibitors, respectively [10].

In bladder cancer, it has been shown that there is a correlation with the levels of MMP-2 and MMP-9 and tumor grade and invasiveness [11]. MMP-2 levels were also found to be strongly associated with tumor stage and prognosis [12]. In a study, serum level of MMP-7 was also found to be associated with the prognosis of the patient. It is reported that in bladder cancer patients treated with radical surgery high MMP-7 plasma levels were significantly associated with poor overall- and disease-specific survival [13].

5. Role of ECM molecules on urinary system

5.1. Upper urinary system

In the renal cortex, the ECM is present in anatomically distinct areas with different functions depending on its molecular components. Glomerular basal membrane, which is thicker, compared to most other basal membranes in the body mainly contains laminin, collagen type IV, nidogen and heparan sulfate proteoglycans [14].

Laminins in glomerular basal membrane form a network required to maintain the basement membrane integrity. A genetic defect in laminin β2 chain will result in Pierson syndrome, which is characterized by congenital nephrotic syndrome and diffuse mesangial sclerosis, muscular hypotonia, distinct ocular abnormalities like microcoria (small pupils) and impairment of vision and neurodevelopment [15]. Leading cause of death, which occurs within first days, or weeks of life in Pierson syndrome is renal failure.

Nidogens bind to collagen type IV and laminin separately. Nidogens have a role in the basement membrane formation but experimental evidence on animal studies showed that they are not essentially required for GBM formation [16]. The most common type of heparan sulfate found in healthy basal membrane is agrin [4].

http://dx.doi.org/10.5772/62807

In normal physiologic conditions, ECM of glomerular mesangium consists of fibronectin, collagen type IV, collagen type V, laminin, chondroitin sulfate, heparan sulfate and nidogen [7, 17]. Mesangial ECM allows larger molecules to pass to the mesangium in contrast to glomerular basal membrane. The small proteoglycans like decorin, biglycan, fibromodulin and lumican are most commonly localized in the tubular interstitium and they are also weakly expressed in mesangium [7].

Normally, the renal tubulointerstitial matrix is composed of collagen types I, III, V, VI, VII and XV, both sulfated and nonsulfated GAGs, glycoproteins and polysaccharides. During fibrosis, decreased degradation and increased synthesis of ECM components result in accumulation of these components leading to the formation of scar tissue in the interstitial space [14, 18].

Increasing evidence suggest that MMPs have a complex role in renal fibrosis [19]. For example, MMP-9 mediates collagen degradation. As a result, collagen fragments were formed, and these fragments mediate neutrophil chemotaxis. Including their action on ECM components, it is also shown that MMPs have functional effect on the modulation of growth factors, their receptors and adhesion molecules [19].

In diabetic nephropathy, glomerular hypertension and hyperfiltration lead to mechanical stress on glomerular cells, resulting increased transcription of transforming growth factor (TGF)-β1 and decreased MMP activity. As a result, in diabetic nephropathy, changes seen on glomerular basal membrane increase in the concentration of laminin, fibronectin, collagen IV and VI; increase in glycation of collagen IV, increase in crosslinking of collagen IV and decrease in the concentration of agrin, perlecan and collagen XVIII [18]. Stokes et al. showed an increase in decorin, collagen type 1 and biglycan levels on mesengial matrix in renal fibrosing disease [20]. Collagen type 4 reported to increase both type 1 and type 2 DM associated with the degree of decline in renal function [14].

There are some conditions in which defects on ECM components affect upper urinary tract. Mutations on the $\alpha5$ chain of collagen type IV result in Alport's syndrome. Genetic defects on the $\alpha3$ and $\alpha4$ chains of collagen type 4 can cause autosomal dominant or recessive Alport's syndrome and thin basement membrane nephropathy. Goodpasture syndrome and Alport posttransplantation disease are two autoimmune conditions in which autoantibodies attacking glomerular basal membrane cause rapidly progressive glomerulonephritis.

Thrombospondin-1 (TSP-1) is a glycoprotein, which has adhesive properties, and it is involved in fibroblast proliferation and migration. TSP-1 is correlated with the degree of tubulointerstitial fibrosis. It is also shown that TSP-1 is transiently expressed at early stages of fibrosis. It is suggested that by the activation of TGF-β, TSP-1 could have a possible role as a mediator of interstitial fibrosis [14].

Matrix molecules such as heparan sulfate, proteoglycans, laminins, integrins and MMPs along with a group of growth factors (e.g., TGF-β) are involved in stimulation or inhibition of growth and branching of the ureteral bud [21]. The important role of ECM components and MMPs on the development of ureters puts these molecules on the scope of most recent studies investigating pathophysiology of congenital ureter-related abnormalities. It is suggested that an increase in ECM components such as collagen 1 may result in ureter-related disorders such as ureteropelvic junction (UPJ) obstruction and ureterovesical junction (UVJ) obstruction [9, 22–24].

5.2. Bladder and lower urinary system

The bladder ECM consists of proteins, proteoglycans and GAGs. ECM in bladder provides support and signaling to the cells of the bladder [25]. ECM components have an important role in the protection of urothelium and the storage of urine. The protective layer of GAGs (predominantly chondroitin and heparan sulfates) that cover urothelial cells forms a barrier against various toxic components [25].

Bladder lamina propria forms a highly effective barrier between epithelial and mesenchymal layers. It consists of mainly connective tissue and it also contains myofibroblasts, nerve fibers, lymphatics and blood vessels [26, 27].

Detrusor muscle is associated with laminin, osteopontin and collagen fibrils (I and III) During physiologic bladder filling and emptying, keratoepithelin is organized in complex folds and facilitates expansion and compaction of the bladder. Further, the ECM composition of the bladder wall, and in particular the type of collagen (type I favored in normally compliant bladders), as well as the collagen-to-elastin ratio, are critical to the maintenance of a low-pressure state in the bladder during normal filling [21].

Studies on bladder cancer show that changes in ECM play a crucial role in the course of the disease. It has been shown that bladder cancer cells cultured in a normal ECM lose their invasiveness or ability to form papillary structures. Instead, they align in either multi- or single-layered formation resembling normal urothelium [28].

Altered distribution of laminin-5 γ2-chain is found to be associated with worse overall survival, higher risk of recurrence and progression; and it is regarded as independent prognostic factor in bladder cancer treated with TUR-B. Studies demonstrated that loss of collagen IV was associated with invasive behavior and worse overall survival [29].

Fibronectin is found at increased levels in lamina propria and in urine in urothelial carcinoma. Increased expression of it is also found to be associated with stage of the cancer but has no prognostic value. Increased value of fibronectin in urine suggested to be used for early detection of the tumor whereas decreased fibronectin level in the urine can be used to assess response to Bacillus Calmette Guérin (BCG) therapy [29].

Increased stromal expression of tenascin C is found to be associated with worse overall survival in bladder cancer; on the contrary tumor cell expression of tenascin C is associated with improved overall survival [29]. It is also found that in patients with decreased expression levels of TSP-1, high rate of recurrence and worse overall survival is seen [29].

In the function and diseases of prostate the noncellular stroma and ECM of the organ play an important role. Prostate basement membrane contains type IV and V collagen meshwork that is laminin rich and supports basal cells, stem cells, transit-amplifying cells and secretory epithelium.

6. ECM on UPJ obstruction

Total or partial blockage at the level where renal pelvis and the ureter are joined is defined as UPJ obstruction. Obstruction can be congenital or acquired. In this case, the passage of urine from the kidney to the ureter was damaged partially or completely, depending on the grade of the obstruction. As a result, deterioration in renal function due to hydronephrosis may occur in untreated cases in the future.

In a normal kidney, the UPJ does not differ histologically from the renal pelvis. However, in an obstructed kidney, the longitudinal muscle fibers are significantly increased with more collagen deposits around the muscle fibers in addition to attenuation of muscle bundles [21].

The role of ECM in the pathogenesis of UPJ obstruction is still unclear. Major pathologic component of obstructive renal injury is tubulointerstitial fibrosis, which results in obstruction-induced renal dysfunction. Tubulointerstitial fibrosis is regarded as the final common pathway for all kidney diseases that lead to chronic renal failure [30]. One of the earliest histologic changes in the obstructed kidney is an increase in inflammatory cell infiltration into the interstitial compartment of the kidney. This results in the secretion of growth factors and cytokines. As a response to increased cytokine and growth factor levels, matrix-producing fibroblasts accumulate in renal interstitium. In response to stimulation from cytokines and growth factors, fibroblasts will secrete collagen, elastin, proteoglycans and fibronectin into the interstitial space. MMPs strictly regulate ECM secretion process in healthy individuals. Tissue inhibitors of MMPs (TIMPs) are produced by both tubular cells and interstitial cells in the kidney, and they function to inhibit the activity of MMPs [21]. An increase in TIMPs expression has been shown as a result of urinary obstruction. Although it is thought that this mechanism could be the result of ECM accumulation the role of MMPs on renal fibrosis is still not clear. Some studies show that inhibition of MMPs results in increased renal fibrosis [31] whereas there is evidence that MMP-9-deficient mice have a dramatic reduction on interstitial fibrosis in response to urinary obstruction [32].

Kaya et al. show that there seems to be increased expression of ECM components in the patients with congenital UPJ obstruction. In their study, surgical specimens of 21 patients who underwent a pyeloplasty surgery were examined immunohistochemically. Their study showed that collagen III and tenascin C expression was significantly higher in patients with UPJ obstruction. Their study also reveals that in UPJ obstruction MMP-2 expression was significantly elevated compared with healthy controls, which represents increased matrix turnover. This study also showed decreased S100 protein expression emphasizing decreased neural structure which helps us to better understand pathophysiology of this condition [9].

Another study performed by Kim et al. [33] in 65 patients demonstrated that the more collagen compared to smooth muscle the worse renal function recovery after surgery. Although this study showed that increased collagen levels are associated with poor prognosis it lacks to investigate relations with collagen subgroups.

Supporting these findings, in 2009, Özel et al. performed a controlled study with 36 patients performing immunohistochemistry and found that fibronectin, type 4 collagen and laminin levels were significantly higher in patients with UPJ obstruction. They also expressed that apoptosis was higher in UPJ obstruction group [34].

Although it is a highly investigated area, role of ECM proteins in the development of UPJ obstruction and their impact on treatment success is still controversial. Current literature lacks a study that compares child and adult patient populations, which could give us a clearer picture for the progression of UPJ obstruction thought the life. Such information could help the physician to decide the timing of the surgery with more objective data.

7. ECM on UVJ

Similar to UPJ studies, UVJ shows decreased muscle density and increased ECM components in diseased patients. In a study published in 2004, 36 UVJ segments were evaluated and MMP1 production was found significantly higher in the group with an obstructed junction. This study also found that an increased level of CD68+ macrophages was found in obstructed junctions and as a result there was an increase in cytokines and growth factors and ECM is secreted at elevated levels [23].

Oswald et al. have shown that markers of smooth muscle structure decrease in UVJ pathologies whereas collagen concentration increases significantly by examining tissue specimens of 29 patients with a refluxing ureter and comparing them with nonrefluxing tissues.

Studies showing changes in ECM composition in UVJ-related disease gives us a picture of what happens after pathological process starts. In most conditions, our knowledge lacks the information of what really starts these changes on subcellular level. There is a need for more detailed and larger studies to get a clear picture of the conditions and develop better treatment strategies.

8. Conclusion

With advancement in the technology and increased data on ECM components, it was realized that many diseases have a close relation with the composition of ECM in the affected tissue.

ECM composition, in some conditions, gives us a view about the prognosis and progression of disease whereas in others it can give information about the cause and pathophysiology of the disease. With better understanding of ECM one can develop new treatment and follow-up models. Also better knowledge of ECM is essential for tissue engineering. Although there is a lot of data on ECM subject there are still needs for well-planned clinical trials, which can change our perspective on this subject.

Author details

Cevdet Kaya[1*] and Bahadır Şahin[2]

*Address all correspondence to: drckaya@hotmail.com

1 Department of Urology, Marmara University School of Medicine, Istanbul, Turkey

2 Resident of Urology, Marmara University School of Medicine, Istanbul, Turkey

References

[1] Frantz C, Stewart KM, Weaver VM. The extracellular matrix at a glance. Journal of Cell Science 2010;123:4195.

[2] Singh B, Fleury C, Jalalvand F, Riesbeck K. Human pathogens utilize host extracellular matrix proteins laminin and collagen for adhesion and invasion of the host. FEMS Microbiology Reviews 2012;36:1122.

[3] Järveläinen H, Sainio A, Koulu M, Wight TN, Penttinen R. Extracellular matrix molecules: potential targets in pharmacotherapy. Pharmacological Reviews 2009;61:198.

[4] Schaefer L, Schaefer RM. Proteoglycans: from structural compounds to signaling molecules. Cell and Tissue Research 2010;339:237.

[5] Lullo GA, Sweeney SM, Körkkö J, Ala-Kokko L, Antonio JD. Mapping the ligand-binding sites and disease-associated mutations on the most abundant protein in the human, type I collagen. Journal of Biological Chemistry 2002;277:4223.

[6] Rozario T, DeSimone DW. The extracellular matrix in development and morphogenesis: a dynamic view. Developmental Biology 2010;341:126.

[7] Chen YM, Miner JH. Glomerular basement membrane and related glomerular disease. Translational Research: The Journal of Laboratory and Clinical Medicine. 2012;160:291.

[8] Verma RP, Hansch C. Matrix metalloproteinases (MMPs): chemical–biological functions and (Q)SARs. Bioorganic & Medicinal Chemistry 2007;15:2223.

[9] Kaya C, Bogaert G, de Ridder D, Schwentner C, Fritsch H, Oswald J, et al. Extracellular matrix degradation and reduced neural density in children with intrinsic ureteropelvic junction obstruction. Urology 2010;76:185.

[10] Faba RO, Palou-Redorta J, Fernández-Gómez JM, Algaba F, Eiró N, Villavicencio H, et al. Matrix metalloproteinases and bladder cancer: what is new? ISRN Urology. 2012;2012:581539.

[11] Davies B, Waxman J, Wasan H, Abel P, Williams G, Krausz T, et al. Levels of matrix metalloproteases in bladder cancer correlate with tumor grade and invasion. Cancer Research 1993;53:5365.

[12] Kanayama Ho, Yokota Ky, Kurokawa Y, Murakami Y, Nishitani M, Kagawa S. Prognostic values of matrix metalloproteinase-2 and tissue inhibitor of metalloproteinase-2 expression in bladder cancer. Cancer. 1998;82:1359.

[13] Szarvas T, Jäger T, Becker M, Tschirdewahn S, Niedworok C, Kovalszky I, et al. Validation of circulating MMP-7 level as an independent prognostic marker of poor survival in urinary bladder cancer. Pathology & Oncology Research 2011;17:325.

[14] Genovese F, Manresa AA, Leeming D, Karsdal M, Boor P. The extracellular matrix in the kidney: a source of novel non-invasive biomarkers of kidney fibrosis? Fibrogenesis & Tissue Repair 2014;7:4.

[15] Matejas V, Hinkes B, Alkandari F, Al-Gazali L, Annexstad E, Aytac MB, et al. Mutations in the human laminin beta2 (LAMB2) gene and the associated phenotypic spectrum. Human Mutation 2010;31:992.

[16] Miner JH. The glomerular basement membrane. Experimental Cell Research 2012;318:973.

[17] Schlöndorff D, Banas B. The mesangial cell revisited: no cell is an island. Journal of the American Society of Nephrology: JASN 2009;20:1179.

[18] Kolset SO, Reinholt FP, Jenssen T. Diabetic nephropathy and extracellular matrix. Journal of Histochemistry and Cytochemistry 2012;60:976.

[19] Catania JM, Chen G, Parrish AR. Role of matrix metalloproteinases in renal pathophysiologies. American Journal of Physiology. Renal Physiology 2007;292:F905.

[20] Stokes MB, Holler S, Cui Y, Hudkins KL, Eitner F, Fogo A, et al. Expression of decorin, biglycan, and collagen type I in human renal fibrosing disease. Kidney International 2000;57:487.

[21] Wein AJ, Kavoussi LR, Partin AW, Peters CA. Campbell-Walsh urology: Elsevier Health Sciences; USA, 2015.

[22] Oswald J, Brenner E, Schwentner C, Deibl M, Rtsch G, Fritsch H, et al. The intravesical ureter in children with vesicoureteral reflux: a morphological and immunohistochemical characterization. The Journal of Urology 2003;170:2423.

[23] Oswald J, Schwentner C, Brenner E, Deibl M, Fritsch H, Bartsch G, et al. Extracellular matrix degradation and reduced nerve supply in refluxing ureteral endings. Journal of Urology 2004;172:1099.

[24] Schwentner C, Oswald J, Lunacek A, Pelzer AE, Fritsch H, Schlenck B, et al. Extracellular microenvironment and cytokine profile of the ureterovesical junction in children with vesicoureteral reflux. Journal of Urology 2008;180:694.

[25] Aitken KJ, Bagli DJ. The bladder extracellular matrix. Part I: architecture, development and disease. Nature Reviews. Urology 2009;6:596.

[26] Chang SL, Chung JS, Yeung MK, Howard PS, Macarak EJ. Roles of the lamina propria and the detrusor in tension transfer during bladder filling. Scandinavian Journal of Urology and Nephrology 1999;33:38.

[27] Brown B, Lindberg K, Reing J, Stolz DB, Badylak SF. The basement membrane component of biologic scaffolds derived from extracellular matrix. Tissue Engineering 2006;12:519.

[28] Dozmorov MG, Kyker KD, Saban R, Knowlton N, Dozmorov I, Centola MB, et al. Analysis of the interaction of extracellular matrix and phenotype of bladder cancer cells. BMC Cancer 2006;6:12.

[29] Brunner A, Tzankov A. The role of structural extracellular matrix proteins in urothelial bladder cancer (review). Biomarker Insights 2007;2:418.

[30] Zeisberg M, Neilson EG. Mechanisms of tubulointerstitial fibrosis. Journal of the American Society of Nephrology 2010;21:1819.

[31] Zeisberg M, Khurana M, Rao VH, Cosgrove D, Rougier J-P, Werner MC, et al. Stage-specific action of matrix metalloproteinases influences progressive hereditary kidney disease. PLoS Medicine 2006;3:e100.

[32] Wang X, Zhou Y, Tan R, Xiong M, He W, Fang L, et al. Mice lacking the matrix metalloproteinase-9 gene reduce renal interstitial fibrosis in obstructive nephropathy. American Journal of Physiology-Renal Physiology 2010;299:F973.

[33] Kim WJ, Yun SJ, Lee TS, Kim CW, Lee HM, Choi H. Collagen-to-smooth muscle ratio helps prediction of prognosis after pyeloplasty. The Journal of Urology 2000;163:1271.

[34] Ozel SK, Emir H, Dervisoglu S, Akpolat N, Senel B, Kazez A, et al. The roles of extracellular matrix proteins, apoptosis and c-kit positive cells in the pathogenesis of ureteropelvic junction obstruction. Journal of Pediatric Urology 2010;6:125.

Chapter 11

Mechanisms of Collagen Network Organization in Response to Tissue/Organ Damage

Takaoki Saneyasu, Saeko Yoshioka and Takao Sakai

Additional information is available at the end of the chapter

http://dx.doi.org/10.5772/63163

Abstract

Fibrosis is a part of the wound-healing response to tissue damage and characterized by excessive accumulation of mainly type I collagen-containing extracellular matrices (ECMs). Transforming growth factor beta (TGF-β) is a profibrogenic master cytokine responsible for promoting differentiation of tissue-resident fibroblasts into myofibroblasts, upregulation of ECM production, and downregulation of ECM degradation. The formation of ECM is an essential response in wound healing. Fibronectin is an ECM glycoprotein substantially expressed during tissue repair. Based on *in vitro* findings, it has been widely accepted that collagen network organization was exclusively fibronectin matrix dependent. Unexpectedly, our fibronectin conditional knockout mouse models have demonstrated a fibronectin-independent mechanism of collagen fibril formation following injury and identified TGF-β signaling and type V collagen as essential elements for collagen fibrillogenesis. Interestingly, the targeting of the TGF-β signaling alone, as proposed in some recent antifibrotic therapies of chronic fibrotic diseases, is not sufficient to completely prevent liver fibrosis. In this chapter, we focus on the present knowledge of the mechanisms of the collagen network organization following tissue/organ damage and pathological processes of chronic fibrotic diseases.

Keywords: collagen, extracellular matrix, fibrogenesis, fibronectin, TGF-β

Abbreviations:

α-SMA, α-smooth muscle actin; BAPN, β-aminopropionitrile; BMP-1, bone morphogenetic protein-1; CCl_4, carbon tetrachloride; Col I, type I collagen; Col III, type III collagen; Col V, type V collagen; Col XI, type XI collagen; CTGF/CCN2, connective tissue growth factor; ECM, extracellular matrix; EGF, epidermal growth factor; HSC, hepatic stellate cell; IL17R, IL-17A receptor; LAP, latency associated protein; LLC, large latent complex; LOX, lysyl oxidase; LTBP, latent TGF-β-binding protein; MMP, matrix metalloproteinase; OB, obliterative bronchiolitis; SLRPs, small leucine-rich proteoglycans/proteins; TGF-β, transforming growth factor-β; TSP-1, Thrombospondin-1; TRI, TGF-β type I receptor; TRII, TGF-β type II receptor; TMLC, mink lung cell line; PAI-1, plasminogen activator inhibitor-1.

1. Introduction

Cells in virtually all tissues are in contact with organized complexes of structural molecules collectively called the extracellular matrix (ECM). ECM induces a variety of signals that regulate the behavior of cells, such as differentiation, adhesion, and migration, and also fundamental physiological processes such as embryonic development and tissue regeneration and remodeling [1]. As a consequence, tissues or organs keep their normal architecture and homeostasis. Aberrations in signal transduction from the ECM cause chronic degenerative and fibrotic disorders.

Considering the adult tissue/organ remodeling following injury, an important unresolved question is how newly deposited ECM contributes to the critical turning point from normal to abnormal healing. Wound healing is a crucial response to maintain tissue/organ structure and integrity after tissue damage, and also tissue/organ homeostasis [2]. Fibrosis is a part of the wound-healing response that maintains organ structure and integrity following tissue damage. However, excessive fibrosis contributes to a number of diseases. Indeed, fibrosis is the common pathological end result of many chronic inflammatory diseases. Fibrosis is an abnormal extension of the wound-healing process that follows tissue damage, characterized by the excessive accumulation of collagenous ECMs. The hallmark of fibrosis is excessive accumulation of mainly type I collagen (Col I) containing ECMs, and therefore involves both wound-healing and fibrotic processes. Fibrosis is recognized as a major cause of morbidity and mortality in most chronic diseases and chronic graft rejection, and also influences tumor invasion and metastasis. Importantly, a critical event in all fibrotic diseases is the activation of myofibroblast, which are the key mediators of fibrotic tissue remodeling [3].

The clarification of regulatory mechanisms underlying excessive accumulation of ECMs in parenchymal organs such as livers during the development of chronic fibrotic diseases is a critical issue. However, currently, the main barrier to designing novel antifibrotic strategies is due to our insufficient understanding of the mechanisms responsible for ECM-network formations following tissue/organ injury. This gap in knowledge translates to lack of experimental models of repair *in vivo*, including gaps in the understanding of the identity and molecular control of factors and cells participating in the repair processes following injury. In this chapter, we will focus on the present knowledge of the mechanisms of the collagen-network organization following tissue/organ damage and pathological processes of chronic fibrotic diseases.

2. Molecules critical for collagen network formation

2.1. Fibronectin

Fibronectin, a dimeric glycoprotein, exists in two isoforms: a soluble isoform in plasma (plasma-type fibronectin, produced solely by hepatocytes) and an insoluble isoform in tissue ECM (cellular-type fibronectin, produced by a variety of cells). Both isoforms are generated from a single gene by alternative splicing [4]. Although considerable *in vitro* functional studies have indicated that fibronectin isoforms play key roles in cell differentiation, proliferation, migration, and survival [5, 6], knowledge of the functional identity of each fibronectin isoform in adult tissue remodeling remains loosely defined due to the complexity and the lack of the systems. Indeed, a prominent expression of fibronectin is often observed during adult tissue repair [7]. In response to adult tissue damage, the initial "provisional matrix" formation between plasma-type fibronectin and fibrinogen stabilizes wounded area, which acts as a nidus for subsequent collagen fibrillogenesis [7, 8]. Little insight into the pathophysiological roles of fibronectin has emerged from studies of genetic changes in humans. There are no documented cases of fibronectin-null patients, the nearest condition being familial glomerulonephritis in which there are mutations in the type III modules of fibronectin [9]. While complete fibronectin-null mice show an embryonic lethal phenotype [10], experimental evidence has documented that skin wounds heal normally in mice lacking plasma-type fibronectin [11]; hence, an absolute requirement for fibronectin in response to adult tissue damage has been speculative.

Based on *in vitro* findings, it has been postulated that collagen network formation depends on the fibronectin matrix [12, 13]. It was, therefore, hypothesized that the removal of fibronectin from the *in vivo* system could abolish extensive ECM network formation following tissue damage. To define the functional identity of fibronectin in adult tissue remodeling, we recently established a null condition for both fibronectin isoforms in adult liver (fibronectin(fl/fl)/Mx-Cre+). We have demonstrated an unexpected finding that the lack of fibronectin does not interfere with reconstruction and resolution of collagen fibril organization after the initial stages of liver injury. We have discovered a fibronectin-independent mechanism of collagen fibrillogenesis and identified transforming growth factor beta (TGF-β) signaling and type V collagen (Col V) as essential elements for collagen fibrillogenesis in response to liver injury [14] (further discussed in Section 3).

2.2. Transforming growth factor-β (TGF-β)

Transforming growth factor (TGF)-β is a profibrogenic master cytokine responsible for promoting differentiation of tissue-resident fibroblasts into myofibroblasts, upregulation of ECM production including fibronectin, and downregulation of ECM degradation [15–18]. TGF-β is secreted as a biologically inactive (latent) form, and importantly, the active TGF-β levels do not often correlate with mRNA and protein levels [19]. Indeed, the activation of latent TGF-β occurs independently of transcription [20], and the bioassay to measure active TGF-β levels has been developed using a mink lung cell line (TMLC) stably transfected with a plasminogen activator inhibitor-1 (PAI-1) promoter fused to luciferase [21].

TGF-β is secreted in a biologically inactive (latent) form in a complex (large latent complex [LLC]) with TGF-β latency associated protein (LAP) [22] and latent TGF-β binding proteins (LTBPs) [23]. Prior to the secretion, TGF-β is synthesized as a precursor and forms dimerized complex through disulfide bonds intracellularly. The dimer precursor is proteolytically cleaved by endopeptidase furin to generate LAP and mature TGF-β peptide from the N-terminal and C-terminal portions, respectively [24, 25]. Interestingly, LAP still associates with mature TGF-β noncovalently, termed as small latent complex. *In vitro* studies demonstrate that LAP is required for the secretion of TGF-β from cells [26, 27]. Furthermore, LAP shields the receptor binding epitope of mature TGF-β, indicating that LAP plays an inhibitory role in binding to its receptors [18]. Small latent complex is further associated with a secreted large glycoprotein LTBP via disulfide bonds. These trimolecular complexes (TGF-β, LAP, and LTBP) termed as LLC are formed intracellularly, then secreted and incorporated into ECM. An *in vitro* observation reveals that LLC and free LTBPs are secreted rapidly from cells as early as 30 min after synthesis, whereas small latent complex is secreted slowly [28]. Another study using TGF-β expressing CHO cells shows that only 50% of recombinant TGF-β precursor is secreted at 6 h after radio labeling with [^{35}S] cysteine and [^{35}S] methionine, and the level of its secretion becomes plateau at 20 h [29]. Secreted LLCs deposit in ECM via LTBPs. LTBPs are extracellular multidomain glycoproteins and share homology with fibrillins, which are the major constituents of connective tissue microfibrillar structure [25, 30–32]. Four different isoforms (LBTP 1-4) have been identified, and each isoform includes four 8-cystein domains and numerous epidermal growth factor (EGF)-like motifs [32]. Three isoforms, LTBP-1, 3, and 4, are known to associate with LAP via the third 8-cystein domain, whereas LTBP-2 does not bind to latent TGF-β [32]. In addition, LTBPs interact with extracellular proteins such as fibrillin-1, fibronectin, heparin, and myostatin [32–36]. Indeed, LTBP-1 colocalizes with both fibrillin-1 and fibronectin *in vitro* [33, 36]. Thus, LTBP-1 plays a central role in secreting and anchoring latent TGF-β into ECM (see Section 3).

In response to injury, the conformation of LLC changes and/or TGF-β is released from LAP, resulting that active TGF-β is exposed to its receptor binding site [37]. Indeed, elevated TGF-β bioavailability is frequently observed in chronic fibrotic diseases, and the inhibition of local TGF-β activation can protect against the progression of fibrosis in several adult chronic fibrotic diseases [38–41]. There are several mechanisms of local TGF-β activation, which mediates αvβ6 and αvβ8 integrins, and thrombospondin-1 (TSP-1). Integrin αvβ6 can directly activate latent TGF-β, which depends on an interaction with RGD amino acid sequence of LAP [25]. In response to injury, integrin αvβ6 induces a conformational change (deformation) of LAP via the interaction between αvβ6 and the cytoskeleton [25, 42, 43]. Consequently, such a deformation makes it possible to release active TGF-β from LAP and then cause mature TGF-β to interact with TGF-β type II receptor (TRII) [25, 42, 43]. Interestingly, this activation process is independent of any proteolysis [25]. In contrast, integrin αvβ8-mediated activation of TGF-β is shown to be dependent on membrane type 1-matrix metalloprotease (MT1-MMP, also known as MMP-14) [44]. Furthermore, this activation does not require β8-cytoplasmic domain [44]. These findings indicate that the mechanism of αvβ8-mediated TGF-β activation is clearly different from that of αvβ6. *In vitro* and *in vivo* studies demonstrate that TSP-1 can also activate latent TGF-β [45, 46]. TSP-1 is a matricellular protein prominently expressed in response to tissue damage and plays a role as a transient component of ECM during tissue repair [47]. TSP-1 directly interacts with the LAP [48], and the interaction is supposed to induce a conformational change of LAP, thereby presumably uncover TGF-β receptor binding site [49, 50]. In addition, MMP-2, 9, 13, bone morphogenetic protein (BMP)-1, and serine proteases (plasmin, thrombin, neutrophil elastase, and kallikrein) have been shown to play an important role in TGF-β activation at least *in vitro* [51, 52]. A very recent study reveals plasma kallikrein-dependent TGF-β activation in fibrotic liver in both animal models and patients [53].

Smads are the central as direct downstream modulators in canonical TGF-β signaling [54–56]. Smads consist of three classes: regulatory/receptor-activated (Smad2 and Smad3), coactivating (Smad4), and inhibitory Smads (Smad6 and Smad7). TGF-β binding to TRII initiates the formation of the complex with the TGF-β type I receptor (TRI) and phosphorylation of TRI. Subsequently, they activate TRI phosphorylates of Smad2 and Smad3, and then they form complexes with Smad4. These complexes bind to specific motifs "Smad-binding element" with transcription factors/coactivator such as Ap-1, Sp1, and CBP/p300, and promote the gene expression [54, 57]. Lines of evidence suggest that TGF-β signaling plays a key role in regulating myofibroblast phenotypes and fibrosis in the heart, lungs, liver, kidneys, and skin [3, 54]. For example, TGF-β directly induces the transdifferentiation of fibroblasts into collagen-secreting active myofibroblasts [3, 14, 58], and overexpression of TGF-β results in the induction of hepatocyte apoptosis and liver fibrosis [18, 59, 60]. In contrast, TGF-β1 knockout mice show remarkable (~80%) decrease of collagen accumulation in response to liver injury [58, 61]. Furthermore, Smad3-null mice show reduction in liver fibrosis with decreased myofibroblast activation and ECM production in response to liver injury, whereas the disruption of inhibitory Smad7 results in an enhancement of damage and fibrogenesis in chronic liver injury [7]. We have previously generated adult mice lacking TRII from livers (TGF-βIIR(fl/fl)/Mx-Cre+); because TRII is the exclusive type II receptor for all TGF-β ligands, lack of this receptor abolishes all TGF-mediated signaling in the liver. Knockout livers actually show significantly lower ECM deposition (~46% compared to controls) in carbon tetrachloride (CCl_4)-induced chronic injury, which is accompanied by the decreased expression of myofibroblast marker alpha-smooth muscle actin (α-SMA). These findings indicate that TGF-β signaling is indeed a dominant pathway in the development of liver fibrosis [62]. However, elimination of TGF-β or TRII does not completely prevent the accumulation of Col I in chronic liver injury, and in particular, TRII knockout livers still remain ~46.4% fibrosis compared to wild type [61, 62]. Therefore, these findings clearly indicate the TGF-β-independent mechanism(s) in the development of liver fibrosis (see Section 3). Indeed, we have found that TRII-null livers significantly upregulated connective tissue growth factors (CTGF/CCN2) following chronic liver injury, suggesting that CTGF/CCN2 can be an alternative mediator in liver fibrosis.

As described above, the local activation of latent TGF-β is a critical step in TGF-β-mediated fibrosis [37, 42, 63, 64]. We have discovered that fibronectin-null livers show elevated local TGF-β bioavailability and upregulate Smad signaling in activated hepatic stellate cells (HSCs) following injury [14, 65]. This novel finding implies that fibronectin regulates the balance between active and inactive (latent) TGF-β, which in turn modulates ECM production and remodeling following injury, and consequently retains adult tissue/organ functions. We further describe this important topic in the following section.

2.3. Type I/type V collagen

Collagens are the most abundant scaffolding ECM in tissue/organ stroma and contribute significantly to tissue/organ integrity [66, 67]. The collagen superfamily is large and heterogeneous, and there are at least 28 different collagen types in vertebrates [66, 68]. A collagen molecule consists of three polypeptide chains called α chains, displays a right-handed supercoil, and also has one or more triple helical regions with common sequence repeats, Gly-X-Y [69]. Collagens are divided into six subfamilies (or subgroups) based on their structure, function, and distribution [66, 67]. The collagens that we focus in this review are Col I and Col V and those belong to fibril-forming collagens. Fibril-forming collagens are synthesized as procollagen and three pro α chains fold a triple helical structure from its C-terminus to N-terminus [67]. To form ordered fibrils, both N- and C-propeptides should be cleaved by procollagen N-proteinases/ADAMTS-2, 3, 14, and procollagen C-proteinases/BMP-1/tolloid proteinases, respectively [68]. An exception to this is the case in Col α1(V) chain. Its C-procollagen is cleaved by furin, and its N-terminal is cleaved by BMP-1 [70]. The peptides cleaved by proteinase are called "propeptides" and noncollagenous peptides remain after the enzyme cleavages called "telopeptides". The telopeptides of Col I contain intermolecular crosslinking sites for fibrillogenesis [71]. Fibril-forming collagens form the 64–67 nm regularly repeated striated fibrils [72]. When collagen I molecules form fibrils, there is a unique space in the collagen fibrils termed "gap zone". The gap zone is present between the N-terminus of one molecule and C-terminus of the next in the triple helix-formed collagen fibrils, and the gap zone is suggested to play a role in a variety of molecular interactions, including lysyl oxidase (LOX)-mediated collagen cross-linking [73, 74].

Collagens are essential for tissue-specific macromolecular structure and organizations in the ECM. Indeed, collagen-mediated ECM networks affect many important biological properties such as matrix/tissue stiffness and tissue/organ structure. Collagens participate in numerous physiological processes such as embryonic development, and tissue regeneration and remodeling [75]. Collagen networks provide the biomechanical scaffold for cell attachment and trap of macromolecules, and regulate cell growth and proliferation, and also the shape and structural integrities of cells and tissues [1, 76]. While collagens exist outside of cell and are composed of ECM structures, cells always sense the alterations of their ECMs, produce new ECMs, and/or degrade their ECMs, and consequently, tissues/organs maintain their homeostasis. More importantly, collagens induce intracellular signaling pathways, and it is mediated by cell surface ECM receptor, integrins. Integrins are transmembrane $\alpha\beta$ heterodimeric receptors that mediate organization of focal contacts, actin-containing cytoskeleton, and ECM. Integrins are a major family of cell-surface-adhesion receptors (composed of 18 α-subunits and 8 β-subunits) [77]. The ligation of integrins by adhesive ligands can induce intracellular signaling events ("outside-in" signaling) and intracellular signaling pathways can control binding avidity of integrins for extracellular ligands ("inside-out" signaling) [77]. Interestingly, there is a "cross-talk" between integrins and receptor tyrosine kinases in certain cell types [78–80].

Several unique features of collagens are spotlighted as targets of medical treatment or biomarkers for diagnosis. For examples, oral administration of Col V induces immunologic tolerance to lung allografts and downregulates lung allograft rejection [81]. Another study shows using a specific monoclonal antibody against Col V C-terminal propeptide that, in the liver fibrosis patients, Col V C-terminal propeptide levels released in the serum have a positive correlation with the total amount of collagens deposited in the fibrotic livers. This finding has a potential to be used as diagnostic and potentially prognostic markers in monitoring liver fibrosis [82].

Considering the mechanisms in collagen fibril network organization, two key molecules, Col V and Col XI, are involved in this process. Although both collagens are fibrillar collagens and minor components in tissues/organs, they play an important role in controlling fibril diameter of assembled collagen [83]. Col V [two $\alpha1(V)$ and one $\alpha2(V)$] is known to form heterotypic fibrils with Col l [76]. Col V has a conserved multidomain structure, N-terminal domain [73]. Col V controls collagen fibril diameter through both triple helical [84] and N-terminal domains [83, 85]. There is evidence that Col V regulates the fibril diameter *in vitro* [73, 84, 86]: Col I alone formed a broad distribution of relatively large diameter fibrils, while Col V alone formed much thinner nonperiodic fibrils. Interestingly, fibrils formed from Col I in the presence of increasing amounts of Col V displayed a significant decrease in the mean fibril diameter. The variance of the fibril population is also decreased as the percentage of Col V increases. Despite of presence of Col I, deletion of *col 5a1* gene causes embryonic lethal in a mouse model due to a virtual lack of fibril formation in the mesenchyme [87]. The heterozygous mice (*col 5a1* [+/–]) are viable, but show the reduction of fibril number and collagen content in the skin compared to wild-type mice. Furthermore, the abnormal collagen fibrils are observed in the deep dermis of *col 5a1* (+/–) mice; the mutant fibrils display larger diameter and broad distribution, and the diameter is inconsistent along the fibril length. These findings show that col V is essential for Col I fibril assembly.

Col XI [$\alpha1(XI)$, $\alpha2(XI)$, $\alpha3(XI)$] is known to form heterotypic fibrils with Col II. Col XI α-chains shares structural homology with Col V and appears to have a similar nucleating function [88]. The first half of the human $\alpha1(V)$ N-propeptide has 73% homology with the human $\alpha1(XI)$ chain [83]. The $\alpha1$ (XI) chain loses its function by chondrodysplasia mutation [89]. Loss of function mutations in the $\alpha1$ (XI) chain leads to the assembly of Col II and Col V fibrils with abnormally large diameters [73, 89]. This interesting observation clearly indicates that Col XI also have a role to regulate the fibril diameter as Col V. To date, very little evidence in Col XI-mediated fibrillogenesis has been identified, and more detailed studies in the functional role of Col XI remain to be elucidated.

Recently, Burlingham et al. have shown an attractive evidence that Col V is involved in immune response to adult human lung disease, obliterative bronchiolitis (OB), which is characterized by narrowing of bronchiole lumens due to the inflammation and fibrosis [90]. Col V is recognized as an antigen by monocyte and presented to $CD4^+$ T cells (called Th-17), and Col V-specific responses are required both IL-17A produced by Th-17 and the monokines

TNF-α and IL-1β, suggesting that IL-17A induced in response to Col V plays a role in fibrosis. Indeed, IL-17A plays a variety of significant roles in neutrophil recruitment, angiogenesis, inflammation, and autoimmune disease [91]. Very recently, Vittal et al. have shown a unique observation about the relationship between Col V and IL-17A [92]. IL-17A can induce epithelial–mesenchymal transition in rat lung epithelial T-antigen negative cells through upregulation of S100A4 and mesenchymal marker α-SMA, and downregulation of epithelial markers ZO-1 and E-cadherin. Mechanistically, IL-17A results in the downregulation of *Smad7*, and upregulation of *TGF-*β and Smad3 activation, clearly indicating that IL-17A is involved in TGF-β pathway. Col V is upregulated by TGF-β during osteogenesis [93], and we have demonstrated TGF-β-induced Col V-mediated *de novo* Col I and Col III network organization even in the absence of fibronectin [14]. Thus, these findings suggest that IL-17A can act as an upstream mediator to regulate the expression of Col V via TGF-β signaling-mediated pathway, and as a consequence, IL-17A could modulate Col V-mediated fibrogenesis. There is another observation showing that IL-17A is involved in liver fibrosis. Mouse primary HSCs are shown to express IL-17A receptor (IL-17R). The treatment of HSC with recombinant IL-17A upregulates α-SMA, collagens, and TGF-β mRNA expression levels. Furthermore, the IL-17RA-null mouse model shows ~50% reduction of liver fibrosis induced by CCl_4 with decreased levels liver damage and inflammations [94], suggesting that IL-17A plays a significant role in HSC transdifferentiation into active myofibroblasts during the development of liver fibrosis. Thus, further analysis of the regulatory mechanisms of Col V by IL-17A could open the new avenue as a drug target for liver fibrosis.

2.4. Other modifying molecules such as lysyl oxidase and small leucine-rich proteoglycans

Collagen contributes significantly to tissue/organ integrity, and collagen cross-linking stiffens the ECM [1]. A recent elegant study has demonstrated that collagen cross-linking leads to cancer progression by enhancing ECM receptor integrin signaling [95, 96]. However, the functional contribution of collagen cross-linking to noncancer pathogenesis remains largely unknown. The LOX family enzymes are copper-dependent amine oxidase and catalyze the post-translational modification of peptidyl lysine to the peptidyl aldehyde, α-aminoadipic-δ-semialdehyde [97]. This chemical change enables the covalent cross-linking of collagen and elastin, resulting in insolubilizing and stabilizing ECM proteins. Collagen cross-linking stiffens the ECM and is accompanied by tissue/organ fibrosis that is mediated by several profibrogenic cytokines [95, 98]. Indeed, LOX inhibitor β-aminopropionitrile (BAPN) reduces organ stiffness following injury and TGF-β1-induced collagen fibril stiffness *in vitro* [65, 99]. Lines of study show that LOX localizes in ECM of several tissue such as skin, aorta, heart, lung, liver, and cartilage [97]. LOX is secreted as inactive proenzyme (proLOX) and then proteolytically cleaved to active enzyme. *In vitro* study shows the possibility that the activation of proLOX occurs on the cell surface in a complex with cellular fibronectin. Indeed, LOX colocalizes well with assembled fibronectin fibrils in cultured fibroblasts and normal human tissues [100]. Furthermore, fibronectin-null mouse embryonic fibroblasts exhibit drastic decrease of the proteolytic processing of proLOX [100], strongly suggesting that fibronectin matrix regulates ECM stiffness via LOX activation.

http://dx.doi.org/10.5772/63163

Small leucine-rich proteoglycans/proteins (SLRPs) such as decorin, biglycan, and fibromodulin are known to contribute significantly to collagen assembly. SLRPs consist of five classes, and almost all SLRPs bind collagen fibrils through their leucine-rich repeat domain [101–103]. Accumulating evidence using SLRP-null mouse models shows that SLRPs regulate the diameter and/or alter structure of collagen fibrils [101–103]. For example, decorin knockout mouse skin exhibits a broader range of collagen fibril diameters [104], and treatment of decorin with decorin-null skin fibroblasts results in the formation of more uniform collagen fibrils [105]. Fibromodulin-deficient tail tendons exhibit thinner collagen fibrils compared to wild-type controls [106, 107]. A dynamic modulus in biglycan-null tendons is significantly increased compared to wild-type tendons [108]. The elasticity of collagen fiber networks in cultured decorin-siRNA-transfected mouse NIH3T3 fibroblasts is declined during the incubation period, whereas it remains unchanged in untransfected cells. It is therefore likely that SLRPs could regulate the physiological properties of ECM (e.g., mechanical strength).

3. Possible mechanisms of collagen network formation

As described in Section 2.1, it has been believed that collagen-network formation depends on the fibronectin matrix in culture [12, 13]. Indeed, the prominent expression of fibronectin is observed during adult tissue repair [11, 109]. Another line of evidence shows that TGF-β plays a central role as a profibrogenic cytokine in the accumulation of ECMs, including fibronectin. Therefore, it would be possible that TGF-β-induced ECM accumulation is dependent on fibronectin. However, to date, fibronectin/TGF-β interdependence in the fibrogenic response to tissue damage has not yet been addressed. Furthermore, it remains to be elucidated how ECM remodeling by myofibroblasts results in changes in mechanical tension and supports the activation of pathogenic signaling pathways during the development of chronic fibrotic diseases. We hypothesized that the removal of fibronectin or TGF-β signaling *in vivo* could prevent extensive ECM network formation following tissue damage. To define the functional identity of fibronectin and TGF-β signaling in adult tissue remodeling, we recently established two animal models lacking fibronectin (both isoforms) or TRII, respectively, in adult liver [14, 62]. Our new findings suggest fibronectin-/TGF-β-independent mechanisms are involved in the development of liver fibrosis.

3.1. Fibronectin-dependent assembly

3.1.1. Fibronectin matrix assembly

Fibronectin matrix assembly consists of multistep process (**Figure 1**) [110–113]. Importantly, fibronectin assembly is cell dependent; binding of fibronectin to cell surface and cellular contractility are required. In the first step, fibronectin binds to cell surface, and integrin plays an important role. Integrins $\alpha5\beta1$ and $\alpha v\beta3$ are characterized as fibronectin receptors [114]. Lines of evidence show that $\alpha5\beta1$ is a primary receptor in fibronectin matrix assembly, whereas $\alpha v\beta3$ dominates the formation of focal contacts [115–117]. However, the binding of fibronectin to the cell is not sufficient for fibronectin assembly. A critical step of this assembly is considered to be the cell-driven exposure of cryptic site for self-association in fibronectin. Although one mechanism for exposing the cryptic site could be conformational changes induced by fibronectin binding to integrin, cellular contractility is necessary for fibronectin fibrillogenesis [88, 111, 118]. Indeed, loss of cellular contractility by RhoA inhibitor prevents fibronectin matrix formation [119]. Furthermore, recent study demonstrates that β1 cytoplasmic domain modulates fibronectin assembly via recruitment of cytoplasmic adaptor protein talin, which links integrin to the actin cytoskeleton [120, 121]. Thus, integrin-mediated association of fibronectin with cytoskeleton is important for fibronectin assembly.

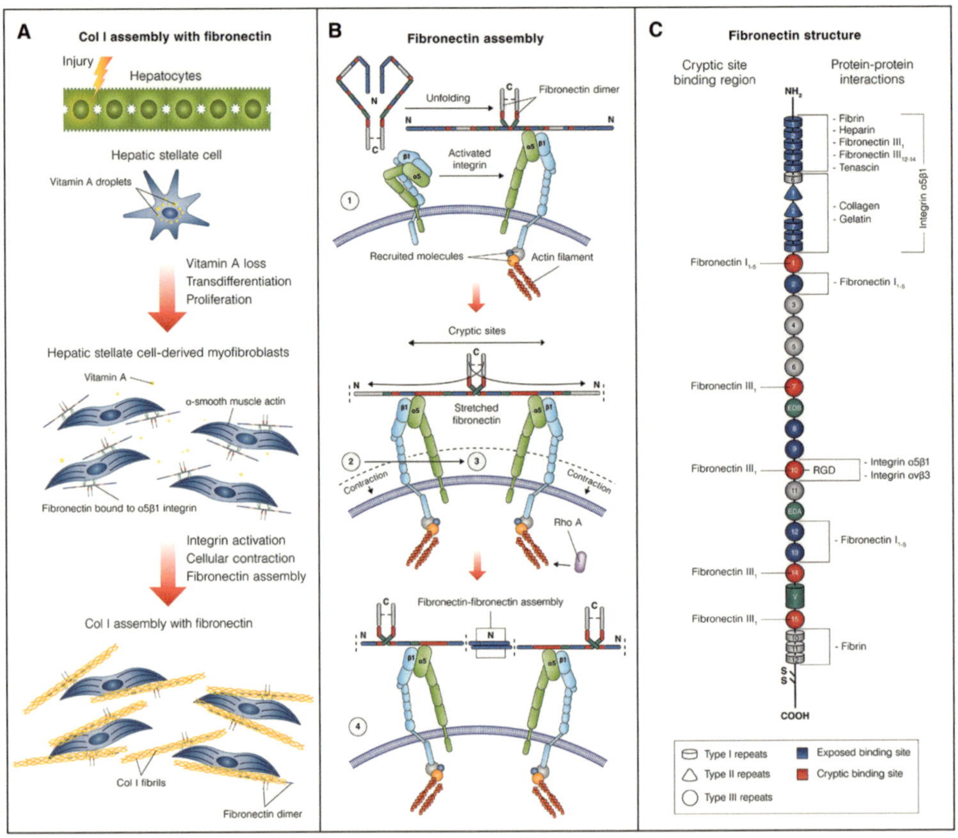

Figure 1. Proposed mechanism of fibronectin-dependent collagen assembly. (A) Type I collagen assembly with fibronectin. (B) Fibronectin assembly. ①Fibronectin binding to activated integrin; ②conformational changes of fibronectin cryptic site by cellular contraction; ③exposure of fibronectin cryptic site, resulting in acquiring ability to associate with other fibronectin molecules; and ④formation and extension of fibronectin fibril assembly. (C) Fibronectin structure. See details in the text.

3.1.2. Collagen assembly with fibronectin matrix

Fibronectin contains collagen-binding domain [122] and directly binds to collagens. *In vitro* studies have shown an extensive codistribution of fibronectin and Col I/III (**Figure 1**) [12,

123]. A very recent study using fibroblasts demonstrates that type I collagen fibrils preferentially colocalize with more-relaxed fibronectin fibrils in the ECM *in vitro* [124]. Fibronectin-null embryonic fibroblastic cells cannot organize collagen fibril networks *in vitro*, but they can form collagen networks when fibronectin is exogenously added [123]. Interestingly, collagen-binding integrins α2β1 and α11β1 are not required for collagen polymerization in fibronectin-null embryonic fibroblastic cells when cultured with exogenous fibronectin [12]. Furthermore, integrin β1-null line GD25 cells, which express fibronectin and its receptor αvβ3, contract collagen gel more forcefully than integrin α2- and β1-transfected GD25 cells that contract collagen gel via α2β1 but not αvβ3 [125]. Other studies show that fibronectin polymerization stimulates collagen gel contraction [126] and that the disruption of fibronectin–collagen association inhibits this contraction [13]. These findings demonstrate that fibronectin matrix is required for collagen assembly and enhances ECM contraction of cells.

3.1.3. Initial incorporation of latent TGF-β complex

Growing evidence suggests a mechanism by which fibronectin plays a role in the initial incorporation of latent TGF-β complex into ECM [18, 33]. LTBP-1 associates with not only fibrillin-1 but also fibronectin [33, 36]. Fibronectin-null fibroblasts fail to incorporate LTBP into ECM [33], and minimally activate latent TGF-β [127]. Furthermore, cells lacking fibronectin receptor integrin α5β1 show defective activity of latent TGF-β by αvβ6 [127]. These findings suggest that fibronectin regulates latent TGF-β activation via deposition of latent TGF-β into the matrix.

3.2. Fibronectin-independent collagen assembly

As described above, based on *in vitro* findings, it has been postulated that collagen network organization and assembly depends on the fibronectin matrix in culture [12, 13]. However, the contribution of fibronectin to these processes remains to be defined *in vivo*. We therefore investigated whether fibronectin is a suitable molecular target for ameliorating the fibrogenic response to liver injury. Since mice with complete inactivation of fibronectin gene die at an early embryonic stage [10], we generated conditional fibronectin-floxed and liver-specific adult fibronectin-null mice (lacking both plasma and cellular isoforms of fibronectin) using *Cre-loxP* technology [11, 14], investigated their phenotypes, and have demonstrated fibronectin-independent mechanisms for collagen network formation following liver injury.

3.2.1. TGF-β and Col V-mediated collagen assembly in fibronectin-null liver

The adult mouse model lacking fibronectin shows no abnormalities in anatomical and histological analyses of the liver and hepatic biochemical markers under standard laboratory conditions. Unexpectedly, the lack of fibronectin did not interfere with the reconstruction of collagen fibril organization in response to both acute liver and chronic liver injuries up to 8 weeks induced by CCl_4 [14]. Fibronectin-null livers show significant increased HSC activation with elevated Smad signaling following injury. To determine whether TGF-β is involved in Col III/I collagen network formation in the absence of fibronectin, we further assessed which factors that regulate activated HSC phenotypes were involved in collagen fibrillogenesis. We have identified TGF-β1-induced Col V as a novel and essential element for Col I/III fibrogenesis in hepatic stellate cells (**Figure 2**). Thus, our study provides compelling evidence that collagen fibrillogenesis in response to adult tissue/organ damage is mediated by both fibronectin and type V collagen.

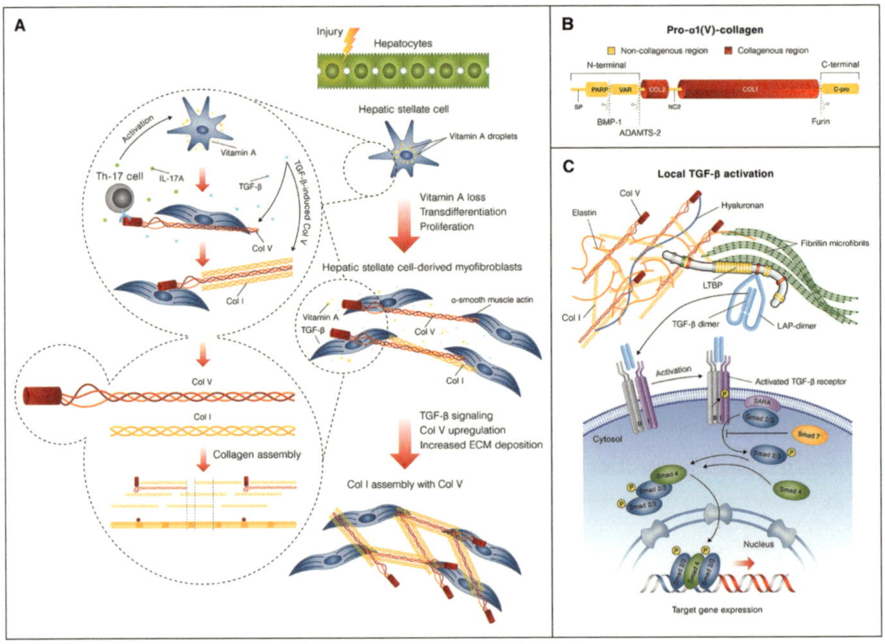

Figure 2. Proposed mechanism of fibronectin-independent collagen assembly mediated by tissue growth factor beta (TGF-β) signaling and type V collagen (Col V). (A) Type I collagen assembly with Col V. (B) Pro-α1(V) collagen structure. PARP, proline- and arginine-rich domain; VAR, variable domain. (C) Local TGF-β activation. See details in the text.

Interestingly, type V collagen–mediated Col III/I fibril assembly following liver injury seems to be specific for adult HSCs because TGF-β1 supports neither Col V nor Col III/I fibril assembly in fibronectin-null embryonic fibroblasts *in vitro* [14]. Furthermore, fibronectin-null livers show substantial depositions of LTBP-1, -3, and -4 with fibrillar structures in the ECM following both acute and chronic liver injuries [14, 65], whereas fibronectin-null embryonic fibroblasts fail to incorporate LTBP-1 into ECM *in vitro* [33]. Therefore, further studies for mechanisms underlying Col V-mediated collagen network formation remain to be elucidated, i.e., the phenotypic differences among myofibroblasts, and the contribution to adult tissue/organ remodeling following injury.

3.2.2. Elevated collagen matrix stiffness in advanced stage of liver fibrosis/cirrhosis in fibronectin-null liver

We showed that collagen network organization can be formed even without fibronectin in response to liver injury [14]. However, it remains unknown whether the initial deposition of fibronectin could contribute to the turning point from normal healing to chronic fibrotic disorders. Furthermore, it remains to be elucidated how ECM remodeling by myofibroblasts affects mechanical tension and supports the activation of pathogenic signaling during the development of chronic fibrotic diseases. We therefore have further investigated whether fibronectin could be a suitable target for ameliorating fibrosis during advanced stages of chronic liver injury [65]. Fibronectin-null livers have exhibited constitutively elevated local TGF-β activity, induced more myofibroblast phenotypes, and accumulated highly disorganized/diffused collagenous ECM networks during chronic liver fibrogenesis induced by CCl_4. The deposition and network formation of Col V are also significantly increased. Consequently, fibronectin-null livers have led to more extensive liver cirrhosis, which is accompanied by significant increased liver matrix stiffness and deteriorated hepatic functions. Mechanistically, fibronectin-null livers have shown elevated LOX expressions, and a significant amount of active LOX is released in fibronectin-null hepatic stellate cells in response to TGF-β1. Furthermore, treatment of fibronectin in fibronectin-null stellate cells recovers collagen fibril stiffness to wild-type levels [65].

We propose that there are the functional links between fibronectin-mediated control of TGF-β bioavailability and collagen fibril stiffness regulated by LOX. All these novel findings strongly suggest that locally activated TGF-β signaling and Col V are essential elements for collagen fibrogenesis without fibronectin in adult tissue remodeling. Although the contribution of Col V-nucleated Col I fibrogenesis in adult tissues is largely undetermined, our finding raises the hypothesis that the accumulation of Col V-mediated ECMs during persistent chronic damage could influence a formation of disorganized ECM architecture.

3.3. TGF-β-independent collagen assembly

3.3.1. CTGF-mediated collagen assembly

TGF-β signaling is the dominant pathway of ECM productions in HSCs [58]. To address whether the elimination of this signaling is sufficient to prevent liver fibrosis, we have generated adult liver-specific TRII-null mice (TGFβIIR[flox/flox]/Mx-Cre+) [62]. Actually, this mutant liver exhibits a significant decrease of ECM deposition and α-SMA expression in CCl_4-induced chronic liver injury. However, elimination of TRII does not completely prevent the collagen accumulation in chronic liver injury, and TRII-null livers still remain ~46.4% fibrosis compared to wild type. Furthermore, we have found that matricellular protein CTGF/CCN2 expression is significantly upregulated (1.94-fold compared to wild type) in mutant livers following chronic liver injury. Therefore, these findings clearly indicate that TGF-β-independent mechanisms play an alternative role in developing liver fibrosis. Since CTGF/CCN2 synergizes with the action of TGF-β, CTGF/CCN2 is considered to act as a TGF-β downstream modulator [128]. Accompanied by the elevated TGF-β activity, CTGF/CCN2 expression is upregulated in several fibrotic tissues, including kidney, lung, heart, liver, pancreas, bowel, and skin [128, 129]. A recent study has demonstrated that overexpression of CTGF/CCN2 in fibroblasts alone is sufficient to cause spontaneous multiorgan fibrosis *in vivo* and that this signal pathway does not involve canonical TGF-β-Smad signaling *in vitro.*

However, the liver does not show significant fibrosis [130]. We have found that TRII-null livers in chronic injury show elevated expression of CTGF/CCN2 in spite of a lack of TGF-β signaling, indicating CTGF/CCN2 as a potent mediator in liver fibrosis [62]. We propose two hits, induction of CTGF/CCN2 and adult tissue/organ injury, for the progression of liver fibrosis. In fibrotic livers, CTGF/CCN2 is known to be synthesized in a diversity of cells such as hepatocytes, myofibroblasts (activated HSCs), endothelial cells, proliferating bile duct epithelial cells, and inflammatory cells [128]. Thus, it remains to be elucidated to identify cellular contribution and mechanisms underlying CTGF/CCN2 production in the progression of tissue/organ fibrosis.

4. Perspective

Novel findings obtained imply that fibronectin regulates the balance between active and inactive (latent) TGF-β, which in turn modulates ECM production and remodeling following injury, and consequently retains adult tissue/organ functions. This regulatory mechanism by fibronectin could be translated for a potential therapeutic target in a broader variety of chronic fibrotic diseases. It is obvious that fibronectin matrix networks play crucial roles in many important biological events and in adult homeostasis. Therefore, the regulation of TGF-β bioavailability by fibronectin with retaining fibronectin matrix networks would be an essential element for long-term antifibrotic strategies in chronic fibrotic diseases to preserve tissue/organ function and homeostasis.

Considering the adult tissue/organ remodeling following injury, important unresolved questions are (1) what is the trigger for fibrosis resolution (regression); and (2) how fibrosis resolution proceeds. Quiescent HSCs are adipocyte-like (vitamin A stock) cells expressing marker genes such as PPARγ, SREBP-1c, and leptin, whereas activated HSCs are proliferative myofibroblasts expressing myogenic marker α-SMA, c-myb, and MEF-2 [131]. CCl_4-induced experimental liver injury models have suggested that elimination of activated HSCs by apoptosis [132] or senescence [133] is a key step in the onset of fibrosis regression [134]. Growing evidence demonstrates that activated HSCs are reverted to quiescent-like state both *in vitro* [135, 136] and *in vivo* [137, 138], and a very recent *in vitro* study using human primary HSCs reveals EGF, fibroblast growth factor 2, fatty acids, and retinol as potential factors in activated HSCs to reverse quiescent-like phenotypes [139]. Interestingly, reverted HSCs show more rapid reactivation into myofibroblasts in response to TGF-β than quiescent HSCs [137]. It is therefore likely that reverted HSCs may not fully revert to a quiescent state [134]. A recent elegant study demonstrates that a history of liver injury is transmitted to offspring via epigenetic modification of PPARγ and TGF-β genes in rats, and consequently, healing response to hepatic injury is suppressed in offspring of CCl_4-injured group compared to injury-inexperienced group [140]. Nevertheless, the clarification of molecular mechanisms underlying excessive accumulation of collagenous ECM during the development of chronic fibrotic diseases could translate basic antifibrotic research into improved clinical therapeutic approaches, which will have a significant benefit in public health impact. Thus, an establishment of novel models/systems, e.g., in which we enable to analyze global alterations of signaling

http://dx.doi.org/10.5772/63163

cascades/metabolisms or epigenetics, would be crucial for challenging these complicated mechanisms in chronic fibrotic diseases.

Acknowledgements

The authors are grateful to Drs Takako Sasaki and Kazunori Mizuno for critical reading of this manuscript and valuable suggestions, and Dr Nikola Kolundžić for his excellent artwork. The authors wish to acknowledge many outstanding contributions of investigators in the field whose work could not be cited because of space constraints. This work was supported in part by National Institutes of Health research grant DK074538 (to T. Sakai). The authors are also grateful to the Institute of Translational Medicine, University of Liverpool, for support (to T. Sakai).

Conflict of interest

The authors disclose no conflicts.

Author details

Takaoki Saneyasu, Saeko Yoshioka and Takao Sakai*

*Address all correspondence to: sakait@liverpool.ac.uk

MRC Centre for Drug Safety Science, Department of Molecular and Clinical Pharmacology, Institute of Translational Medicine, Liverpool, UK

References

[1] Kolacna, L., Bakesova, J., Varga, F., Kostakova, E., Planka, L., Necas, A., Lukas, D., Amler, E., and Pelouch, V. (2007) Biochemical and biophysical aspects of collagen nanostructure in the extracellular matrix. Physiol. Res. 56(Suppl 1):S51–S60.

[2] Shaw, T. J., and Martin, P. (2009) Wound repair at a glance. J. Cell Sci. 122, 3209–3213.

[3] Wynn, T. A., and Ramalingam, T. R. (2012) Mechanisms of fibrosis: therapeutic translation for fibrotic disease. Nat. Med. 18, 1028–1040.

[4] Schwarzbauer, J. E. (1991) Fibronectin: from gene to protein. Curr. Opin. Cell Biol. 3, 786–791.

[5] Mosher, D. F. (1989) Fibronectin. Academic Press, San Diego.

[6] Hynes, R. O. (1990) Fibronectins. Springer-Verlag, New York.

[7] Hayashi, H., and Sakai, T. (2011) Animal models for the study of liver fibrosis: new insights from knockout mouse models. Am. J. Physiol. Gastrointest. Liver Physiol. 300, G729–G738.

[8] Clark, R. A. F. (1996) The molecular and cellular biology of wound repair, 2nd ed., Plenum Press, New York.

[9] Castelletti, F., Donadelli, R., Banterla, F., Hildebrandt, F., Zipfel, P. F., Bresin, E., Otto, E., Skerka, C., Renieri, A., Todeschini, M., Caprioli, J., Caruso, R. M., Artuso, R., Remuzzi, G., and Noris, M. (2008) Mutations in FN1 cause glomerulopathy with fibronectin deposits. Proc. Natl. Acad. Sci. U. S. A. 105, 2538–2543.

[10] George, E. L., Georges-Labouesse, E. N., Patel-King, R. S., Rayburn, H., and Hynes, R. O. (1993) Defects in mesoderm, neural tube and vascular development in mouse embryos lacking fibronectin. Development. 119, 1079–1091.

[11] Sakai, T., Johnson, K. J., Murozono, M., Sakai, K., Magnuson, M. A., Wieloch, T., Cronberg, T., Isshiki, A., Erickson, H. P., and Fassler, R. (2001) Plasma fibronectin supports neuronal survival and reduces brain injury following transient focal cerebral ischemia but is not essential for skin-wound healing and hemostasis. Nat. Med. 7, 324–330.

[12] Velling, T., Risteli, J., Wennerberg, K., Mosher, D. F., and Johansson, S. (2002) Polymerization of type I and III collagens is dependent on fibronectin and enhanced by integrins alpha 11beta 1 and alpha 2beta 1. J. Biol. Chem. 277, 37377–37381.

[13] Sottile, J., Shi, F., Rublyevska, I., Chiang, H. Y., Lust, J., and Chandler, J. (2007) Fibronectin-dependent collagen I deposition modulates the cell response to fibronectin. Am. J. Physiol. Cell Physiol. 293, C1934–C1946.

[14] Moriya, K., Bae, E., Honda, K., Sakai, K., Sakaguchi, T., Tsujimoto, I., Kamisoyama, H., Keene, D. R., Sasaki, T., and Sakai, T. (2011) A fibronectin-independent mechanism of collagen fibrillogenesis in adult liver remodeling. Gastroenterology. 140, 1653–1663.

[15] Feng, X. H., and Derynck, R. (2005) Specificity and versatility in TGF-beta signaling through Smads. Annu. Rev. Cell Dev. Biol. 21, 659–693.

[16] tenDijke, P., and Arthur, H. M. (2007) Extracellular control of TGFbeta signalling in vascular development and disease. Nat Rev Mol. Cell Biol. 8, 857–869.

[17] Wakefield, L. M., and Stuelten, C. (2007) Keeping order in the neighborhood: new roles for TGFbeta in maintaining epithelial homeostasis. Cancer Cell. 12, 293–295.

[18] Hayashi, H., and Sakai, T. (2012) Biological significance of local TGF-beta activation in liver diseases. Front Physiol. 3, 12.

[19] Theodorescu, D., Bergsma, D., Man, M. S., Elshourbagy, N., Sheehan, C., Rieman, D., and Kerbel, R. S. (1991) Cloning and overexpression of TGF-beta 1 cDNA in a mammary adenocarcinoma: *in vitro* and *in vivo* effects. Growth Factors. 5, 305–316.

[20] Boulanger, J., Reyes-Moreno, C., and Koutsilieris, M. (1995) Mediation of glucocorticoid receptor function by the activation of latent transforming growth factor beta 1 in MG-63 human osteosarcoma cells. Int. J. Cancer. 61, 692–697.

[21] Abe, M., Harpel, J. G., Metz, C. N., Nunes, I., Loskutoff, D. J., and Rifkin, D. B. (1994) An assay for transforming growth factor-beta using cells transfected with a plasminogen activator inhibitor-1 promoter-luciferase construct. Anal. Biochem. 216, 276–284.

[22] Ju, W., Ogawa, A., Heyer, J., Nierhof, D., Yu, L., Kucherlapati, R., Shafritz, D. A., and Bottinger, E. P. (2006) Deletion of Smad2 in mouse liver reveals novel functions in hepatocyte growth and differentiation. Mol. Cell. Biol. 26, 654–667.

[23] Rifkin, D. B. (2005) Latent transforming growth factor-beta (TGF-beta) binding proteins: orchestrators of TGF-beta availability. J. Biol. Chem. 280, 7409–7412.

[24] Dubois, C. M., Laprise, M. H., Blanchette, F., Gentry, L. E., and Leduc, R. (1995) Processing of transforming growth factor beta 1 precursor by human furin convertase. J. Biol. Chem. 270, 10618–10624.

[25] Munger, J. S., Harpel, J. G., Gleizes, P. E., Mazzieri, R, Nunes, I, and Rifkin, D. B. (1997) Latent transforming growth factor-beta: structural features and mechanisms of activation. Kidney Int. 51, 1376–1382.

[26] Lopez, A. R., Cook, J., Deininger, P. L., and Derynck, R. (1992) Dominant negative mutants of transforming growth factor-beta1 inhibit the secretion of different transforming growth factor-beta isoforms. Mol. Cell. Biol. 12, 1674–1679.

[27] Gray, A. M., and Mason, A. J. (1990) Requirement for activin A and transforming growth factor-beta1 proregions in homodimer assembly. Science. 247, 1328–1330.

[28] Miyazono, K., Olofsson, A., Colosetti, P., and Heldin, C. H. (1991) A role of the latent TGF-beta1-binding protein in the assembly and secretion of TGF-beta1. EMBO J. 10, 1091–1101.

[29] Sha, X., Brunner, A. M., Purchio, A. F., and Gentry, L. E. (1989) Transforming growth factor β1: importance of glycosylation and acidic proteases for processing and secretion. Mol. Endocrinol. 3, 1090–1098.

[30] Kanzaki, T., Olofsson, A., Moren, A., Wernstedt, C., Hellman, U., Miyazono, K., Claesson-Welsh, L., and Heldin, C. H. (1990) TGF-beta1 binding protein: a component of the large latent complex of TGF-beta1 with multiple repeat sequences. Cell. 61, 1051–1061.

[31] Robertson, I. B., Horiguchi, M., Zilberberg, L., Dabovic, B., Hadjiolova, K., and Rifkin, D. B. (2015) Latent TGF-beta-binding proteins. Matrix Biol. 47, 44–53.

[32] Todorovic, V., and Rifkin, D. B. (2012) LTBPs, more than just an escort service. J. Cell. Biochem. 113, 410–418.

[33] Dallas, S. L., Sivakumar, P., Jones, C. J., Chen, Q., Peters, D. M., Mosher, D. F., Humphries, M. J., and Kielty, C. M. (2005) Fibronectin regulates latent transforming growth

factor-beta (TGF beta) by controlling matrix assembly of latent TGF beta-binding protein-1. J. Biol. Chem. 280, 18871–18880.

[34] Nunes, I., Munger, J. S., Harpel, J. G., Nagano, Y., Shapiro, R. L., Gleizes, P. E., and Rifkin, D. B. (1996) Structure and activation of the large latent transforming growth factor-beta complex. Int. J. Obes. Relat. Metab. Disord. 20(Suppl 3), S4–S8.

[35] Isogai, Z., Ono, R. N., Ushiro, S., Keene, D. R., Chen, Y., Mazzieri, R., Charbonneau, N. L., Reinhardt, D. P., Rifkin, D. B., and Sakai, L. Y. (2003) Latent transforming growth factor beta-binding protein 1 interacts with fibrillin and is a microfibril-associated protein. J. Biol. Chem. 278, 2750–2757.

[36] Taipale, J., Miyazono, K., Heldin, C. H., and Keski-Oja, J. (1994) Latent transforming growth factor-beta1 associates to fibroblast extracellular matrix via latent TGF-beta binding protein. J. Cell. Biol. 124, 171–181.

[37] Annes, J. P., Munger, J. S., and Rifkin, D. B. (2003) Making sense of latent TGFbeta activation. J. Cell. Sci. 116, 217–224.

[38] Margadant, C., and Sonnenberg, A. (2010) Integrin-TGF-beta crosstalk in fibrosis, cancer and wound healing. EMBO Rep. 11, 97–105.

[39] Sheppard, D. (2006) Transforming growth factor beta: a central modulator of pulmonary and airway inflammation and fibrosis. Proc. Am. Thorac. Soc. 3, 413–417.

[40] Varga, J., and Pasche, B. (2009) Transforming growth factor beta as a therapeutic target in systemic sclerosis. Nat. Rev. Rheumatol. 5, 200–206.

[41] Nishimura, S. L. (2009) Integrin-mediated transforming growth factor-beta activation, a potential therapeutic target in fibrogenic disorders. Am. J. Pathol. 175, 1362–1370.

[42] Wipff, P. J., and Hinz, B. (2008) Integrins and the activation of latent transforming growth factor beta1—an intimate relationship. Eur. J. Cell. Biol. 87, 601–615.

[43] Shi, M., Zhu, J., Wang, R., Chen, X., Mi, L., Walz, T., and Springer, T. A. (2011) Latent TGF-β structure and activation. Nature. 474, 343–349.

[44] Mu, D., Cambier, S., Fjellbirkeland, L., Baron, J. L., Munger, J. S., Kawakatsu, H., Sheppard, D., Broaddus, V. C., and Nishimura, S. L. (2002) The integrin alpha(v)beta8 mediates epithelial homeostasis through MT1-MMP-dependent activation of TGF-beta1. J. Cell. Biol. 157, 493–507.

[45] Crawford, S. E., Stellmach, V., Murphy-Ullrich, J. E., Ribeiro, S. M., Lawler, J., Hynes, R. O., Boivin, G. P., and Bouck, N. (1998) Thrombospondin-1 is a major activator of TGF-beta1 *in vivo*. Cell. 93, 1159–1170.

[46] Schultz-Cherry, S., Chen, H., Mosher, D. F., Misenheimer, T. M., Krutzsch, H. C., Roberts, D. D., and Murphy-Ullrich, J. E. (1995) Regulation of transforming growth factor-beta activation by discrete sequences of thrombospondin 1. J. Biol. Chem. 270, 7304–7310.

[47] Adams, J. C. (2001) Thrombospondins: multifunctional regulators of cell interactions. Annu. Rev. Cell Dev. Biol. 17, 25–51.

[48] Schultz-Cherry, S., Ribeiro, S., Gentry, L., and Murphy-Ullrich, J. E. (1994) Thrombospondin binds and activates the small and large forms of latent transforming growth factor-beta in a chemically defined system. J. Biol. Chem. 269, 26775–26782.

[49] Murphy-Ullrich, J. E., and Poczatek, M. (2000) Activation of latent TGF-beta by thrombospondin-1: mechanisms and physiology. Cytokine Growth Factor Rev. 11, 59–69.

[50] Sweetwyne, M. T., and Murphy-Ullrich, J. E. (2012) Thrombospondin1 in tissue repair and fibrosis: TGF-beta-dependent and independent mechanisms. Matrix Biol. 31, 178–186.

[51] Jenkins, G. (2008) The role of proteases in transforming growth factor-beta activation. Int. J. Biochem. Cell Biol. 40, 1068–1078.

[52] Robertson, I. B., and Rifkin, D. B. (2013) Unchaining the beast; insights from structural and evolutionary studies on TGFbeta secretion, sequestration, and activation. Cytokine Growth Factor Rev. 24, 355–372.

[53] Hara, M., Kirita, A., Kondo, W., Matsuura, T., Nagatsuma, K., Dohmae, N., Ogawa, S., Imajoh-Ohmi, S., Friedman, S. L., Rifkin, D. B., and Kojima, S. (2014) LAP degradation product reflects plasma kallikrein-dependent TGF-beta activation in patients with hepatic fibrosis. Springerplus. 3, 221.

[54] Piershma, B., Bank, R. A., and Boersema, M. (2015) Signaling in fibrosis: TGF-β, WNT, and YAP/TAZ converge. Front. Med. 2, 59.

[55] Heldin, C. H., Miyazono, K., and ten Dijke, P. (1997) TGF-beta signalling from cell membrane to nucleus through SMAD proteins. Nature. 390, 465–471.

[56] Massagué, J., and Wotton, D. (2000) Transcriptional control by the TGF-beta/Smad signaling system. EMBO J. 19, 1745–1754.

[57] Oreffo, R. O., Mundy, G. R., Seyedin, S. M., and Bonewald, L. F. (1989) Activation of the bone-derived latent TGF beta complex by isolated osteoclasts. Biochem. Biophys. Res. Commun. 158, 817–823.

[58] Friedman, S. L. (2000) Molecular regulation of hepatic fibrosis, an integrated cellular response to tissue injury. J. Biol. Chem. 275, 2247–2250.

[59] Sanderson, N., Factor, V., Nagy, P., Kopp, J., Kondaiah, P., Wakefield, L., Roberts, A. B., Sporn, M. B., and Thorgeirsson, S. S. (1995) Hepatic expression of mature transforming growth factor beta 1 in transgenic mice results in multiple tissue lesions. Proc. Natl. Acad. Sci. U. S. A. 92, 2572–2576.

[60] Hayashi, H., Sakai, K., Baba, H., and Sakai, T. (2012) Thrombospondin-1 is a novel negative regulator of liver regeneration after partial hepatectomy through transforming growth factor-beta1 activation in mice. Hepatology. 55, 1562–1573.

[61] Hellerbrand, C., Stefanovic, B., Giordano, F., Burchardt, E. R., and Brenner, D. A. (1999) The role of TGFbeta1 in initiating hepatic stellate cell activation *in vivo*. J. Hepatol. 30, 77–87.

[62] Sakai, K., Jawaid, S., Sasaki, T., Bou-Gharios, G., and Sakai, T. (2014) Transforming growth factor-beta-independent role of connective tissue growth factor in the development of liver fibrosis. Am. J. Pathol. 184, 2611–2617.

[63] Koli, K., Saharinen, J., Hyytiainen, M., Penttinen, C., and Keski-Oja, J. (2001) Latency, activation, and binding proteins of TGF-beta. Microsc. Res. Tech. 52, 354–362.

[64] Hyytiainen, M., Penttinen, C., and Keski-Oja, J. (2004) Latent TGF-beta binding proteins: extracellular matrix association and roles in TGF-beta activation. Crit. Rev. Clin. Lab. Sci. 41, 233–264.

[65] Iwasaki, A., Sakai, K., Moriya, K., Sasaki, T., Keene, D. R., Akhtar, R., Miyazono, T., Yasumura, S., Watanabe, M., Morishita, S., and Sakai, T. (2016) Molecular mechanism responsible for fibronectin-controlled alterations in matrix stiffness in advanced chronic liver fibrogenesis. J. Biol. Chem. 291, 72–88.

[66] Ricard-Blum, S. (2010) The collagen family. Cold Spring Harb. Perspect. Biol. 3, a004978–a004978.

[67] Mander, L. N., and Liu, H.-w. (2010) Comprehensive natural products II: chemistry and biology, Elsevier Science, Amsterdam.

[68] Kadler, K. E., Baldock, C., Bella, J., and Boot-Handford, R. P. (2007) Collagens at a glance. J. Cell. Sci. 120, 1955–1958.

[69] Brodsky, B., and Persikov, A. V. (2005) Molecular structure of the collagen triple helix. Adv. Protein Chem. 70, 301–339.

[70] Unsold, C., Pappano, W. N., Imamura, Y., Steiglitz, B. M., and Greenspan, D. S. (2002) Biosynthetic processing of the pro-alpha 1(V)2pro-alpha 2(V) collagen heterotrimer by bone morphogenetic protein-1 and furin-like proprotein convertases. J. Biol. Chem. 277, 5596–5602.

[71] Prockop, D. J., and Fertala, A. (1998) Inhibition of the self-assembly of collagen I into fibrils with synthetic peptides. Demonstration that assembly is driven by specific binding sites on the monomers. J. Biol. Chem. 273, 15598–15604.

[72] Bruckner, P. (2010) Suprastructures of extracellular matrices: paradigms of functions controlled by aggregates rather than molecules. Cell Tissue Res. 339, 7–18.

[73] Birk, D. E. (2001) Type V collagen: heterotypic type I/V collagen interactions in the regulation of fibril assembly. Micron. 32, 223–237.

[74] Siegel, R. C. (1974) Biosynthesis of collagen crosslinks: increased activity of purified lysyl oxidase with reconstituted collagen fibrils. Proc. Natl. Acad. Sci. U. S. A. 71, 4826–4830.

[75] Prockop, D. J., and Kivirikko, K. I. (1995) COLLAGENS: molecular biology, diseases, and potentials for therapy. Annu. Rev. Biochem. 64, 403–434.

[76] Kadler, K. E., Holmes, D. F., Trotter, J. A., and Chapman, J. A. (1996) Collagen fibril formation. Biochem. J. 316(Pt 1), 1–11.

[77] Hynes, R. O. (2002) Integrins: bidirectional, allosteric signaling machines. Cell. 110, 673–687.

[78] Sakai, T., de la Pena, J. M., and Mosher, D. F. (1999) Synergism among lysophosphatidic acid, beta1A integrins, and epidermal growth factor or platelet-derived growth factor in mediation of cell migration. J. Biol. Chem. 274, 15480–15486.

[79] Sundberg, C., and Rubin, K. (1996) Stimulation of beta1 integrins on fibroblasts induces PDGF independent tyrosine phosphorylation of PDGF beta-receptors. J. Cell. Biol. 132, 741–752.

[80] Moro, L., Venturino, M., Bozzo, C., Silengo, L., Altruda, F., Beguinot, L., Tarone, G., and Defilippi, P. (1998) Integrins induce activation of EGF receptor: role in MAP kinase induction and adhesion-dependent cell survival. EMBO J. 17, 6622–6632.

[81] Yasufuku, K., Heidler, K. M., O'Donnell, P. W., Smith, G.N., Jr., Cummings, O. W., Foresman, B. H., Fujisawa, T., and Wilkes, D. S. (2001) Oral tolerance induction by type V collagen downregulates lung allograft rejection. ATS J. 25, 26–34.

[82] Vassiliadis, E., Veidal, S. S., Simonsen, H., Larsen, D. V., Vainer, B., Chen, X., Zheng, Q., Karsdal, M. A., and Leeming, D. J. (2011) Immunological detection of the type V collagen propeptide fragment, PVCP-1230, in connective tissue remodeling associated with liver fibrosis. Biomarkers. 16, 426–433.

[83] Smith, S. M., and Birk, D. E. (2012) Focus on molecules: collagens V and XI. Exp. Eye Res. 98, 205–206.

[84] Adachi, E., and Hayashi, T. (1986) *In vitro* formation of hybrid fibrils of type V collagen and type I collagen. Limited growth of type I collagen into thick fibrils by type V collagen. Connect. Tissue Res. 14, 257–266.

[85] Symons, S., Renard, M., Bonod-Bidaud, C., Syx, D., Vagany, E., and Malfait, F. (2011) Identification of binding partners interacting with the α1-N-propeptide of type V collagen. Biochem. Soc. 433, 371–381.

[86] Birk, D. E., Fitch, J. M., Babiarz, J. P., Doane, K. J., and Linsenmayer, T. F. (1990) Collagen fibrillogenesis *in vitro*: interaction of types I and V collagen regulates fibril diameter. J. Cell. Sci. 95(Pt 4), 649–657.

[87] Wenstrup, R. J., Florer, J. B., Brunskill, E. W., Bell, S. M., Chervoneva, I., and Birk, D. E. (2004) Type V collagen controls the initiation of collagen fibril assembly. J. Biol. Chem. 279, 53331–53337.

[88] Kadler, K. E., Hill, A., and Canty-Laird, E. G. (2008) Collagen fibrillogenesis: fibronectin, integrins, and minor collagens as organizers and nucleators. Curr. Opin. Cell Biol. 20, 495–501.

[89] Li, Y., Lacerda, D. A., Warman, M. L., Beier, D. R., Yoshioka, H., Ninomiya, Y., Oxford, J. T., Morris, N. P., Andrikopoulos, K., Ramirez, F., Wardell, B. B., Lifferth, G. D., Teuscher, C., Woodward, S. R., Taylor, B. A., Seegmiller, R. E., and Olsen, B. R. (1995) A fibrillar collagen gene, Col11a1, is essential for skeletal morphogenesis. Cell. 80, 423–430.

[90] Burlingham, W. J., Love, R. B., Jankowska-Gan, E., Haynes, L. D., Xu, Q., Bobadilla, J. L., Meyer, K. C., Hayney, M. S., Braun, R. K., Greenspan, D. S., Gopalakrishnan, B., Cai, J., Brand, D. D., Yoshida, S., Cummings, O. W., and Wilkes, D. S. (2007) IL-17–dependent cellular immunity to collagen type V predisposes to obliterative bronchiolitis in human lung transplants. J. Clin. Invest. 117, 3498–3506.

[91] Kolls, J. K., and Linden, A. (2004) Interleukin-17 family members and inflammation. Immunity. 21, 467–476.

[92] Vittal, R., Fan, L., Greenspan, D. S., Mickler, E. A., Gopalakrishnan, B., Gu, H., Benson, H. L., Zhang, C., Burlingham, W., Cummings, O. W., and Wilkes, D. S. (2012) IL-17 induces type V collagen overexpression and EMT via TGF-β-dependent pathways in obliterative bronchiolitis. Am. Physiol. Soc. 304, 401–414.

[93] Kahai, S., Vary, C. P., Gao, Y., and Seth, A. (2004) Collagen, type V, alpha1 (COL5A1) is regulated by TGF-beta in osteoblasts. Matrix Biol. 23, 445–455.

[94] Tan, Z., Qian, X., Jiang, R., Liu, Q., Wang, Y., Chen, C., Wang, X., Ryffel, B., and Sun, B. (2013) IL-17A plays a critical role in the pathogenesis of liver fibrosis through hepatic stellate cell activation. J. Immunol. 191, 1835–1844.

[95] Levental, K. R., Yu, H., Kass, L., Lakins, J. N., Egeblad, M., Erler, J. T., Fong, S. F., Csiszar, K., Giaccia, A., Weninger, W., Yamauchi, M., Gasser, D. L., and Weaver, V. M. (2009) Matrix crosslinking forces tumor progression by enhancing integrin signaling. Cell. 139, 891–906.

[96] Chen, Y., Terajima, M., Yang, Y., Sun, L., Ahn, Y. H., Pankova, D., Puperi, D. S., Watanabe, T., Kim, M. P., Blackmon, S. H., Rodriguez, J., Liu, H., Behrens, C., Wistuba, II, Minelli, R., Scott, K. L., Sanchez-Adams, J., Guilak, F., Pati, D., Thilaganathan, N., Burns, A. R., Creighton, C. J., Martinez, E. D., Zal, T., Grande-Allen, K. J., Yamauchi, M., and Kurie, J. M. (2015) Lysyl hydroxylase 2 induces a collagen cross-link switch in tumor stroma. J. Clin. Invest. 125, 1147–1162.

[97] Lucero, H. A., and Kagan, H. M. (2006) Lysyl oxidase: an oxidative enzyme and effector of cell function. Cell Mol. Life Sci. 63, 2304–2316.

[98] van der Slot, A. J., van Dura, E. A., de Wit, E. C., De Groot, J., Huizinga, T. W., Bank, R. A., and Zuurmond, A. M. (2005) Elevated formation of pyridinoline cross-links by profibrotic cytokines is associated with enhanced lysyl hydroxylase 2b levels. Biochim. Biophys. Acta. 1741, 95–102.

[99] Georges, P. C., Hui, J. J., Gombos, Z., McCormick, M. E., Wang, A. Y., Uemura, M., Mick, R., Janmey, P. A., Furth, E. E., and Wells, R. G. (2007) Increased stiffness of the rat liver precedes matrix deposition: implications for fibrosis. Am. J. Physiol. Gastrointest. Liver Physiol. 293, G1147–1154.

[100] Fogelgren, B., Polgar, N., Szauter, K. M., Ujfaludi, Z., Laczko, R., Fong, K. S., and Csiszar, K. (2005) Cellular fibronectin binds to lysyl oxidase with high affinity and is critical for its proteolytic activation. J. Biol. Chem. 280, 24690–24697.

[101] Chen, S., and Birk, D. E. (2013) The regulatory roles of small leucine-rich proteoglycans in extracellular matrix assembly. FEBS J. 280, 2120–2137.

[102] Kalamajski, S., and Oldberg, A. (2010) The role of small leucine-rich proteoglycans in collagen fibrillogenesis. Matrix Biol. 29, 248–253.

[103] Heinegård, D. (2009) Proteoglycans and more-from molecules to biology. Int. J. Exp. Pathol. 90, 575–586.

[104] Danielson, K. G., Baribault, H., Holmes, D. F., Graham, H., Kadler, K. E., and Iozzo, R. V. (1997) Targeted disruption of decorin leads to abnormal collagen fibril morphology and skin fragility. J. Cell. Biol. 136, 729–743.

[105] Seidler, D. G., Schaefer, L., Robenek, H., Iozzo, R. V., Kresse, H., and Schönherr, E. (2005) A physiologic three-dimensional cell culture system to investigate the role of decorin in matrix organisation and cell survival. Biochem. Biophys. Res. Commun. 332, 1162–1170.

[106] Svensson, L., Aszódi, A., Reinholt, F. P., Fässler, R., Heinegård, D., and Oldberg, A. (1999) Fibromodulin-null mice have abnormal collagen fibrils, tissue organization, and altered lumican deposition in tendon. J. Biol. Chem. 274, 9636–9647.

[107] Iwasaki, S., Hosaka, Y., Iwasaki, T., Yamamoto, K., Nagayasu, A., Ueda, H., Kokai, Y., and Takehana, K. (2008) The modulation of collagen fibril assembly and its structure by decorin: an electron microscopic study. Arch. Histol. Cytol. 71, 37–44.

[108] Dourte, L. M., Pathmanathan, L., Mienaltowski, M. J., Jawad, A. F., Birk, D. E., and Soslowsky, L. J. (2013) Mechanical, compositional, and structural properties of the mouse patellar tendon with changes in biglycan gene expression. J. Orthop. Res. 31, 1430–1437.

[109] Moriya, K., Sakai, K., Yan, M. H., and Sakai, T. (2012) Fibronectin is essential for survival but is dispensable for proliferation of hepatocytes in acute liver injury in mice. Hepatology. 56, 311–321.

[110] Mosher, D. F. (1995) Organization of the provisional fibronectin matrix: control by products of blood coagulation. Thromb. Haemost. 74, 529–533.

[111] Geiger, B., Bershadsky, A., Pankov, R., and Yamada, K. M. (2001) Transmembrane crosstalk between the extracellular matrix–cytoskeleton crosstalk. Nat. Rev. Mol. Cell. Biol. 2, 793–805.

[112] Singh, P., Carraher, C., and Schwarzbauer, J. E. (2010) Assembly of fibronectin extracellular matrix. Annu. Rev. Cell. Dev. Biol. 26, 397–419.

[113] Mouw, J. K., Ou, G., and Weaver, V. M. (2014) Extracellular matrix assembly: a multiscale deconstruction. Nat. Rev. Mol. Cell. Biol. 15, 771–785.

[114] Barczyk, M., Carracedo, S., and Gullberg, D. (2010) Integrins. Cell Tissue Res. 339, 269–280.

[115] Wennerberg, K., Lohikangas, L., Gullberg, D., Pfaff, M., Johansson, S., and Fässler, R. (1996) Beta 1 integrin-dependent and -independent polymerization of fibronectin. J. Cell. Biol. 132, 227–238.

[116] Akiyama, S. K., Yamada, S. S., Chen, W. T., and Yamada, K. M. (1989) Analysis of fibronectin receptor function with monoclonal antibodies: roles in cell adhesion, migration, matrix assembly, and cytoskeletal organization. J. Cell. Biol. 109, 863–875.

[117] Fogerty, F. J., Akiyama, S. K., Yamada, K. M., and Mosher, D. F. (1990) Inhibition of binding of fibronectin to matrix assembly sites by anti-integrin (alpha 5 beta 1) antibodies. J. Cell. Biol. 111, 699–708.

[118] Schwarzbauer, J. E., and Sechler, J. L. (1999) Fibronectin fibrillogenesis: a paradigm for extracellular matrix assembly. Curr. Opin. Cell. Biol. 11, 622–627.

[119] Zhong, C., Chrzanowska-Wodnicka, M., Brown, J., Shaub, A., Belkin, A. M., and Burridge, K. (1998) Rho-mediated contractility exposes a cryptic site in fibronectin and induces fibronectin matrix assembly. J. Cell. Biol. 141, 539–551.

[120] Green, J. A., Berrier, A. L., Pankov, R., and Yamada, K. M. (2009) Beta1 integrin cytoplasmic domain residues selectively modulate fibronectin matrix assembly and cell spreading through talin and Akt-1. J. Biol. Chem. 284, 8148–8159.

[121] Calderwood, D. A., Campbell, I. D., and Critchley, D. R. (2013) Talins and kindlins: partners in integrin-mediated adhesion. Nat. Rev. Mol. Cell. Biol. 14, 503–517.

[122] Pankov, R., and Yamada, K. M. (2002) Fibronectin at a glance. J. Cell. Sci. 115, 3861–3863.

[123] Sottile, J., and Hocking, D. C. (2002) Fibronectin polymerization regulates the composition and stability of extracellular matrix fibrils and cell-matrix adhesions. Mol. Biol. Cell. 13, 3546–3559.

[124] Kubow, K. E., Vukmirovic, R., Zhe, L., Klotzsch, E., Smith, M. L., Gourdon, D., Luna, S., and Vogel, V. (2015) Mechanical forces regulate the interactions of fibronectin and collagen I in extracellular matrix. Nat. Commun. 6, 8026.

[125] Cooke, M. E., Sakai, T., and Mosher, D. F. (2000) Contraction of collagen matrices mediated by alpha2beta1A and alpha(v)beta3 integrins. J. Cell. Sci. 113(Pt 13), 2375–2383.

[126] Hocking, D. C., Sottile, J., and Langenbach, K. J. (2000) Stimulation of integrin-mediated cell contractility by fibronectin polymerization. J. Biol. Chem. 275, 10673–10682.

[127] Fontana, L., Chen, Y., Prijatelj, P., Sakai, T., Fassler, R., Sakai, L. Y., and Rifkin, D. B. (2005) Fibronectin is required for integrin alphavbeta6-mediated activation of latent TGF-beta complexes containing LTBP-1. FASEB J. 19, 1798–1808.

[128] Gressner, O. A., and Gressner, A. M. (2008) Connective tissue growth factor: a fibrogenic master switch in fibrotic liver diseases. Liver Int. 28, 1065–1079.

[129] Mason, R. M. (2013) Fell-Muir lecture: Connective tissue growth factor (CCN2)—a pernicious and pleiotropic player in the development of kidney fibrosis. Int. J. Exp. Pathol. 94, 1–16.

[130] Sonnylal, S., Shi-Wen, X., Leoni, P., Naff, K., Van Pelt, C. S., Nakamura, H., Leask, A., Abraham, D., Bou-Gharios, G., and de Crombrugghe, B. (2010) Selective expression of connective tissue growth factor in fibroblasts *in vivo* promotes systemic tissue fibrosis. Arth. Rheum. 62, 1523–1532.

[131] Bataller, R., and Brenner, D. A. (2005) Liver fibrosis. J. Clin. Invest. 115, 209–218.

[132] Iredale, J. P., Benyon, R. C., Pickering, J., McCullen, M., Northrop, M., Pawley, S., Hovell, C., and Arthur, M. J. (1998) Mechanisms of spontaneous resolution of rat liver fibrosis. Hepatic stellate cell apoptosis and reduced hepatic expression of metalloproteinase inhibitors. J. Clin. Invest. 102, 538–549.

[133] Krizhanovsky, V., Yon, M., Dickins, R. A., Hearn, S., Simon, J., Miething, C., Yee, H., Zender, L., and Lowe, S. W. (2008) Senescence of activated stellate cells limits liver fibrosis. Cell. 134, 657–667.

[134] Mallat, A., and Lotersztajn, S. (2013) Reversion of hepatic stellate cell to a quiescent phenotype: from myth to reality? J. Hepatol. 59, 383–386.

[135] Hazra, S., Xiong, S., Wang, J., Rippe, R. A., Krishna, V., Chatterjee, K., and Tsukamoto, H. (2004) Peroxisome proliferator-activated receptor gamma induces a phenotypic switch from activated to quiescent hepatic stellate cells. J. Biol. Chem. 279, 11392–11401.

[136] She, H., Xiong, S., Hazra, S., and Tsukamoto, H. (2005) Adipogenic transcriptional regulation of hepatic stellate cells. J. Biol. Chem. 280, 4959–4967.

[137] Kisseleva, T., Cong, M., Paik, Y., Scholten, D., Jiang, C., Benner, C., Iwaisako, K., Moore-Morris, T., Scott, B., Tsukamoto, H., Evans, S. M., Dillmann, W., Glass, C. K., and Brenner, D. A. (2012) Myofibroblasts revert to an inactive phenotype during regression of liver fibrosis. Proc. Natl. Acad. Sci. U. S. A. 109, 9448–9453.

[138] Troeger, J. S., Mederacke, I., Gwak, G. Y., Dapito, D. H., Mu, X., Hsu, C. C., Pradere, J. P., Friedman, R. A., and Schwabe, R. F. (2012) Deactivation of hepatic stellate cells during liver fibrosis resolution in mice. Gastroenterology. 143, 1073–1083.

[139] El Taghdouini, A., Najimi, M., Sancho-Bru, P., Sokal, E., and van Grunsven, L. A. (2015) *In vitro* reversion of activated primary human hepatic stellate cells. Fibrogenesis Tissue Repair. 8, 14.

[140] Zeybel, M., Hardy, T., Wong, Y. K., Mathers, J. C., Fox, C. R., Gackowska, A., Oakley, F., Burt, A. D., Wilson, C. L., Anstee, Q. M., Barter, M. J., Masson, S., Elsharkawy, A. M., Mann, D. A., and Mann, J. (2012) Multigenerational epigenetic adaptation of the hepatic wound-healing response. Nat. Med. 18, 1369–1377.

Chapter 12

Tumor Microenvironment Heterogeneity: A Review of the Biology Masterpiece, Evaluation Systems, and Therapeutic Implications

Irene Tadeo, Tomás Álvaro, Samuel Navarro and
Rosa Noguera

Additional information is available at the end of the chapter

http://dx.doi.org/10.5772/62479

Abstract

A tumor can be considered as a highly heterogeneous functional tissue, connected and dependent on the microenvironment, which sends and receives signals to and from the tumor tissue itself. Tumor cells alter the mechanical properties of the microenvironment in order to create favorable conditions for their proliferation. Stromal cells and non-cellular elements of the extracellular matrix (ECM), including the host immune system, the fibrous scaffolding, the fundamental substance, and blood vascularization can determine tumoral cell morphologies, functions, aggressiveness, and response to treatment, as well as an accurate assessment of prognosis of the patients. Robust morphometric digital pathology techniques that are able to standardize measurements and analyse whole sets of immunohistochemical images are called for to identify, describe, and quantify the elements of the ECM. The computer-automated segmentation algorithms are therefore required to increase the knowledge on the tumor microenvironment heterogeneity and to provide new therapeutic targets.

Keywords: Extracellular matrix, digital pathology, tumor microenvironment, heterogeneity, reticulin fibers, vascularization

1. Introduction

Many processes relevant to cancer, such as differentiation, maturation, and the malignant potential of tumor cells have all been shown to be influenced by extracellular matrix (ECM) stiffness [1]. In the general context of cancer, within the tumor ECM, various support cells (fibroblasts, Schwann cells), tumor-associated immune cells, and vascular (blood and lymph) endothelial cells are found, lying among a network of various reticulin, collagen, and elastic fibers merged within the interstitial fluids (glycosaminoglycans, proteoglycans, and glycoproteins) and gradients of several chemical species, which constantly interplay with cells and provide much of the structural support available to parenchymal cells in tissues, providing tensile strength and flexibility [2, 3].

Numerous studies have demonstrated that the tumor ECM not only responds to and supports carcinogenesis, but actively contributes to tumor initiation, progression, and metastasis, and must not only be understood as a reactive neighboring, but also as an active contributor [4]. In fact, it has been published that chronic growth stimulation, ECM remodeling, alteration of cell mechanics, and disruption of tissue architecture are a non-genetic basis, influencing cancer progression [5–7]. Given the complexity both within and outside the cancer cell, and the interactions between cancer cells and the surrounding stroma, it is not surprising that a single perturbation within a tumor can create a cascade of changes in multiple pathways and networks, some of which may have lethal repercussions [8].

Tumor malignancy is driven, among other factors, by the remodeling of contiguous stromal tissue to foster growth, metastasis, and therapy resistance. Tumor cells alter the mechanical properties of the microenvironment to create favorable conditions for their proliferation and/or dissemination [9]. They promote stiffening of their environment, which feeds back to increase malignant behaviors such as loss of tissue architecture and invasion [10]. Matrix remodeling by tumor-associated stromal cells entails the assembly of a highly dense ECM, whose physicochemical attributes enhance malignancy through morphogenic deregulation, tumor cell proliferation, vascular recruitment, and stromal cell differentiation [11–13]. Both tumor and stromal cells produce and assemble a matrix of collagens, proteoglycans, and other molecules such as cytokines that hinder the transport of macromolecules and stimulate the otherwise quiescent host cells to initiate a variety of processes, including desmoplasia and angiogenesis [14, 15]. Mediated by increased deposition, unfolding, and crosslinking of fibrilar adhesion proteins, stiffening increases cell contractility which, in turn, can directly and indirectly modify gene expression via altering transcription factor activity and the release of matrix-bound pro-tumorigenic growth factors [16–19]. Similarly, changes in the microstructure, elasticity, distribution of pore sizes, chemical composition, and fiber arrangement due to ECM remodeling can control aspects of tumor cell phenotype such as adhesion, mechanics, and motility [20–24]. The speed of malignant cells *in vitro* is also affected by the geometry of the ECM. Human glioma cells move faster through narrow channels than through wide channels or in non-stretched 2D surfaces. This is thought to be triggered by an increase in the polarity of the traction forces between cell and ECM [25]. Recent publications describe that not only neoplastic ECM stiffness, but also the firmness of tumor cells play a significant role in tumor progression. Steadiness of tumor cells, especially the metastatic cells, is lower than that of the normal cells of the same sample, and is currently caused by the loss of actin filaments and/or microtubules, and the subsequent lower density of the scaffold [26, 27].

The study of ECM elements composing the architectural scaffolding and the blood vessels can be performed by means of automated or semi-automated quantification. This technique can be used on histologic sections of all malignancies, providing important tumor characteristics such as the morphometric features and the spatial context of tumor and stromal cells at single-cell resolution. Differences in the ECM of different cancer subtypes can contribute to a more comprehensive understanding of the highly heterogeneous tumor microenvironment, and can provide abundant and novel targets for therapy.

2. Masterpiece of biology

2.1. Glycosaminoglycans

Since glycosaminoglycans (GAGs) have been recognized as essential players in critical biological processes regulating cellular properties, owing to their particular biofiltering, scaffolding, and cell anchoring properties, they have been related to diverse malignancies and they are now understood as key elements involved in cancer cell biology and novel therapeutical agents [28, 29].

GAGs are long, non-branched polymers of several disaccharides (up to 200 repeated saccharides), consisting of one uronic acid (almost always glucuronic acid and sometimes iduronic acid) and one hexosamine (glucosamine or galactosamine), presenting variable degrees of sulfation, and constitute the main components of the fundamental substance of the ECM [2]. According to their chemical composition, GAGs can be divided in acid and sulfated: chondroitin sulfate (CS), dermatan sulfate (DS), and heparan sulfate or heparin (HS, Hep); acid and non-sulfated: hyaluronan (HA); and non-acid and sulfated: keratan sulfate (KS). GAGs can form proteoglycans by means of a linkage tetrasaccharide, linking GAGs to a central protein (core protein) through a serine residue and catalyzed by four specific enzymes. The formation of the linkage tetrasaccharide is key for GAG synthesis to start in *GAGosomes* [30–32]. Proteoglycans can be located at the cell surface (syndecan, glypican), in the intracellular compartment (serglycin), secreted to the extracellular medium (small leucine-rich proteoglycans, hyalectans), or within the basement membrane (agrin, collagen XVIII, perlecan) [33, 34].

Diagnostic methods have typically centered on the analysis of GAG structure and concentration. Hyaluronan has the capacity to bind large amounts of water to form viscous gels with special filtering properties and it is capable of forming polyvalent connections with other ECM proteins. This GAG, as well as the rest, has the capacity of modulating host-tumor interaction, lymphangiogenesis, angiogenesis, and multidrug resistance. It can therefore be used as a drug carrier [35]. Hyaluronan accumulates in the stroma of various human tumors and modulates intracellular signaling pathways, cell proliferation, motility, and invasive properties of malignant cells: high stromal hyaluronic acid content is associated with poor differentiation and aggressive clinical behavior in human adenocarcinomas. On the contrary, squamous cell carcinomas and malignant melanomas tend to have a reduced HA content [35]. When tumors produce hyaluronan, this fact is associated with invasion, host-tumor interactions, lymphangiogenesis/angiogenesis, epithelial-mesenchymal transition and with local and distant metastases in glioma, pancreatic adenocarcinoma, breast cancer and prostate cancer, respectively [35]. Other studies in prostate carcinoma have found a link between poor prognosis and under-sulfation or overexpression of chondroitin sulphate. Aggressive breast cancer shows an increase of approximately 15% on GAG content with longer chains compared to non-lethal breast cancer tissue [36].

2.2. Fibrous component

The architectural role of the fibrous component of the ECM is clear and central for tissue homeostasis. In fact, scaffold architecture has been found to have a significant impact on cell growth [37].

The primary proteins present in the ECM are the collagens. They are secreted by connective tissue cells, as well as by a variety of other cell types. As a major component of skin and bone, they represent the most abundant proteins in mammals, constituting 25% of the total protein mass in these animals. A typical collagen molecule is extremely rich in proline and glycine, long, stiff, and presents a triple-stranded helical structure, in which three collagen polypeptide chains called α chain are wound around one another in a ropelike superhelix [38]. Reticulin fibers, or type III collagen, are fine fibers forming an extensive branching network in certain organs. Their distribution is rather restricted: they are usually found mainly in the basement of epithelial tissues, the surface of adipose cells, muscle cells and Schwann cells, outside the endothelium of the hepatic sinusoids, and in the fibrous reticulum of lymphoid tissues. These fibers have a diameter of less than 2 μm and support not only the physical structure of the cell, but also various biological functions, largely through their ability to bind multiple interacting partners such as other ECM proteins, growth factors, signal receptors, and adhesion molecules [3, 39]. Collagen type I fibers account for 90% of the body collagen and usually form thick bundles in bone, skin, tendons, ligaments, cornea, and internal organs. These collagen strands measure up to 100 μm thick and usually follow a wavy course without branching in normal tissues.

The architecture of the collagen scaffolds in tumors is severely altered [18, 40]. It has been found that an aligned fiber matrix enhances differentiation of human neural crest stem cells towards the Schwann cell lineage [41] and evidence has pointed to collagen crosslinking as a significant contributor to the changes in cellular mechanical microenvironment that accompanies tumor progression [42]. Reticulin fibers are considered to play an important role in the adherence of cells and constitute a skeletal framework suitable for individual cells and tissues [43]. They are known to increase in amount and disorganize with aging and stress in physiological conditions [44]. Desmoplasia (collagen type I and reticulin fibers accumulation) is associated with several malignancies. The deregulation and disorganization of the tumor stroma alter the composition, structure, and stiffness of the ECM, leading to the creation of niches within tissues and organs that offer a proper environment for tumors to successfully establish metastasis and activate therapy resistance programs [40, 45, 46]. In primary breast tumors, collagen type I density is associated with breast cancer metastasis, and may serve as an imaging biomarker of metastasis. The expression of COL11A1 gene continuously increases during ovarian cancer disease progression, with the highest expression in recurrent metastases. Knockdown of COL11A1 decreases in vitro cell migration and invasion and tumor progression in mice [47]. The tumor-stromal interface of breast primary tumors in 3D culture was studied with second harmonic generation, showing that randomly organized matrix realigned the collagen fibers allowing individual cells to migrate out along radially aligned fibers [48].

Elastic fibers are generally twisted or straight strands of 0.2–1.5 μm which sometimes branch to form a coarse network in loose connective tissue or form flattened sheets in dense elastic tissues such as the aorta [39]. As basic structural elements, elastic fibers aberrations trigger severe pathologies such as Marfan's syndrome, emphysema, hypertension, actinic elastosis, and aortic aneurysms [49]. Nevertheless, little evidence has been reported about its role in neoplastic processes. It has been found that elastic fiber increase, together with a decrease in collagen fibers is associated with oral squamous cell carcinoma and lymph node metastasis [50]. Elastin-rich lung ECM is largely remodeled during tumor invasion. The degradation of elastin produces peptides displaying a wide range of biological activities and increase invasive capacities of lung cancer cells by post-transcriptional regulation of metalloproteinase-2 [51]. This mechanism has also been found to act in melanoma progression, another cancer associated with rich elastin microenvironment [52].

2.3. Vascular component

Depending on their size and composition, blood vessels can be divided in capillaries (5–15μm), post-capillaries/metarterioles (15–20μm), sinusoids (20–50μm), venules/ arterioles (50–200μm), and veins/ arteries (>200μm). In order to grow over the limit of oxygen and nutrients diffusion, tumors have developed different strategies to provide blood supply. These mechanisms are sprouting angiogenesis: the growth of new capillary vessels out of preexisting ones; intussusceptive angiogenesis: the division of preexisting vessels in two new vessels by the formation of transvascular tissue in the lumen of the vessels; recruitment of endothelial progenitor cells or angioblasts; vessel co-option: tumor cells can grow along existing vessels without evoking an angiogenic response; vasculogenic mimicry: tumor cells dedifferentiate to an endothelial phenotype and make tube-like structures, providing tumors with a secondary circulation system; mosaic vessels: both endothelial cells and tumor cells form the luminal surface of the vessels [53–55].

In many aspects, tumor vessels are different from normal vessels. They are dilated, tortuous, and poorly covered by pericytes [56, 57]. The fact that tumor growth is dependent on angiogenesis has given rise to anti-angiogenic therapies targeting different pro-angiogenic molecules [58]. The tumor microenvironment comprises numerous signaling molecules and pathways that influence the angiogenic response. Understanding how these components functionally interact as angiogenic stimuli or as repressors and how mechanisms of resistance arise is required for the identification of new therapeutic strategies [59].

In some malignancies, different studies have shown conflicting results, some indicating a prognostic value of angiogenesis and others rejecting such conclusions [60–64]. This may arise from the fact that tissue vascularization has been quantified following different methods [62, 63], all based on the detection of the differential staining of the vessels with more or less specific immunostaining (anti-factor VIII, von Willebrand factor, CD34, CD31, caveolin or CD105) on image sections [65], hot spots [60], whole sections [62], or in tissue microarrays sections [64], thereby indicating the need for robust morphometric techniques which may standardize the measurement of angiogenesis.

2.4. Immune system

The immune system is composed of different lineages of immune cells with different functions that protect every organism against infection. Among its tasks, immunosurveillance impedes transformed cells to become neoplastic cells. Indeed, increasing literature support the hypothesis that cancer development is influenced by the host immune system [66]. In fact, specialized blood vessels, the high endothelial venules (HEV), allowing the extravasation of lymphocytes, are present in primary human solid cancers [67, 68]. Therefore, it has been postulated that analyzing the composition, distribution, and architecture of the immune infiltrate for each tumor type, will offer new prognostic or predictive biomarkers [68, 69].

Immune infiltrates are heterogeneous between tumor types, and are diverse from patient to patient. All immune cell types may be found in a tumor, including macrophages, dendritic cells (DC), mast cells, natural killer (NK) cells, naïve and memory lymphocytes, B cells, and T lymphocytes (which include various subsets of T cell: TH1, TH2, TH17, regulatory T cells (TREGS), T follicular helper cells (TFH), and cytotoxic T cells). The analysis of the location, density, and functional orientation of different immune cell populations (termed the immune contexture) in large collections of annotated human tumors has allowed the identification of components that are beneficial for patients and those that are deleterious [70–72]. The prognostic impact of immune cells such as B cells, NK cells, myeloid derived suppressor cells (MDSC), macrophages, and subset of Thelper populations (TH2, TH17, TREG cells) may differ depending on the type of cancer, and on the cancer stage [72]. In contrast, cytotoxic T cells, TH1 cells, and memory T cells were strongly associated with good clinical outcome for all cancer types [66, 72]. Notably, two large studies have shown that tumor immune infiltrate patterns and subsets in colorectal cancer are significant prognostic biomarkers [73, 74]. A potential clinical translation of these observations is the establishment of an Immunoscore, based on the numeration of two lymphocyte populations (CD3/CD45RO, CD3/CD8 or CD8/CD45RO), both in the core and in the invasive margin of tumors, as a clinically useful prognostic marker [75]. This immunoscore sheds light into the prognostic role of the tumoral immune infiltrate, but still needs to be validated in colorectal cancer. Its utility in other malignancies should be tested as well.

3. Methods for the quantification of ECM elements above mentioned

The methods explained are based on the experience of the group in the study of the ECM in neuroblastic tumor samples.

3.1. Samples

The study of tissue microarrays (TMAs) have emerged as a tool for rapid analysis for diagnostic and prognostic studies because several markers can be tested in huge amounts of samples [76–

79]. The main advantage of TMAs consists in the standardization of the methods, given that all samples contained in a single TMA are subject to the same protocols. TMA are mainly used to validate biologic markers with potential diagnostic value, which can be used to develop screening programs or enhance a subclassification of a disease [80]. The steps to follow are summarized in **Figure 1**. Alternatively, whole slides for individual samples can be used.

Figure 1. Steps followed to construct a TMA. **A)** Hematoxylin & eosin of a sample with two representative areas selected and their corresponding location in the paraffin block (already perforated). **B)** Beecher Instrument (Silver Springs, MD) and the different parts composing it (1 and 2: fixed plate and receptor block; 3: small needle to perforate the receptor block; 4 and 5: mobile plate with one donor block and its hematoxylin & eosin to be able to locate the representative regions selected by the pathologist; 6: big needle to extract cylinders from donor blocks; 7: fingerwheel to move in the X axis; 8: fingerwheel to move in the Y axis. **C)** TMA obtained where each sample is represented by 2 cylinders and 2 control cylinders are placed asymmetrically.

3.2. Stainings

Several serial sections of 3μm can be made and stained with histochemistry (HC) for GAGs, reticulin fibers and collagen type I fibers, and with immunohistochemistry (IHC) anti-CD31 for blood vessels.

Alcian blue pH 2.5 Stain Kit (Artisan™, Dako) stains acid GAGs (HS, CS, DS, and HA), sulfomucins and sialomucins in blue, the nucleus in red and the cytoplasm in pale pink (**Figure 2**).

Reticulin stains are silver stains based on the argyrophilic properties of reticulin fibers and a slightly modified Gomori can be used, which stains reticulin fibers in black (**Figure 3**). The first step in the staining procedure consists of oxidation of the hexose sugars in reticulin fibers to yield aldehydes. The second step is called sensitization in which a metallic compound such as ammonium sulfate is deposited around the reticulin fibers, followed by silver impregnation in which an ammonical or diamine silver solution is reduced by the exposed aldehyde groups to metallic silver. Further reduction of the diamine silver is achieved by transferring the sections to formaldehyde; this step is called revealing. The last step consists of toning by gold chloride in which the silver is replaced by metallic gold and the color of the reticulin fibers changes from brown to black. Masson's trichrome stain consists in the sequential staining with Harris hematoxylin which stains nuclei dark red, aniline blue which stains collagen and reticulin blue, and molybdic orange G which stains erythrocytes dark orange (**Figure 4**). Orcein is a natural dye obtained from lichens which are found to stain elastic fibers dark brown (**Figure 5**).

CD31 is a single chain type 1 transmembrane protein with a molecular mass of approximately 135 kDa, belonging to the immunoglobulin superfamily. CD34 can also be applied to a subgroup, but this marker also stains cells other than endothelial. CD31 is expressed on endothelial cells of epithelial origin (all continuous endothelia, including those of arteries, arterioles, venules, veins, and capillaries, but it is not completely expressed on discontinuous endothelium in, for example, splenic red pulp). In addition, CD31 is expressed diffusely on the surfaces of megakaryocytes, platelets, myeloid cells, natural killer cells, and some subsets of T cells, as well as on B-cell precursors. Cells labeled by the antibody predominantly display membrane staining with weaker cytoplasmic staining (**Figure 6**).

Finally, IHC is used to detect diverse subpopulations of lymphocytes in tumoral tissues. For example, CD45 is a transmembrane glycoprotein expressed on most nucleated cells of hematopoetic origin, *i.e.*, all human leucocytes; CD20 reacts with an epitope located on the surface of B cells and appears early during B-cell maturation; CD3 is a pan-T cell marker for identification of T cells. It is well-suited for labeling reactive T cells in tissue with lymphoid infiltrates, and for classification of T-cell neoplasms; CD7 is expressed by the majority of peripheral blood T cells, NK cells, and all thymocytes. It is one of the earliest surface antigens on T and NK-cell lineages; CD4 is a transmembrane glycoprotein, expressed on normal thymocytes, T-helper cells, majority of mature peripheral T cells, and a subset of suppressor or cytotoxic T cells; CD8 is a 68 kDa transmembrane glycoprotein expressed as a heterodimer by a majority of thymocytes, and by class I major histocompatibility complex restricted, mature, suppressor/cytotoxic T cells; CD68 labels human monocytes and macrophages, but not myeloid cells; CD163 has been shown to mark cells of monocyte/macrophage lineage; CD11b is expressed on the surface of many leukocytes including monocytes, neutrophils, natural killer cells, granulocytes and macrophages, as well as on 8% of spleen cells and 44% of bone marrow cells; and CD11c is expressed prominently on the plasma membranes of monocytes, tissue macrophages, NK cells, and most dendritic cells (**Figure 7**).

3.3. Evaluation of the samples

3.3.1. Subjective assessment

The samples contained in TMAs or in whole slides can subjectively be analyzed by a pathologist to assess the amount of each ECM element as non-informative: artefact, scant material, lost cylinder; negative: no expression or <5% of stained area is detected; positive 1+: mild staining, 5–10% of the area; positive 2+: moderate staining, 10–50% of the area.; and positive 3+: strong staining, >50% of the area.

3.3.2. Automated quantification

The growing size and number of medical diagnostic images requires the use of computer-automated segmentation algorithms for the delineation of ECM structures of interest, which play a vital role in the research of new biomedical-imaging markers [81]. This is critical to analyzing whole sets of HC and IHC images to identify, describe, and quantify tissue alterations of the ECM, and require digitation of the samples. Morphometric techniques based on computer-automated segmentation algorithms attempt to decrease human error, increase efficiency, assess large areas or huge amounts of tumor samples, create reproducible results, and help to standardize the measurements [81].

3.3.2.1. Digitization of the sections

The digitization of the samples is required to perform morphometric analysis. The method available in most laboratories is the capture of several images per sample with a photomicroscope. Sequential photos can be done at 20x or 40x magnifications with a photomicroscope and then carefully merged with Adobe Photoshop to reconstruct a single whole cylinder image (**Figure 2**). Following our experience, this method needs approximately two weeks to digitize 30 samples (2 cylinders per sample) for a single staining at 20x magnification, or approximately 6 to 7 weeks at 40x magnification.

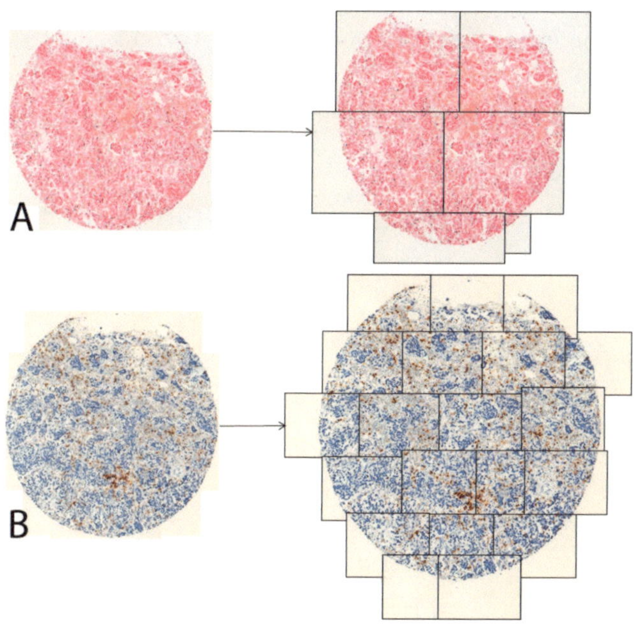

Figure 2. Process of hematoxylin & eosin and CD31 immunostaining TMA image capture. **A)** At 20x magnification, 6 individual images of the cylinder must be captured to reconstruct the whole cylinder image. **B)** The whole 1mm cylinder image at 40x magnification must be reconstructed from 20 individual images.

The use of a slide scanner is the only method that permits the digitization of a high number of cases for a reasonable time when developing a routine image biobank [82, 83] or a research project. Additionally, the use of slide scanners provides the possibility to standardize the image quality using preserved light conditions.

In order to enhance the standardization of the image capture conditions and therefore, the quality of the measurements and to save time, whole slide scanning is advised, as described next. Our group used the ScanScope XT scanner, Aperio technologies, but increasing alternatives such as Panoramic Midi from3D Histech and the Ventana iScann Coreo Au from Roche, among others can also be considered [82, 84]. ScanScope XT scanner is a brightfield scanner that digitizes whole sections at 20x or 40x magnification providing high resolution images in approximately 15 to 30 minutes per slide, depending on the magnification and the size of the tissue. 40x magnification was used, originating images with a resolution of 0.25 μm/pixel. Given to the enormous amount of pixels scanned, the images were compressed in JPEG2000 to 100–200 megabytes for the average size of a TMA, and saved to a proprietary TIFF format (SVS).

The sections were placed in a mobile plate one by one. An option for TMAs digitization provided by the scanner driver was used. This option recognizes the tissue cylinders and places several points per sample where the objective adjusts the focus to obtain clear and focused images. The mobile plate with the section moves in consecutive stripes, until it sweeps the whole section while the objective scans. Individual scanned stripes are originated and stitched together automatically to reconstruct the whole image, which can be visualized up to 40x magnifications with the Image Scope viewer software (Aperio technologies). The process is briefly shown in **Figure 3**.

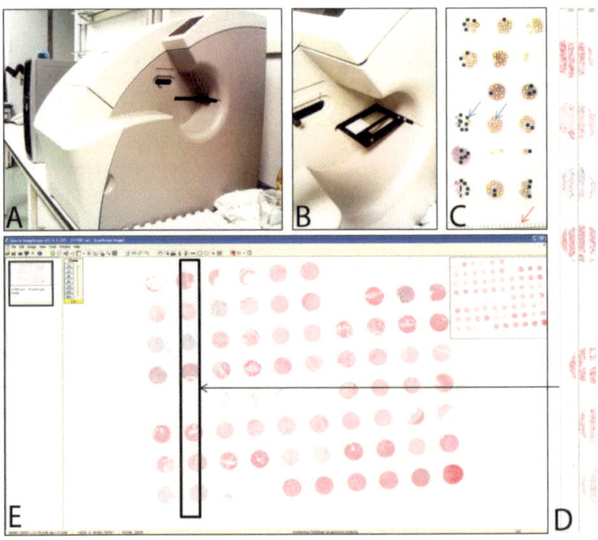

Figure 3. Digitization process. **A)** ScanScope XT, Aperio technologies. **B)** Mobile plate with one section on it. The plate is introduced in the scanner and placed under the objective to start the process. **C)** Preview of an area of the section. Blue arrows show the pre-set points where the scanner shall readjust the focus. Green points have already been properly focused and yellow points still have to be focused. The red arrow shows different marks corresponding to the different horizontal stripes which are going to be scanned individually and then stitched together. **D)** 4 single consecutive stripes are shown corresponding to the section in E. **E)** All the stripes are stitched to form a single image of the whole section. The image is opened in the free viewer ImageScope, Aperio technologies.

3.3.2.2. Design of automated image analysis algorithms

Depending on the staining to be measured and the availability of morphometric systems, different methods for image analysis can be used, all following a common workflow (**Figure 4**).

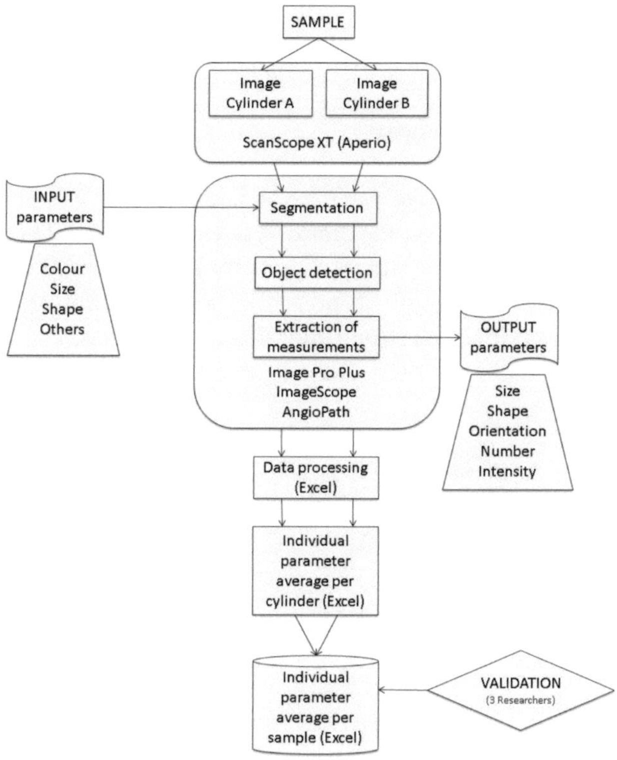

Figure 4. Flowchart showing the multi-resolution image analysis system for TMA with two cylinders per sample. Images belonging to different samples stained with different markers have been quantified by image analysis following a common process including segmentation (differential recognition of the staining) with specific input parameters for each marker and method and extraction of some given parameters. Adapted from Tadeo *et al.* [85].

OPTION 1: To use commercial software for image analysis with established protocols which the user can adapt to the marker specific color.

Example 1: Algorithm for the fundamental substance (GAGs). Each cylinder was analyzed individually with Aperio Positive Pixel Count Algorithm on Aperio ScanScope software by selecting each cylinder outline in the whole section image. This algorithm was used with the default parameters and counted the number of pixels belonging to a given staining intensity, being the intensity of each pixel the average between the values of red, green, and blue (RGB) intensities. It recognizes Alcian blue stained pixels which have average intensities over 221. Nuclear fast red stained pixels were detected with weak, middle, and strong intensities painted in yellow, orange, and brown following default parameters (**Figure 5**). The remaining customizable parameters were set by default.

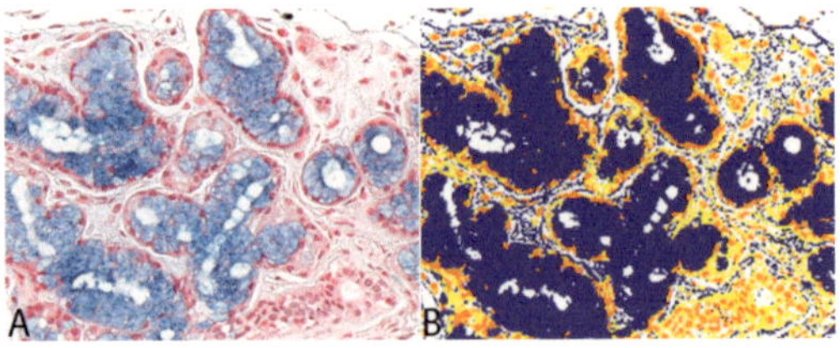

Figure 5. Segmentation obtained with the positive pixel count algorithm on a salivary gland. **A)** Original image. **B)** Mark-up image after segmentation. Alcian blue staining is marked in blue and nuclear fast red is marked in yellow, orange or maroon, depending on the intensity of staining.

Example 2: Algorithm for the immune system cells. Each cylinder was analyzed individually with the Nuclear Quant algorithm on Panoramic Viewer software (3D Histech) by selecting each cylinder outline in the whole section image. This algorithm was used with the default parameters and counted the number of cells presenting a given staining intensity. Stained cells were detected with weak, middle, and strong intensities painted in yellow, orange, and maroon (**Figure 6**).

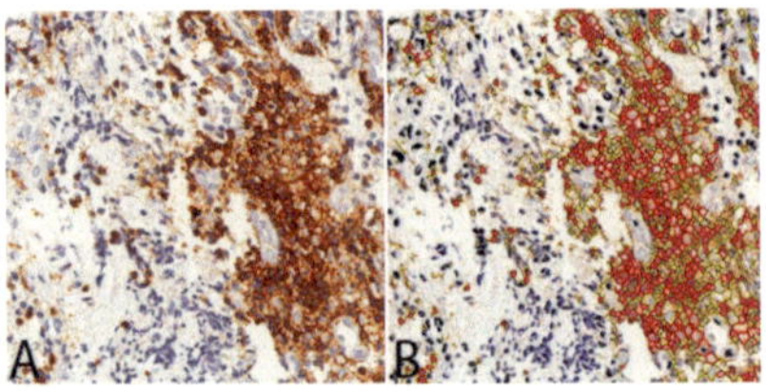

Figure 6. Segmentation of a neuroblastic tumor sample stained with IHC anti-CD45. **A)** Original image. **B)** Mark-up image after segmentation. Immune system cells are marked in yellow, orange or maroon, depending on the intensity of staining blue and hematoxylin is marked in blue.

OPTION 2: To use commercial software for image analysis which allows the user to configure personal image analysis protocols or macros capable of recognizing and describing the specific color and shape of the element of interest.

Example: Algorithm for the reticulin fibers. The grading of fibrosis in general and namely of reticulin fibers is of main interest in bone marrow pathologies [86] and fibrous diseases of the liver [87]. For this reason, different methods for the quantification of reticular fibrosis have been developed, some of them consisting in automated morphometry [88–93]. However, these methods quantify the percentage of stained area and we considered that not only the amount of fibers, but also the morphometric features were relevant, given that these are usually subjectively assessed. Image Pro-Plus software (Media Cybernetics), which enables the design of specific algorithms, was used to analyze the fibrous component. An image of every cylinder was extracted in a separate JPEG-quality 80 image from the whole section scan. An algorithm capable of specifically detecting the reticulin fibers and measure their amount, size, and shape was customized. Image J is a free software providing similar options. An example of the segmentation process in a control tissue (kidney) is provided by **Figure 7**.

http://dx.doi.org/10.5772/62479

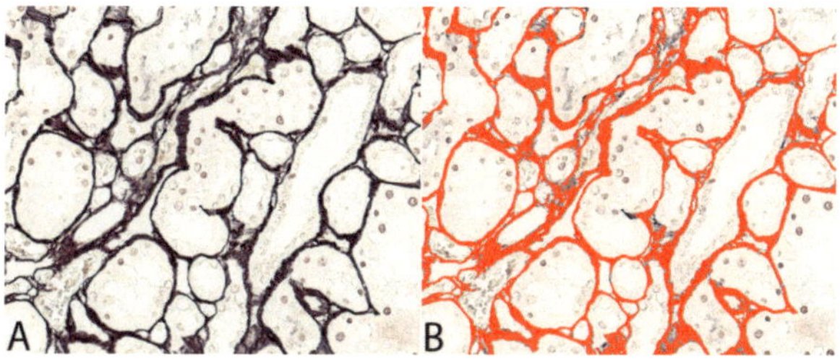

Figure 7. Segmentation process on a kidney tissue. **A)** Original image. **B)** Image after segmentation. The reticulin fibers recognized by the algorithm are marked-up in red.

OPTION 3: To design a personal application capable of solving commercial softwares lacks.

Example: Algorithm for the vascular component. A common feature of tumor vessels studies is that the researchers focus on microvessel density overlooking other parameters that might be significant, such as the size and shape of the microvessels [94]. Studies have revealed the importance of the size and shape of blood vessels in, for instance, laryngeal tumors [95]. The morphometric tool Angiopath closes blood vessels with discontinuous endothelial layer, recognizes all blood vessels and classifies them in six categories corresponding to different types of vessels, differentiated by their largest diameter or length. Density (density and occupied area), size (area, width, length, and perimeter), and shape (perimeter-ratio, shape index, branching, aspect, roundness, and deformity) parameters are extracted [96, 97]. An example of the segmentation process is provided in **Figure 8**.

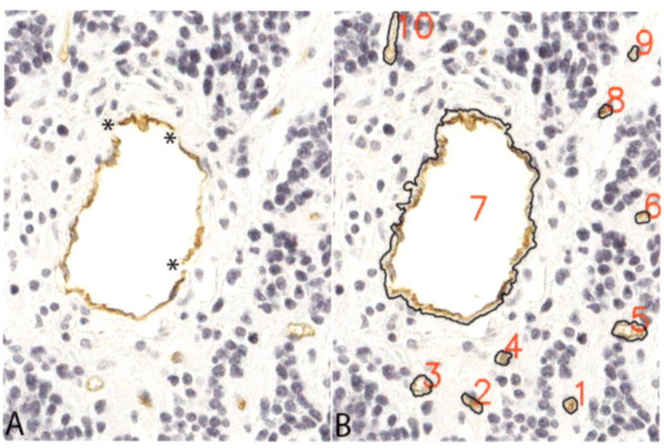

Figure 8. Segmentation process on a neuroblastic tumor sample immunostained with CD31. **A)** Original image. **B)** Image after segmentation. Note that the big blood vessel with an interrupted staining of the endothelial cells surrounding the vascular lumen (asterisk) has been closed, thus providing morphometric measures.

3.3.2.3. Spatial distribution of microenvironmental components

Topological network analysis and the graph theory in combination with Vonoroi tessellations [98] have recently been found to be useful in the diagnosis of muscular dystrophies and neurogenic atrophies, in the classification of neuromuscular disease or to model the progression of dementia [99–102]. All the generated information is subject to capture relevant information about the organization of different tissue markers.

3.3.2.4. Texture analysis

Another novel approach to cancer research is the texture analysis of different tissue images and machine-learning methods to train automated algorithms which find different patterns in the tissue capable of discriminating between prognostic groups, among other variables, resulting in computer-aided diagnosis tools [103–106].

4. Therapeutic implications of the study of ECM elements

The accumulation of non-fluid elements in the tumor ECM, understood as an increased density of fibrous elements with huge crosslinking, carries the establishment of a solid pressure within the tumor tissue, which, together with other pressure sources, leads to an elevated tumor interstitial pressure [107]. Elevated tumor pressure alters gene expression, cancer cell proliferation, apoptosis and invasiveness, stromal cell function, and ECM synthesis and organization [108–110]. Additionally, it has a negative impact on cancer treatment and is known to reduce the efficacy of cancer therapy through a reduced uptake and heterogeneous distribution of drugs [111, 112]. The direct intratumoral injection of anticancer agents has been evaluated extensively in the past decades, as one approach to counteract such intratumoral pressure [113]. For example, is has been shown that intratumoral injection of immunostimulatory oligonucleotides induces differentiation and reduces immunosuppressive activity of myeloid-derived suppressor cells in mice inoculated with colon cancer cell lines [114]. Regarding ECM fibers, therapies that attack some of the major physical changes themselves, such as inhibiting matrix crosslinking, could be developed to revert pressure within the tumor tissue [115]. Most of the therapeutic strategies under study are based on altering the fibroblastic pathways of collagen synthesis and inhibiting fibroblast growth factor [116–118]. Other strategies focusing on the destruction of existing fibrosis are being developed. In this sense, collagenase has been shown to reduce tumor interstitial pressure by cleaving collagen and consequently destroying the collagen crosslinked networks in human osteosarcoma xenografts [119]. Moreover, this enzyme could indirectly affect tumor ECM by degrading the reticulin fibers associated with blood vessels. Another therapeutic approach not affecting the amount of fibers but the morphology consists in impeding an increased crosslinking. The reduction of lysyl-oxydase-like-2, catalyzing the production of a crosslinked collagenous matrix, by the monoclonal antibody AB0023, was efficacious in reducing disease in different models of cancer and fibrosis [115]. Incorporating biomechanical forces into our current picture of the tumor microenvironment, through the study of the interaction and the status of different ECM elements, fundamentally changes the context in which we study the basic cell and molecular biology of cancer. If there is one core message from the last two decades of research into the tumor ECM, it is that context matters [120].

Solid stresses affect tumor pathophysiology directly by compressing cancer and stromal cells but also indirectly by deforming blood and lymphatic vessels as described in pancreatic cancer, where compressed vessels are observable at the histological level [111]. The compression of tumor vasculature also leads to a reduction in tumor blood flow and subsequently reduces the anticancer agent transport to tumor cells [121]. Therapeutic chemicals are normally transported through the interstitial space by convection, a transport process that is dependent on pressure gradients, but increased tumor vessel permeability creates a less steep gradient along the vessels. This leads to a reduction in convective transport across the vessels in tumor bulk [122]. Additionally, an increase in vessels permeability enables the extravasation of stem and immune cells, among others, and the entrance of malignant cells into the bloodstream to promote metastasis [123]. Moreover, abnormal and tortuous tumor vasculature causes blood stasis, which leads to the reduction of oxygen and blood flow in tumors causing hypoxia and subsequently ischemia and necrosis [121]. This effect increases anabolic metabolism which produce an acid ECM that degrades or deactivates some therapeutic drugs and renders them ineffective [124]. Hypoxia also can cause resistance to radiotherapy and can help tumor cells to escape immunosurveillance [45]. Because of its importance to the tumor, tumor vasculature represents an "Achilles' heel" that can be used for cancer therapy [125]. Indeed, multiple antiangiogenic therapeutic strategies have been developed in the last decades for many different malignancies [57]. Several direct angiogenic inhibitors (angiostatin, endostatin, and thrombospondin) targeting endothelial cells and indirect anti-antiangiogenic agents blocking the production or activity of pro-angiogenic molecules, such as vascular endothelial growth factor (VEGF), have been developed [126]. In this case, again intratumoral injections of adenoviral vectors encoding tumor-targeted immunoconjugates induce a cytotoxic immune response against the neovasculature endothelial cells and tumor cells in mice xenografted with human melanoma cell lines [127]. Recent studies have focused on the value of normalizing tumor vasculature to improve responses to conventional anticancer therapies rather than destroying tumor vessels to starve primary tumors from oxygen and induce tumor shrinkage [128]. Vessel normalization results in reduced vessel diameter, increased pericyte coverage, and normalized basement membrane, accompanied by normalization of its function. This normalized tumor vasculature becomes less permeable and tortuous and leads to reduced fluid and protein extravasation into the interstitium, resulting in the decrease tumor interstitial pressure [129, 130]. For example, various preclinical studies have shown that the high levels of VEGF observed in tumors induce vessel abnormalities [131, 132]. Therefore, targeting VEGF using anti-VEGF antibodies and inhibitors such as bevacizumab has been shown to reduce tumor interstitial pressure [133, 134]. Another protein playing an important role in tumor angiogenesis is PDGF. Imatinib, a PDGFR-inhibitor has been shown to reduce tumor interstitial pressure and enhance therapeutic efficacy [135]. Other drugs such as hydralazine and cachectin are vasodilators that cause a decrease in vascular resistance followed by an increase in tumor blood flow, which can potentially improve intratumoral transport of macromolecules and that have been shown to reduce intratumoral pressure [136, 137]. In addition to these therapeutic agents, some physical approaches have been tested to decrease tumor ECM pressure that can be summarized as follows: irradiation reduces the vascular wall permeability to fluid and leads to a reduction of intratumoral pressure [138]; induced hyperthermia or hypothermia have shown similar results regarding the reduction of intratumoral pressure as for irradiation [139]; ultrasounds due to mechanical (cavitation) and thermal pressure cause damage to tumor cells and ECM, which increase the interstitial hydraulic conductivity, reduces matrix tension and enhances tumor blood flow [140]; hyperbaric oxygen increases oxygen tension and delivery to tissues [141]; finally, photodynamic therapy impaired microcirculation in tumors [142]. All these strategies are possible by the knowledge of the physiopathology of vascularity in tumors ECM. Most of them are in the experimental phases and others are already in clinical trials.

In the specific case of neuroblastic tumors, many relevant processes, such as stem cell differentiation, neuronal maturation, neurite extension, MYCN expression and the malignant potential and phenotype of tumor cells have all been shown to be influenced by ECM stiffness [1]. In this respect, our group performed a preliminary study where some differences regarding ECM elements could be described between favorable and unfavorable neuroblastic tumors [85]. This study was extended to a cohort of more than 500 samples where a stiffer, crosslinked and less porous ECM was mainly found in unfavorable tumors [143, 144]. Additionally, unfavorable tumors presented a highly vascularized ECM with mostly sinusoid vessels with abnormal morphologies [145]. Some ECM elements morphometric features helped the description of an ultra-high risk subgroup which could beneficiate of novel therapies [146].

Cancer cells often express a variety of abnormal proteins that can serve as targets for an immune response. Although spontaneous immune responses to these antigens can occur, these reactions are rarely sufficient to cause tumor regression; however, the local administration of immune-activating agents can induce tumor-associated inflammation and protective immunity. By binding to their targets, antibodies exercise their functions through several effector mechanisms, including steric inhibition and neutralization, complement activation, and activation of cell-mediated cytotoxicity. Each of these mechanisms may play a role in the antitumor activity of monoclonal antibodies; however, at present the relative importance of these mechanisms is not completely clear. In general, immune adjuvant–based therapies have only proven effective against early stage tumors; yet in this context they can be remarkably effective with minimal risk of serious adverse reactions [147]. Advances in immunotherapy have changed the management of several malignancies, being monoclonal antibodies the most specific targeted therapies currently in use [148]. Therefore, understanding how tumor cells develop immune escape mechanisms and create microenvironments that improve their survival and dissemination and to reach assay normalization seems crucial to achieving optimal treatment [69, 149, 150]. Most notable has been the ability of the anti-CTLA4 antibody, ipilimumab, to achieve a significant increase in survival for patients with metastatic melanoma, for which conventional therapies have failed. In the context of advances in the understanding of how tolerance, immunity and immunosuppression regulate antitumor immune responses together with the advent of targeted therapies, suggesting that active immunotherapy represents a path to obtain a durable and long-lasting response in cancer patients [151]. Five monoclonal antibodies are clinically approved for the treatment of hematologic tumors such as chronic lymphocytic leukemia (alemtuzumab), acute myelogenous leukemia (gemtuzumab), and non-Hodgkin's lymphoma (rituximab, ibritumomab tiuxetan, and tositumomab)

[147]. Cetuximab and panitumumab increase progression-free survival in patients with metastatic colorectal cancer who have previously failed standard chemotherapy and are associated with 10–20% and 10% response rates, respectively [152, 153]. Trastuzumab was the second monoclonal antibody approved for cancer therapy and is used, either alone or in combination with paclitaxel, for the treatment of invasive, HER2/*neu*-positive breast cancer, which represents approximately 20–30% of invasive breast cancers with a response rate of 50% when used in combination with chemotherapy [154]. Bevacizumab is the only monoclonal antibody with anticancer activity that does not directly target malignant cells. Instead, bevacizumab binds VEGF, a critical mediator of tumor angiogenesis. Inhibiting angiogenesis slows the delivery of nutrients and oxygen to tumors, inhibiting growth without severely compromising normal tissue function [155, 156].

5. Conclusions

In cancer biology, tumors are described as complex tissues comprised of heterogeneous neoplastic cells interwoven with tumor associated stroma. The characterization of the ECM elements associated with the tumoral and stromal cells presents opportunities for targeted therapeutic intervention. However, the heterogeneity of tumoral elements dictates that, in order to achieve successful clinical treatment, it is necessary to employ a combination of targeted therapies.

Novel and accurate image analysis algorithms must be developed and used to further investigate about tumor ECM composition and conformation and its relationship with different malignancies prognosis. Digital pathology approaches based on automated image analysis are necessary to obtain standardization, reproducibility, and eliminate observers' biases.

The definition of favorable and unfavorable ECM characteristics, together with the understanding of the fact that biomechanical forces affect cell-cell and cell-matrix crosstalk and alter tumor cells fate, suggests the possibility of developing new therapies that can target the behaviors that arise from these complex interactions and from the heterogeneity of tumor tissues.

Acknowledgements

The authors want to thank Marcial García Rojo, Gloria Bueno and Jose Benavent for technical support on digital pathology. This study was supported by the FIS contracts PI10/00015 and PI14/01008 and RTICC contract RD12/0036/0020, grants from the ISCIII & FEDER (European Regional Development Fund), Spain. The authors declare no conflict of interest.

Author details

Irene Tadeo[1*], Tomás Álvaro[2], Samuel Navarro[1] and Rosa Noguera[1]

*Address all correspondence to: irenetadeo@hotmail.com

1 Pathology Department, Medical School, University of Valencia-INCLIVA, Avda. Blasco Ibañez 15, Valencia, Spain

2 Department of Pathology, Hospital de Tortosa, Verge de la Cinta, IISPV, URV, Tortosa, Spain

References

[1] Lam WA, Cao L, Umesh V, Keung AJ, Sen S, Kumar S. Extracellular matrix rigidity modulates neuroblastoma cell differentiation and N-myc expression. Mol Cancer. 2010;9:35.

[2] Noguera R, Nieto OA, Tadeo I, Farinas F, Alvaro T. Extracellular matrix, biotensegrity and tumor microenvironment. An update and overview. Histology and histopathology. 2012;27(6):693–705.

[3] Kim SH, Turnbull J, Guimond S. Extracellular matrix and cell signalling: the dynamic cooperation of integrin, proteoglycan and growth factor receptor. J Endocrinol. 2011;209(2):139–51.

[4] Hu M, Polyak K. Microenvironmental regulation of cancer development. Curr Opin Genet Dev. 2008;18(1):27–34.

[5] Huang S, Ingber DE. A non-genetic basis for cancer progression and metastasis: self-organizing attractors in cell regulatory networks. Breast Dis. 2006;26:27–54.

[6] Tadeo I, Berbegall AP, Escudero LM, Alvaro T, Noguera R. Biotensegrity of the extracellular matrix: physiology, dynamic mechanical balance, and implications in oncology and mechanotherapy. Front Oncol. 2014;4:39.

[7] Tlsty TD, Coussens LM. Tumor stroma and regulation of cancer development. Annu Rev Pathol. 2006;1:119–50.

[8] Stasinopoulos I, Penet MF, Chen Z, Kakkad S, Glunde K, Bhujwalla ZM. Exploiting the tumor microenvironment for theranostic imaging. NMR Biomed. 2011;24(6):636–47.

[9] Aguilar-Cuenca R, Juanes-Garcia A, Vicente-Manzanares M. Myosin II in mechanotransduction: master and commander of cell migration, morphogenesis, and cancer. Cell Mol Life Sci. 2013.

[10] DuFort CC, Paszek MJ, Weaver VM. Balancing forces: architectural control of mechanotransduction. Nat Rev Mol Cell Biol. 2010;12(5):308–19.

[11] Chandler EM, Seo BR, Califano JP, Andresen Eguiluz RC, Lee JS, Yoon CJ, et al. Implanted adipose progenitor cells as physicochemical regulators of breast cancer. Proc Natl Acad Sci U S A. 2012;109(25):9786–91.

[12] Kalluri R, Zeisberg M. Fibroblasts in cancer. Nat Rev Cancer. 2006;6(5):392–401.

[13] Otranto M, Sarrazy V, Bonte F, Hinz B, Gabbiani G, Desmouliere A. The role of the myofibroblast in tumor stroma remodeling. Cell Adh Migr. 2012;6(3):203–19.

[14] Delnero P, Song YH, Fischbach C. Microengineered tumor models: insights & opportunities from a physical sciences-oncology perspective. Biomed Microdevices. 2013.

[15] Netti PA, Berk DA, Swartz MA, Grodzinsky AJ, Jain RK. Role of extracellular matrix assembly in interstitial transport in solid tumors. Cancer Res. 2000;60(9):2497–503.

[16] Chandler EM, Saunders MP, Yoon CJ, Gourdon D, Fischbach C. Adipose progenitor cells increase fibronectin matrix strain and unfolding in breast tumors. Phys Biol. 2011;8(1):015008.

[17] Hinz B. Tissue stiffness, latent TGF-beta1 activation, and mechanical signal transduction: implications for the pathogenesis and treatment of fibrosis. Curr Rheumatol Rep. 2009;11(2):120-6.

[18] Levental KR, Yu H, Kass L, Lakins JN, Egeblad M, Erler JT, et al. Matrix crosslinking forces tumor progression by enhancing integrin signaling. Cell. 2009;139(5):891–906.

[19] Mammoto A, Connor KM, Mammoto T, Yung CW, Huh D, Aderman CM, et al. A mechanosensitive transcriptional mechanism that controls angiogenesis. Nature. 2009;457(7233):1103-8.

[20] Willis AL, Sabeh F, Li XY, Weiss SJ. Extracellular matrix determinants and the regulation of cancer cell invasion stratagems. J Microsc. 2013;251(3):250–60.

[21] Pathak A, Kumar S. From molecular signal activation to locomotion: an integrated, multiscale analysis of cell motility on defined matrices. PloS one. 2011;6(3):e18423.

[22] Zaman MH, Trapani LM, Sieminski AL, Mackellar D, Gong H, Kamm RD, et al. Migration of tumor cells in 3D matrices is governed by matrix stiffness along with cell-matrix adhesion and proteolysis. Proc Natl Acad Sci U S A. 2006;103(29):10889–94.

[23] Carey SP, Kraning-Rush CM, Williams RM, Reinhart-King CA. Biophysical control of invasive tumor cell behavior by extracellular matrix microarchitecture. Biomaterials. 2012;33(16):4157–65.

[24] Delnero P, Song YH, Fischbach C. Microengineered tumor models: insights & opportunities from a physical sciences-oncology perspective. Biomed Microdevices.

[25] Pathak A, Kumar S. Independent regulation of tumor cell migration by matrix stiffness and confinement. Proc Natl Acad Sci U S A. 2012;109(26):10334-9.

[26] Cross SE, Jin YS, Rao J, Gimzewski JK. Nanomechanical analysis of cells from cancer patients. Nat Nanotechnol. 2007;2(12):780-3.

[27] Lekka M, Laidler P, Gil D, Lekki J, Stachura Z, Hrynkiewicz AZ. Elasticity of normal and cancerous human bladder cells studied by scanning force microscopy. Eur Biophys J. 1999;28(4):312-6.

[28] Karamanos NK, Tzanakakis GN. Glycosaminoglycans: from "cellular glue" to novel therapeutical agents. Curr Opin Pharmacol. 2012.

[29] Afratis N, Gialeli C, Nikitovic D, Tsegenidis T, Karousou E, Theocharis AD, et al. Glycosaminoglycans: key players in cancer cell biology and treatment#. FEBS J. 2012.

[30] Bai X, Zhou D, Brown JR, Crawford BE, Hennet T, Esko JD. Biosynthesis of the linkage region of glycosaminoglycans: cloning and activity of galactosyltransferase II, the sixth member of the beta 1,3-galactosyltransferase family (beta 3GalT6). J Biol Chem. 2001;276(51):48189–95.

[31] Zhang L. Glycosaminoglycan (GAG) biosynthesis and GAG-binding proteins. Prog Mol Biol Transl Sci. 2010;93:1–17.

[32] Victor XV, Nguyen TK, Ethirajan M, Tran VM, Nguyen KV, Kuberan B. Investigating the elusive mechanism of glycosaminoglycan biosynthesis. J Biol Chem. 2009;284(38): 25842–53.

[33] Theocharis AD, Skandalis SS, Tzanakakis GN, Karamanos NK. Proteoglycans in health and disease: novel roles for proteoglycans in malignancy and their pharmacological targeting. FEBS J. 2010;277(19):3904–23.

[34] Edwards IJ. Proteoglycans in prostate cancer. Nat Rev Urol. 2012;9(4):196–206.

[35] Sironen RK, Tammi M, Tammi R, Auvinen PK, Anttila M, Kosma VM. Hyaluronan in human malignancies. Exp Cell Res. 2011;317(4):383–91.

[36] Weyers A, Yang B, Yoon DS, Park JH, Zhang F, Lee KB, et al. A structural analysis of glycosaminoglycans from lethal and nonlethal breast cancer tissues: toward a novel class of theragnostics for personalized medicine in oncology? OMICS. 2012;16(3):79–89.

[37] Lowery JL, Datta N, Rutledge GC. Effect of fiber diameter, pore size and seeding method on growth of human dermal fibroblasts in electrospun poly(epsilon-caprolactone) fibrous mats. Biomaterials. 2010;31(3):491–504.

[38] Alberts B, Johnson A, Lewis J, Raff M, Roberts K, Walter P. Molecular Biology of the Cell. 5th ed. New York: Garland Science; 2008.

[39] Ushiki T. Collagen fibers, reticular fibers and elastic fibers. A comprehensive understanding from a morphological viewpoint. Arch Histol Cytol. 2002;65(2):109–26.

[40] Egeblad M, Rasch MG, Weaver VM. Dynamic interplay between the collagen scaffold and tumor evolution. Curr Opin Cell Biol. 2010;22(5):697–706.

[41] Ren YJ, Zhang S, Mi R, Liu Q, Zeng X, Rao M, et al. Enhanced differentiation of human neural crest stem cells towards the Schwann cell lineage by aligned electrospun fiber matrix. Acta Biomater. 2013;9(8):7727–36.

[42] Ng MR, Brugge JS. A stiff blow from the stroma: collagen crosslinking drives tumor progression. Cancer Cell. 2009;16(6):455-7.

[43] Fukuda T, Tsuneyoshi M. Adhesion proteins, cellular morphology and fibrous components around the cell/extracellular-matrix interface in myxoid liposarcomas. J Cancer Res Clin Oncol. 2000;126(6):320-4.

[44] Yu E, Lee I. Reticular network of the human thymus. J Korean Med Sci. 1993;8(6):431-6.

[45] Choi IK, Strauss R, Richter M, Yun CO, Lieber A. Strategies to increase drug penetration in solid tumors. Front Oncol. 2013;3:193.

[46] Sun Y, Nelson PS. Molecular pathways: involving microenvironment damage responses in cancer therapy resistance. Clinical cancer research : an official journal of the American Association for Cancer Research. 2012;18(15):4019–25.

[47] Cheon DJ, Tong Y, Sim MS, Dering J, Berel D, Cui X, et al. A collagen-remodeling gene signature regulated by TGFbeta signaling is associated with metastasis and poor survival in serous ovarian cancer. Clinical cancer research : an official journal of the American Association for Cancer Research. 2013.

[48] Provenzano PP, Eliceiri KW, Campbell JM, Inman DR, White JG, Keely PJ. Collagen reorganization at the tumor-stromal interface facilitates local invasion. BMC Med. 2006;4(1):38.

[49] Urban Z, Boyd CD. Elastic-fiber pathologies: primary defects in assembly-and secondary disorders in transport and delivery. Am J Hum Genet. 2000;67(1):4-7.

[50] Agrawal U, Rai H, Jain AK. Morphological and ultrastructural characteristics of extracellular matrix changes in oral squamous cell carcinoma. Indian J Dent Res. 2011;22(1):16–21.

[51] Toupance S, Brassart B, Rabenoelina F, Ghoneim C, Vallar L, Polette M, et al. Elastin-derived peptides increase invasive capacities of lung cancer cells by post-transcriptional regulation of MMP-2 and uPA. Clin Exp Metastasis. 2012;29(5):511–22.

[52] Devy J, Duca L, Cantarelli B, Joseph-Pietras D, Scandolera A, Rusciani A, et al. Elastin-derived peptides enhance melanoma growth in vivo by upregulating the activation of Mcol-A (MMP-1) collagenase. British journal of cancer. 2012;103(10):1562–70.

[53] Chang YS, di Tomaso E, McDonald DM, Jones R, Jain RK, Munn LL. Mosaic blood vessels in tumors: frequency of cancer cells in contact with flowing blood. Proc Natl Acad Sci U S A. 2000;97(26):14608–13.

[54] Hillen F, Griffioen AW. Tumour vascularization: sprouting angiogenesis and beyond. Cancer Metastasis Rev. 2007;26(3-4):489–502.

[55] Styp-Rekowska B, Hlushchuk R, Pries AR, Djonov V. Intussusceptive angiogenesis: pillars against the blood flow. Acta Physiol (Oxf). 2011;202(3):213–23.

[56] Bergers G, Benjamin LE. Tumorigenesis and the angiogenic switch. Nat Rev Cancer. 2003;3(6):401–10.

[57] Carmeliet P, Jain RK. Angiogenesis in cancer and other diseases. Nature. 2000;407(6801):249–57.

[58] Ichihara E, Kiura K, Tanimoto M. Targeting angiogenesis in cancer therapy. Acta Med Okayama. 2011;65(6):353–62.

[59] Weis SM, Cheresh DA. Tumor angiogenesis: molecular pathways and therapeutic targets. Nat Med. 2011;17(11):1359–70.

[60] Canete A, Navarro S, Bermudez J, Pellin A, Castel V, Llombart-Bosch A. Angiogenesis in neuroblastoma: relationship to survival and other prognostic factors in a cohort of neuroblastoma patients. Journal of clinical oncology : official journal of the American Society of Clinical Oncology. 2000;18(1):27–34.

[61] Jakovljevic G, Culic S, Stepan J, Kosuta I, Seiwerth S. Relationship between tumor vascularity and vascular endothelial growth factor as prognostic factors for patients with neuroblastoma. Coll Antropol. 2011;35(4):1071-9.

[62] Meitar D, Crawford SE, Rademaker AW, Cohn SL. Tumor angiogenesis correlates with metastatic disease, N-myc amplification, and poor outcome in human neuroblastoma. Journal of clinical oncology : official journal of the American Society of Clinical Oncology. 1996;14(2):405–14.

[63] Ozer E, Altungoz O, Unlu M, Aygun N, Tumer S, Olgun N. Association of MYCN amplification and 1p deletion in neuroblastomas with high tumor vascularity. Appl Immunohistochem Mol Morphol. 2007;15(2):181-6.

[64] Peddinti R, Zeine R, Luca D, Seshadri R, Chlenski A, Cole K, et al. Prominent microvascular proliferation in clinically aggressive neuroblastoma. Clinical cancer research : an official journal of the American Association for Cancer Research. 2007;13(12):3499–506.

[65] Ribatti D, Surico G, Vacca A, De Leonardis F, Lastilla G, Montaldo PG, et al. Angiogenesis extent and expression of matrix metalloproteinase-2 and -9 correlate with progression in human neuroblastoma. Life Sci. 2001;68(10):1161-8.

[66] Galon J, Pages F, Marincola FM, Angell HK, Thurin M, Lugli A, et al. Cancer classification using the Immunoscore: a worldwide task force. Journal of translational medicine. 2012;10:205.

[67] Martinet L, Garrido I, Filleron T, Le Guellec S, Bellard E, Fournie JJ, et al. Human solid tumors contain high endothelial venules: association with T- and B-lymphocyte infiltration and favorable prognosis in breast cancer. Cancer Res. 2011;71(17):5678–87.

[68] Rahir G, Moser M. Tumor microenvironment and lymphocyte infiltration. Cancer Immunol Immunother. 2012;61(6):751-9.

[69] Fridman WH, Galon J, Pages F, Tartour E, Sautes-Fridman C, Kroemer G. Prognostic and predictive impact of intra- and peritumoral immune infiltrates. Cancer Res. 2011;71(17):5601-5.

[70] Galon J, Fridman WH, Pages F. The adaptive immunologic microenvironment in colorectal cancer: a novel perspective. Cancer Res. 2007;67(5):1883-6.

[71] Wang E, Miller LD, Ohnmacht GA, Mocellin S, Perez-Diez A, Petersen D, et al. Prospective molecular profiling of melanoma metastases suggests classifiers of immune responsiveness. Cancer Res. 2002;62(13):3581-6.

[72] Fridman WH, Pages F, Sautes-Fridman C, Galon J. The immune contexture in human tumours: impact on clinical outcome. Nat Rev Cancer. 2012;12(4):298–306.

[73] Nosho K, Baba Y, Tanaka N, Shima K, Hayashi M, Meyerhardt JA, et al. Tumour-infiltrating T-cell subsets, molecular changes in colorectal cancer, and prognosis: cohort study and literature review. The Journal of pathology. 2010;222(4):350–66.

[74] Ogino S, Nosho K, Irahara N, Meyerhardt JA, Baba Y, Shima K, et al. Lymphocytic reaction to colorectal cancer is associated with longer survival, independent of lymph node count, microsatellite instability, and CpG island methylator phenotype. Clinical cancer research : an official journal of the American Association for Cancer Research. 2009;15(20):6412–20.

[75] Pages F, Kirilovsky A, Mlecnik B, Asslaber M, Tosolini M, Bindea G, et al. In situ cytotoxic and memory T cells predict outcome in patients with early-stage colorectal cancer. Journal of clinical oncology : official journal of the American Society of Clinical Oncology. 2009;27(35):5944–51.

[76] Wang H, Zhang W, Fuller GN. Tissue microarrays: applications in neuropathology research, diagnosis, and education. Brain Pathol. 2002;12(1):95–107.

[77] Henshall S. Tissue microarrays. J Mammary Gland Biol Neoplasia. 2003;8(3):347–58.

[78] Avninder S, Ylaya K, Hewitt SM. Tissue microarray: A simple technology that has revolutionized research in pathology. J Postgrad Med. 2008;54(2):158–62.

[79] Brown LA, Huntsman D. Fluorescent in situ hybridization on tissue microarrays: challenges and solutions. J Mol Histol. 2007;38(2):151-7.

[80] Hassan S, Ferrario C, Mamo A, Basik M. Tissue microarrays: emerging standard for biomarker validation. Curr Opin Biotechnol. 2008;19(1):19–25.

[81] Pham DL, Xu C, Prince JL. Current methods in medical image segmentation. Annu Rev Biomed Eng. 2000;2:315–37.

[82] Peces C, Garcia-Rojo M, Sacristan J, Gallardo AJ, Rodriguez A. Serendipia: Castilla-La Mancha telepathology network. Diagn Pathol. 2008;3 Suppl 1:S5.

[83] Slodkowska J, Garcia-Rojo M. Digital pathology in personalized cancer therapy. Studies in health technology and informatics. 2012;179:143–54.

[84] Rojo MG, Bueno G, Slodkowska J. Review of imaging solutions for integrated quantitative immunohistochemistry in the Pathology daily practice. Folia histochemica et cytobiologica / Polish Academy of Sciences, Polish Histochemical and Cytochemical Society. 2009;47(3):349–54.

[85] Tadeo I, Piqueras M, Montaner D, Villamon E, Berbegall AP, Canete A, et al. Quantitative modeling of clinical, cellular, and extracellular matrix variables suggest prognostic indicators in cancer: a model in neuroblastoma. Pediatric research. 2014;75(2): 302–14.

[86] Thiele J, Kvasnicka HM, Facchetti F, Franco V, van der Walt J, Orazi A. European consensus on grading bone marrow fibrosis and assessment of cellularity. Haematologica. 2005;90(8):1128–32.

[87] Chen LB, Huang HH, Shu X, Xu QH, Chen N, Zhang K, et al. [Pathological study of liver biopsy from 156 patients clinically diagnosed with mild chronic hepatitis B based on current guideline]. Zhonghua Shi Yan He Lin Chuang Bing Du Xue Za Zhi. 2009;23(2):138–40.

[88] Caballero T, Perez-Milena A, Masseroli M, O'Valle F, Salmeron FJ, Del Moral RM, et al. Liver fibrosis assessment with semiquantitative indexes and image analysis quantification in sustained-responder and non-responder interferon-treated patients with chronic hepatitis C. J Hepatol. 2001;34(5):740-7.

[89] Duregon E, Fassina A, Volante M, Nesi G, Santi R, Gatti G, et al. The reticulin algorithm for adrenocortical tumor diagnosis: a multicentric validation study on 245 unpublished cases. Am J Surg Pathol. 2013;37(9):1433–40.

[90] Huss S, Schmitz J, Goltz D, Fischer HP, Buttner R, Weiskirchen R. Development and evaluation of an open source Delphi-based software for morphometric quantification of liver fibrosis. Fibrogenesis Tissue Repair. 2010;3(1):10.

[91] Vertemati M, Moscheni C, Petrella D, Lamperti L, Cossa M, Gambacorta M, et al. Morphometric analysis of hepatocellular nodular lesions in HCV cirrhosis. Pathol Res Pract. 2012;208(4):240-4.

[92] Vertemati M, Vizzotto L, Moscheni C, Dhillon A, Quaglia A. A morphometric model to minimize subjectivity in the histological assessment of hepatocellular carcinoma and its precursors in cirrhosis. Microsc Res Tech. 2008;71(8):606–13.

[93] Teman CJ, Wilson AR, Perkins SL, Hickman K, Prchal JT, Salama ME. Quantification of fibrosis and osteosclerosis in myeloproliferative neoplasms: a computer-assisted image study. Leuk Res. 2010;34(7):871-6.

[94] Korkolopoulou P, Patsouris E, Kavantzas N, Konstantinidou AE, Christodoulou P, Thomas-Tsagli E, et al. Prognostic implications of microvessel morphometry in diffuse astrocytic neoplasms. Neuropathol Appl Neurobiol. 2002;28(1):57–66.

[95] Laitakari J, Nayha V, Stenback F. Size, shape, structure, and direction of angiogenesis in laryngeal tumour development. J Clin Pathol. 2004;57(4):394–401.

[96] Fernández-Carrobles MM, Tadeo I, Bueno G, Noguera R, Navarro S, Déniz O, et al. TMA vessel segmentation based on color and morphological features. Application to angiogenesis research. Journal of Biomedicine and Biotechnology. 2013:Aceptado para publicación.

[97] Fernández-Carrobles MM, Tadeo I, Noguera R, Navarro S, García-Rojo M, Déniz O, et al., editors. A morphometric tool applied to angiogenesis research based on vessel segmentation. Diagnostic Pathology; 2013.

[98] Voronoi GF. Nouvelles applications des paramètres continus à la théorie de formes quadratiques. Journal für die reine und angewandte Mathematik 1908;134:198–287.

[99] Barabasi AL, Oltvai ZN. Network biology: understanding the cell's functional organization. Nat Rev Genet. 2004;5(2):101–13.

[100] Raj A, Kuceyeski A, Weiner M. A network diffusion model of disease progression in dementia. Neuron. 2012;73(6):1204–15.

[101] Saez A, Acha B, Montero-Sanchez A, Rivas E, Escudero LM, Serrano C. Neuromuscular disease classification system. J Biomed Opt. 2013;18(6):066017.

[102] Saez A, Rivas E, Montero-Sanchez A, Paradas C, Acha B, Pascual A, et al. Quantifiable diagnosis of muscular dystrophies and neurogenic atrophies through network analysis. BMC Med. 2013;11:77.

[103] Ganeshan B, Miles KA. Quantifying tumour heterogeneity with CT. Cancer Imaging. 2013;13:140-9.

[104] Kalinli A, Sarikoc F, Akgun H, Ozturk F. Performance comparison of machine learning methods for prognosis of hormone receptor status in breast cancer tissue samples. Comput Methods Programs Biomed. 2013;110(3):298–307.

[105] Kowal M, Filipczuk P, Obuchowicz A, Korbicz J, Monczak R. Computer-aided diagnosis of breast cancer based on fine needle biopsy microscopic images. Computers in biology and medicine. 2013;43(10):1563–72.

[106] Ninos K, Kostopoulos S, Sidiropoulos K, Kalatzis I, Glotsos D, Athanasiadis E, et al. Computer-based image analysis system designed to differentiate between low-grade and high-grade laryngeal cancer cases. Anal Quant Cytol Histol. 2013;35(5):261–72.

[107] Stylianopoulos T, Martin JD, Snuderl M, Mpekris F, Jain SR, Jain RK. Coevolution of solid stress and interstitial fluid pressure in tumors during progression: implications for vascular collapse. Cancer Res. 2013;73(13):3833–41.

[108] Demou ZN. Gene expression profiles in 3D tumor analogs indicate compressive strain differentially enhances metastatic potential. Annals of biomedical engineering. 2010;38(11):3509–20.

[109] Paszek MJ, Weaver VM. The tension mounts: mechanics meets morphogenesis and malignancy. J Mammary Gland Biol Neoplasia. 2004;9(4):325–42.

[110] Tse JM, Cheng G, Tyrrell JA, Wilcox-Adelman SA, Boucher Y, Jain RK, et al. Mechanical compression drives cancer cells toward invasive phenotype. Proc Natl Acad Sci U S A. 2012;109(3):911-6.

[111] Jain RK, Baxter LT. Mechanisms of heterogeneous distribution of monoclonal antibodies and other macromolecules in tumors: significance of elevated interstitial pressure. Cancer Res. 1988;48(24 Pt 1):7022–32.

[112] Salmon H, Franciszkiewicz K, Damotte D, Dieu-Nosjean MC, Validire P, Trautmann A, et al. Matrix architecture defines the preferential localization and migration of T cells into the stroma of human lung tumors. J Clin Invest. 2012;122(3):899–910.

[113] Lammers T, Peschke P, Kuhnlein R, Subr V, Ulbrich K, Huber P, et al. Effect of intratumoral injection on the biodistribution and the therapeutic potential of HPMA copolymer-based drug delivery systems. Neoplasia. 2006;8(10):788–95.

[114] Shirota Y, Shirota H, Klinman DM. Intratumoral injection of CpG oligonucleotides induces the differentiation and reduces the immunosuppressive activity of myeloid-derived suppressor cells. Journal of immunology. 2012;188(4):1592-9.

[115] Barry-Hamilton V, Spangler R, Marshall D, McCauley S, Rodriguez HM, Oyasu M, et al. Allosteric inhibition of lysyl oxidase-like-2 impedes the development of a pathologic microenvironment. Nat Med. 2010;16(9):1009–17.

[116] Kendall RT, Feghali-Bostwick CA. Fibroblasts in fibrosis: novel roles and mediators. Front Pharmacol. 2014;5:123.

[117] Loeffler M, Kruger JA, Niethammer AG, Reisfeld RA. Targeting tumor-associated fibroblasts improves cancer chemotherapy by increasing intratumoral drug uptake. J Clin Invest. 2006;116(7):1955–62.

[118] Ostapoff KT, Kutluk Cenik B, Wang M, Ye R, Xu X, Nugent D, et al. Neutralizing murine TGFbetaR2 promotes a differentiated tumor cell phenotype and inhibits pancreatic cancer metastasis. Cancer Res. 2014.

[119] Eikenes L, Bruland OS, Brekken C, Davies Cde L. Collagenase increases the transcapillary pressure gradient and improves the uptake and distribution of monoclonal antibodies in human osteosarcoma xenografts. Cancer Res. 2004;64(14):4768–73.

[120] Shieh AC. Biomechanical forces shape the tumor microenvironment. Annals of biomedical engineering. 2011;39(5):1379–89.

[121] Padera TP, Stoll BR, Tooredman JB, Capen D, di Tomaso E, Jain RK. Pathology: cancer cells compress intratumour vessels. Nature. 2004;427(6976):695.

[122] Ariffin AB, Forde PF, Jahangeer S, Soden DM, Hinchion J. Releasing pressure in tumors: what do we know so far and where do we go from here? A review. Cancer Res. 2014;74(10):2655–62.

[123] Reymond N, d'Agua BB, Ridley AJ. Crossing the endothelial barrier during metastasis. Nat Rev Cancer. 2013;13(12):858–70.

[124] Vaupel P, Kallinowski F, Okunieff P. Blood flow, oxygen and nutrient supply, and metabolic microenvironment of human tumors: a review. Cancer Res. 1989;49(23): 6449–65.

[125] Johannessen TC, Wagner M, Straume O, Bjerkvig R, Eikesdal HP. Tumor vasculature: the Achilles' heel of cancer? Expert opinion on therapeutic targets. 2013;17(1):7–20.

[126] Rossler J, Taylor M, Geoerger B, Farace F, Lagodny J, Peschka-Suss R, et al. Angiogenesis as a target in neuroblastoma. Eur J Cancer. 2008;44(12):1645–56.

[127] Hu Z, Garen A. Intratumoral injection of adenoviral vectors encoding tumor-targeted immunoconjugates for cancer immunotherapy. Proc Natl Acad Sci U S A. 2000;97(16): 9221-5.

[128] Carmeliet P, Jain RK. Principles and mechanisms of vessel normalization for cancer and other angiogenic diseases. Nat Rev Drug Discov. 2011;10(6):417–27.

[129] Goel S, Wong AH, Jain RK. Vascular normalization as a therapeutic strategy for malignant and nonmalignant disease. Cold Spring Harb Perspect Med. 2012;2(3):a006486.

[130] Jain RK, Tong RT, Munn LL. Effect of vascular normalization by antiangiogenic therapy on interstitial hypertension, peritumor edema, and lymphatic metastasis: insights from a mathematical model. Cancer Res. 2007;67(6):2729–35.

[131] Baffert F, Le T, Sennino B, Thurston G, Kuo CJ, Hu-Lowe D, et al. Cellular changes in normal blood capillaries undergoing regression after inhibition of VEGF signaling. Am J Physiol Heart Circ Physiol. 2006;290(2):H547–59.

[132] Jain RK. Normalization of tumor vasculature: an emerging concept in antiangiogenic therapy. Science. 2005;307(5706):58–62.

[133] Salnikov AV, Heldin NE, Stuhr LB, Wiig H, Gerber H, Reed RK, et al. Inhibition of carcinoma cell-derived VEGF reduces inflammatory characteristics in xenograft carcinoma. Int J Cancer. 2006;119(12):2795–802.

[134] Tong RT, Boucher Y, Kozin SV, Winkler F, Hicklin DJ, Jain RK. Vascular normalization by vascular endothelial growth factor receptor 2 blockade induces a pressure gradient across the vasculature and improves drug penetration in tumors. Cancer Res. 2004;64(11):3731-6.

[135] Fan Y, Du W, He B, Fu F, Yuan L, Wu H, et al. The reduction of tumor interstitial fluid pressure by liposomal imatinib and its effect on combination therapy with liposomal doxorubicin. Biomaterials. 2013;34(9):2277–88.

[136] Jarm T, Podobnik B, Sersa G, Miklavcic D. Effect of hydralazine on blood flow, oxygenation, and interstitial fluid pressure in subcutaneous tumors. Adv Exp Med Biol. 2003;510:25-9.

[137] Kristensen CA, Nozue M, Boucher Y, Jain RK. Reduction of interstitial fluid pressure after TNF-alpha treatment of three human melanoma xenografts. British journal of cancer. 1996;74(4):533-6.

[138] Multhoff G, Vaupel P. Radiation-induced changes in microcirculation and interstitial fluid pressure affecting the delivery of macromolecules and nanotherapeutics to tumors. Front Oncol. 2012;2:165.

[139] Sen A, Capitano ML, Spernyak JA, Schueckler JT, Thomas S, Singh AK, et al. Mild elevation of body temperature reduces tumor interstitial fluid pressure and hypoxia and enhances efficacy of radiotherapy in murine tumor models. Cancer Res. 2011;71(11):3872–80.

[140] Watson KD, Lai CY, Qin S, Kruse DE, Lin YC, Seo JW, et al. Ultrasound increases nanoparticle delivery by reducing intratumoral pressure and increasing transport in epithelial and epithelial-mesenchymal transition tumors. Cancer Res. 2012;72(6):1485–93.

[141] Stuhr LE, Raa A, Oyan AM, Kalland KH, Sakariassen PO, Petersen K, et al. Hyperoxia retards growth and induces apoptosis, changes in vascular density and gene expression in transplanted gliomas in nude rats. J Neurooncol. 2007;85(2):191–202.

[142] Kleemann B, Loos B, Scriba TJ, Lang D, Davids LM. St John's Wort (Hypericum perforatum L.) Photomedicine: Hypericin-Photodynamic Therapy Induces Metastatic Melanoma Cell Death. PloS one. 2014;9(7):e103762.

[143] Tadeo I. Study of the architectural scaffolding and the vascular system of neuroblastic tumors. Valencia, Spain: University of Valencia; 2015.

[144] Tadeo I, Berbegall AP, Navarro S, Castel V, Noguera R. Extracellular matrix histological patterns are associated to malignancy in Neuroblastic tumors. Manuscript in preparation. 2016.

[145] Tadeo I, Bueno G, Berbegall AP, Fernández-Carrobles MM, Castel V, Garcia-Rojo M, et al. Vascular patterns provide therapeutic targets in aggressive neuroblastic tumors. Under consideration in Oncotarget. 2016.

[146] Tadeo I, Berbegall AP, Castel V, García-Miguel P, Callaghan R, Påhlman S, et al. Extracellular Matrix Composition defines an Ultra-High Risk Group of Neuroblastoma within the High Risk Patient Cohort. Under consideration in British Journal of Cancer. 2016.

[147] Dougan M, Dranoff G. Immune therapy for cancer. Annual review of immunology. 2009;27:83–117.

[148] Neves H, Kwok HF. Recent advances in the field of anti-cancer immunotherapy. BBA clinical. 2015;3:280-8.

[149] Seeger RC. Immunology and immunotherapy of neuroblastoma. Semin Cancer Biol. 2011;21(4):229–37.

[150] Kerkar SP, Restifo NP. Cellular constituents of immune escape within the tumor microenvironment. Cancer Res. 2012;72(13):3125–30.

[151] Mellman I, Coukos G, Dranoff G. Cancer immunotherapy comes of age. Nature. 2011;480(7378):480-9.

[152] Meyerhardt JA, Mayer RJ. Systemic therapy for colorectal cancer. N Engl J Med. 2005;352(5):476–87.

[153] Van Cutsem E, Peeters M, Siena S, Humblet Y, Hendlisz A, Neyns B, et al. Open-label phase III trial of panitumumab plus best supportive care compared with best supportive care alone in patients with chemotherapy-refractory metastatic colorectal cancer. Journal of clinical oncology : official journal of the American Society of Clinical Oncology. 2007;25(13):1658–64.

[154] Hudis CA. Trastuzumab--mechanism of action and use in clinical practice. N Engl J Med. 2007;357(1):39–51.

[155] Cohen MH, Gootenberg J, Keegan P, Pazdur R. FDA drug approval summary: bevacizumab (Avastin) plus Carboplatin and Paclitaxel as first-line treatment of advanced/ metastatic recurrent nonsquamous non-small cell lung cancer. The oncologist. 2007;12(6):713-8.

[156] Cohen MH, Gootenberg J, Keegan P, Pazdur R. FDA drug approval summary: bevacizumab plus FOLFOX4 as second-line treatment of colorectal cancer. The oncologist. 2007;12(3):356–61.

Section 4

Biomaterials, Stem Cells and Tissue Engineering

Chapter 13

Exploring the Extracellular Matrix to Create Biomaterials

Sylvain Vigier and Tamas Fülöp

Additional information is available at the end of the chapter

http://dx.doi.org/10.5772/62979

Abstract

The extracellular matrix (ECM) represents the framework of tissues and organs and is involved in cell differentiation and function. The study of ECM is challenging and required a combination of identification and imaging techniques to give a valuable scheme of its composition, organization, and finally function. The study of ECM enables to culture cells *ex vivo*, but cultures are restricted to two-dimensional surfaces, whereas in the meantime, material sciences were developing devices able to bring cell culture in a three-dimensional (3D) environment. This chapter presents basic techniques to investigate extracellular matrices composition and organization. Basic knowledge on ECM composition and organization should inspire material scientists to propose more biologically relevant materials. In a second time, we present strategies available to create *ex vivo* models of ECM and a series of examples of 3D materials that were engineered to investigate cell adhesion, phenotype, and differentiation in a biologically relevant microenvironment. The production of a gold-standard material is possible for a specific biological question, and it might be developed from an intelligible dialogue between material scientists, that bring engineering strategies, and cell biologists who implement the material design to meet the biological process that has to be investigated *ex vivo*.

Keywords: extracellular matrix, tissue organization, 3D materials, in vitro cell culture models, tissue engineering

1. Introduction

Multicellular organisms require a framework to delineate functional territories and to provide a shelf where the cells can attach to perform their specific functions. The extracellular matrix (ECM) represents this framework for tissues and organs and as such it is an important actor of organisms' physiology. The most known examples of ECM-related tissues are the skin, where ECM act as a barrier against the outside environment, and the bones where ECM is strengthened by a mineral phase which allows the body to stand and to move. However, its apparent structural and mechanical properties have hidden more subtle roles of ECM in cell differentiation and function as ECMs are not restricted to load-bearing organs but are present and required in all types of tissues and organs. During the development of the embryo, neural crest cells lose their cell-cell adhesion properties toward cell–ECM interactions that allow them to move along the dorsal part of the embryo and reach their specific site of function and give birth to the future skeleton. Again, tissue remodeling, as observed during the healing processes, can release messenger molecules that were entrapped in the ECM, waiting for the right moment to trigger their signal and healing functions [1]. Some lack of knowledge on ECM functions remains mainly because of the challenge represented by its comprehensive study. Indeed, ECM is made of several high molecular weight proteins, proteoglycans, and polysaccharides molecules self-arranged into fibers and networks difficult to solubilize and individualize. Basic biochemistry techniques have led to the identification of the major components of ECMs such as collagens or laminins, but as the investigations are progressing, this results in the constant growing of the constituent members of collagen and laminin families and in the discovery of new ECM components with unknown functions [2]. Moreover, understanding the ECM not only means discovering new molecules but also to unravel their organization in the ECM network. So the study of ECM requires the combination of identification and imaging techniques to give a valuable scheme of its composition, organization, and finally function. Interestingly, unraveling ECM complexity meets one of the fundamental questions for biologists: how to recreate and maintain life outside a living organism (literally *ex vivo* but commonly referred as *in vitro*)?

The beginning of the 20th century aroused the possibility to dissociate cells from living tissues and to culture them *ex vivo*. This new technique has triggered the emergence of the new discipline of cell biology which has brought most of the knowledge that we possess today on cell proliferation, differentiation, metabolism, cell fate, and death. However, *ex vivo* cell cultures were restricted to two-dimensional (2-D) culture systems, originally on glass and subsequently on plastic dishes, occasionally supplemented by the coating of ECM molecules to favor cell adhesion. Parallel to the development of cell biology, the broad field of materials science was creating polymers and devices able to bring *ex vivo* cell culture to the third dimension, and to the 21st century. Dedicated to materials that interact with living tissues, the field of biomaterials encompasses several scientific disciplines, from physics and chemistry to biochemistry and medicine. Several types of three-dimensional (3D) materials have been engineered which may represent valuable tools for fundamental cell research, but a lack of knowledge on ECM structures have undermined their use for cell biology. On the other hand, cell biologists are not necessarily aware of the development and possibilities created by extensive research in the field of 3D biomaterials, and this partly compromises the expansion of 3D cell culture models.

In this chapter, we will present basic techniques involved in the investigation of extracellular matrices and data generated by their use to understand ECM composition and organization. Basic knowledge on ECM composition and organization should be useful for biomaterial scientists to propose more biologically relevant materials. Such methodologies are fully transposable for the characterization of biomaterials and 3D models of ECMs. In a second section, we will present a series of biomaterials that were engineered based on the investigation of ECMs composition and organization *in vivo* and could become suitable 3D cell culture models for mechanobiology, aging, cell migration, cell differentiation, and studies on pathologies and their treatments.

2. Exploring the ECM

Extracellular matrices are multimolecular three-dimensional (3D) networks made of a large variety of ECM-specific molecules and their compositions and organizations are tissue-specific. Exploring the ECM means (1) the determination of its distribution within the tissue and its relation to the cell content, (2) the identification and quantification of its composition, and (3) the characterization of the 3-D architecture of the ECM network [2]. ECMs contain similar biomolecules which can be organized in two main classes (1) proteins and glycoproteins and (2) proteoglycans and polysaccharides. Variation in the composition or in the amount of certain ECM molecules will change dramatically the physical properties of the ECM such as the tensile strength observed in the hard mineralized ECM in bones, the elasticity observed in dermis of the skin, or even the transparency in the cornea of the eye. The biochemistry of ECM components strongly influences the techniques used to investigate them. Light microscopy associated with histological staining is based on the differences of biochemical features of tissues (i.e., hydrophobicity, electrical charge, and molecular weight). Proteomics associated with mass spectrometry is a powerful tool to exhaustively identify proteins in a complex sample, but biochemistry of ECM proteins is particularly unfavorable to this method that need significant adaptation to be effective with ECM samples. Finally, electron microscopy is the ideal method to investigate the molecular and fibrillary organization of the ECM network.

2.1. Biochemistry of the main ECM components

2.1.1. Proteins and glycoproteins

A large diversity of proteins is found in ECMs where they are the principal component. They are classified either in structural proteins that are directly involved in the overall architecture of the ECM or in soluble factors that are globular proteins entrapped in the ECM network. Structural proteins are mainly fibrous, insoluble, and high molecular weight molecules, including collagens, elastin, laminins, and fibronectins. They are direct actors of the shape and the mechanical properties of tissues and organs and further possess the ability to auto-assemble among themselves as well as to interact with each other to form fibrillary network and complex 3-D architectures. Most of the ECM proteins have sequences recognized by cells for adhesions and some of them can bind specifically soluble growth factors or cytokines. These molecules present several posttranslational modifications like hydroxylation at Proline and

Lysine residues in collagens and O-glycosylation and N-glycosylation in laminins and fibronectin.

Collagens are found in all types of ECMs and are the main constituent of connective tissues like skin, bone, and tendons [3]. They belong to a large family of molecules with to date 28 members identified (numbered from collagen type I to type XXVIII). Collagens are trimeric proteins, made of the association of three alpha-chains specific to each type of collagens that assemble together to form a super-helix structure. For some collagen types several alpha-chains exist, leading to multiple isoforms of the same collagen molecule and raising the diversity and the complexity of the collagen family. In ECMs, collagens are organized in different supramolecular assemblies inherited from the specificity for each collagen types taking into account their amino-acid sequences and the 3-D folding of their tertiary structure [4]. Fibril-forming collagens include collagen type I, II, III, V, and XI. They assemble in large fibrils (up to 500 nm in diameter) that can merge to form collagen fibers of micrometric size. All ECMs contain fibrillary collagens. Connective tissues are characterized by an abundant ECM content made mainly by fibrils of collagen type I in dermis and bone, or of collagen type II fibrils in cartilage. Basement membranes (BM) are a specialized form of ECM mainly found in epithelial tissues and contain heterotypic fibrils combining collagen I and III or V [5]. Size and diameters of collagen fibrils are regulated by other ECM molecules like fibril-associated collagens or proteoglycans. Collagen fibrils and fibers are finally stabilized by covalent cross-links making these structures highly resistant to mechanical load and stresses. Network-forming collagens are mostly found in BM where collagen type IV is the most abundant. Collagen IV molecules assemble in a hexameric superstructure that propagate to form finally a 2-D network that is maintained by covalent crosslinks with methionine and lysine residues [6].

Laminins are large molecular weight (from 400 to 900 kDa), heterotrimeric glycoproteins and, along with collagen type IV, they are the main constituent of BM [7]. Even found in every BM, laminin is a large family of molecules, and their distribution among BM is tissue-specific. A laminin molecule consists of the association of one alpha, one beta, and one gamma chain. To date, 5 alpha, 3 beta, and 3 gamma chains have been identified which may be assembled in 16 different laminin molecules. All laminins share common structural features: a cross-shaped 3-D structure with one long and two shorts arms, di-sulfide bridges in-between the chains that maintain their association and the presence of several N-glycosylation on asparagine residues. Laminins auto-assemble in a network interlaced with the collagen type IV network. Directed toward the cells, laminins gives cues for cell adhesions through integrin receptors.

Elastin is organized in fibers closely linked to fibrillar collagens where it gives the elasticity to tissues and compensate the tensile strength of collagen fibers [8]. Elastin is secreted by cells as a 60–70 kDa monomeric soluble precursor, tropoelastin, which contains intermittent hydrophobic domains. Tropoelastin monomers auto-assemble to form elastin fibers that are stabilized by enzymatic cross-linking through Lysine residues and rendering the elastin network highly insoluble. Stacks of hydrophobic domains in the elastin network are responsible for its elastic properties and make elastin highly resistant to enzymatic degradation and solubilization in aqueous solutions.

http://dx.doi.org/10.5772/62979

Fibronectin is a large (500 kDa) dimeric glycoprotein made of the association of two nonidentical monomers linked by two disulfides bounds at their C-terminal extremities [9]. Diversity of the monomers is due to alternate splicing of the fibronectin mRNA, as fibronectin is encoded by only one gene. Fibronectin is expressed by several cell types and found in most of ECMs. It assembles through disulfide bridges in oligomers and finally in insoluble fibers possessing various diameters ranging from 10 nm to microns [10]. A soluble form made of the dimeric monomer may be also found to circulate in the blood. Fibronectin primary structure is arranged in several domains that specifically interact with collagens or with cells via integrins.

There are globular, soluble proteins associated with the ECM network of structural proteins. Among the globular proteins there are growth factors, cytokines, and ECM-specific proteolytic enzymes like matrix metalloproteinases (MMP). They play an important role in cell signaling and in the remodeling of the ECM network and finally in the overall biological activity of ECMs. They can be linked to structural proteins by labile interactions at specific biding sites or be trapped in the high molecular weight chains of the structural proteins and proteoglycan. However, they are not core proteins of the ECM network, and their biochemistry is similar to most of the other globular proteins.

2.1.2. Proteoglycans and polysaccharides

Polysaccharides found in ECMs of vertebrae are glycosaminoglycans (GAG) and are covalently linked to a core protein to form proteoglycans, except for hyaluronan representing the only "pure" polysaccharide of ECMs [11]. Even if this chapter focuses on mammalian ECMs, it has to be mentioned that polysaccharides are the main ECM components of invertebrates and plants represented by chitin and cellulose, respectively. Hyaluronan, equally called hyaluronic acid, has the particularity to be synthesized at the plasma membrane by three different Hyaluronan synthase enzymes and not inside the Golgi apparatus like all the other proteoglycans [12]. GAGs are linear, unbranched polysaccharides composed by tens to hundreds of disaccharides units. The combination of disaccharide units is highly heterogeneous, but can be specific for each individual chain. The disaccharide unit is made of glucosamine or galactosamine linked to another modified hexose, the most often to glucuronic acid, iduronic acid, or galactose. These monosaccharides are mainly modified by N-acetylation and N-sulfatation. The nature of the disaccharide unit and the types of modifications lead to the formation of different types of GAG, including chondroitin sulfate, dermatan sulfate, keratan sulfate, heparan sulfate, and hyaluronan. At physiological pH, GAGs chains are highly negatively charged due to the sulfate and carboxylic acid functions carried on modified hexoses. The net negative charge of GAGs make them highly hydrophilic, and thus, they play an important role in the hydration of ECMs [13]. High amounts of water associated with GAGs ensure some mechanical properties to ECMs, especially the resistance to compression as in the cartilage. Proteoglycans are abundant within ECM, but may be also found at the cell membrane or intracellularly. The most active part of the proteoglycans is the GAGs chain which can interact with growth factors, cytokines, cell receptors, and other constituents of the ECM. However, their core proteins also possess interaction sites that make proteoglycans highly versatile molecules inside ECMs [14]. Due to their interactions with ECM components, they play a role in ECM organization, but their important role is to be a reservoir for growth factors and to anchor signal molecules that are released through specific enzymes in particular after injuries and favor wound healing.

2.2. Exploring the organization and the composition ECMs within tissues by histology

Histology is an old, but still a powerful technique to image the organization of tissues at the scale of the cells. Organs or tissues have to be fixed to stabilize their organization after removing them from the living body and to be embedded in a hard material for being sectioned into thin slices to allow light to pass through the tissue and so be visible by the lenses of an optical microscope. Usually, tissues are fixed in paraformaldehyde and embedded in wax (paraffin). Tissues can also be frozen-fixed in liquid nitrogen. This approach may help to avoid some prejudicial effects of chemical fixations and do not require an embedding step to slice the tissue. However, frozen sections give pictures with less contrast and sharp edges, therefore formalin-fixed paraffin embedded (FFPE) samples are preferred for routine examination. To be interpretable under light microscopic examination, the histological sections have to be stained to give some contrast to the different tissue and cells structures [15]. Numerous special stains have been developed to give a precise contrast to the organization and to the components of ECM, and this approach still keeps all its strength and efficiency to describe and accurately decorticate a tissue [16]. A significant further step in histology is the use of antibodies by the mean of immunohistochemistry (IHC) to specifically identify a target molecule, and at the same time, visualize its exact localization within the tissue [17]. However, IHC relies on the quality of the antibody, and it appears more difficult to obtain commercially reliable antibodies against ECM molecules than for intracellular molecules. A model of the accuracy and the potentials of histological techniques for investigation of ECMs are reported in a recent and thorough study on tooth root cementum using exclusively light microscopy techniques, with a combination of special stains, polarized light, and immunodetection [18].

2.2.1. Special stains for histology of ECMs

Masson's and Gomori's trichrome– they are ideal for connective tissues and in particular to stain fibrillar collagens. Trichrome staining indicates that three different dyes are used to discriminate the ECM materials from the cell cytoplasm and the nucleus. Selectivity of the dye for the different tissue structures is based mainly on size exclusion: a small dye will penetrate into low porosity elements while larger dye will penetrate preferably into the more porous structures. Gomori's trichrome stains collagen-based ECM in green, cell cytoplasm and noncollagenous tissue in pink to purple and cells nuclei in black to blue. Same discrimination of connective tissue comportments but with different colors is obtained with Masson's trichrome (collagen in blue, cytoplasm in red, and nucleus in black). This trichrome staining is longer and maybe more delicate to perform than the classical hematoxylin-eosin staining, however, this staining should be used each time when collagen-based samples (tissue or biomaterial) are investigated because of the high degree of contrast it enables compared to H&E [19].

http://dx.doi.org/10.5772/62979

Von Kossa staining— It is used to reveal mineral deposits within tissues. Mineralisation of ECM is observed in bones or teeth but can also appear pathologically in blood vessels, skin, or cartilage. Von Kossa staining detects calcium phosphate deposits by its substitution with silver nitrate to form black precipitates. This method is robust and useful to follow mineralization of bone substitute materials or progressive differentiation of naive cells into the osteoblastic lineage. It can also discriminate calcium–phosphate based materials from organic compounds.

Picro-sirius—This staining method is specific of fibrillar collagens (type I–III), although it can slightly stain collagen type IV in BM. Sirius-red dye is an anionic molecule that arrange parallel to collagen molecules by interaction with basic (cationic) amino acids [20]. Collagenous tissue is stained in red while remaining tissue is pale yellow, and nuclei can be counterstained in black or deep blue if necessary. The parallel organization of picro-sirius is used to enhance the birefringence signal of collagen fibers observed under a polarized microscope. With the combination of polarized light and picro-sirius, it is possible to address the 3D orientation of the collagen fibers as the intensity of the birefringence depends on fibers orientation, with maximum birefringence intensity for fibers organized at 45°(+/–90°) toward the transmitted light axis, while no birefringence signal is observed with fibers organized parallel or orthogonal to the transmission axis [21,22]. Picro-sirius is also used to make quantitative analysis of the collagen content in tissues, in particular to address the degrees of fibrosis.

Verhoeff–Van Gieson—This coloration method stains preferentially elastic fibers which have a higher affinity for the dye than the rest of the tissue due to hydrophobic interactions. The elastic fibers are stained black or dark brown while collagen fibers are red and the rest of the tissue appears pale yellow. This coloration is particularly appropriate to reveal atrophy of elastic tissues in case of vascular diseases and to discriminate arteries from veins due to differences in the elastic fibers organization and content [23].

Alcian blue—This dye is a cationic molecule and interacts specifically with negatively charged polysaccharides like sulfated and carboxylated GAGs. Because the selectivity of the dye is based on the charge of GAGs it is of importance to ascertain the pH of the alcian blue solution, as the carboxylic acid groups are in their acidic neutral form at $pH < 2$ and therefore will not interact with the dye. This specificity allows the discrimination between the carboxylated and the sulfated GAGs, as the latter are negatively charged at $pH < 2$. Polysaccharide's rich tissues are stained in light blue, counterstaining with nuclear fast red, making cells cytoplasm and nuclei colored in pink-red. Alcian blue is routinely used to stain cartilaginous ECM due to their large content of GAGs [24].

Periodic Acid–Schiff (PAS)—This staining is also sensitive to polysaccharides and is used to stain BM due to their high content in glycoproteins and proteoglycans. Unlike alcian blue, PAS staining reveals both charged and neutral polysaccharides because this coloration is based on the oxidation of monosaccharides with periodic acid and subsequent reaction of newly formed aldehyde groups with the Schiff reagent dye. BM are finely stained in pink-purple color, whereas the cytoplasm and the nuclei are colored in shades of blue when counterstained with Harris' hematoxylin. This staining method is also used to reveal other polysaccharides-rich compounds, such as mucins and glycogen [15].

2.2.2. *Immunohistochemistry of the extracellular matrix*

IHC enables the identification of a specific component of the ECM and to image its distribution within the tissue [25]. The target molecule is recognized by an antibody that reacts to a specific epitope and then is visualized by light microscopy through a chromogenic enzymatic reaction (alkaline phosphatase or horseradish peroxidase) or through a fluorescent dye with a fluorescence microscope. The antibody is observed directly if the dye or the enzyme is linked to it, but most of the time it is detected indirectly by a labeled (by a fluorophore or an enzyme) secondary antibody which reacts to the first one through its Fc fragment. Frozen sections are more appropriate for Immunohistochemistry because they avoid the use of fixative that may alter the epitope, but frozen sections cannot be counterstained and so keeps the tissue organization around the epitope not visible [26]. In contrast, FFPE samples are well preserved and can be counterstained with different dyes after antibody incubation and detection. However, if the fixative (generally 4% paraformaldehyde in neutral buffer) preserves the morphology of the tissue, it can severely compromise the antigenicity of the target molecule, and then make immunodetection inefficient or inoperative. Paraformaldehyde fixative triggers intra- and intermolecular cross-linking of proteins by the formation of methylene bridges between amino acids residues [25]. It may also alter the molecular structure of polysaccharides, lipids, and nucleic acids. The degree of cross-linking will depend on the concentration and the pH of the fixative solution, as well as on the time and the temperature at what the fixation is performed. The formation of intra- and intermolecular cross-linking modifies the secondary and tertiary structures of proteins that lower the detection by antibodies because of the modification of the target epitopes [27]. In the early 1990s, an antigen retrieval (AR) method was introduced to recover the antigenicity of FFPE tissue sections impaired by the fixation treatment [28]. The AR method originally refers to the high-temperature processing of FFPE sections, but with the development of other methods it is nowadays a generic term for any kind of treatment used to recover the original antigenicity of the FFPE sections [29]. The rationale of AR is the breaking of fixative-induced cross-links and methylene bridges that enable a renaturation of the proteins and a partial recovery of the epitopes. However, it has to be noticed that the true mechanism of AR is not yet understood, and it remains an empirical technique that requires several positive and negative controls to avoid true- or false-positive reactions [30]. AR is performed with the use of heat (called heat-induced antigen or epitope retrieval) or enzymes (referred-to PIER for proteolytic enzyme-induced epitope retrieval) to break fixative cross-links.

In heat-Induced epitope retrieval (HIER), three parameters appear essential in the outcome of the AR: temperature and pH of the solution and time of incubation of the sections [29,31]. Classically, sections are incubated for 10–20 minutes at 95°C in a water bath. Microwave and steam-cookers are also used to heat sections and have shown good AR properties, although the control of the temperature is more delicate. The pH of the solution is a critical factor because some epitopes will be revealed only in acidic or in alkaline buffer. The most common acidic buffer is citrate used in a pH range of 3–6. The most used alkaline buffer is Tris supplemented or not with EDTA at pH 8–10 [29,31]. All pH, temperature, and time have to be checked carefully because extreme conditions will damage the tissue sections.

Enzyme treatment is thought to break some of the fixative methylene bridges and to elicit the reconstitution of epitopes after a moderate digestion of proteins. It is generally performed with proteolytic enzymes such as pepsin, trypsin and proteinase K at a concentration of 0.05–1% for 10–30 minutes. Glycosidases, such as hyaluronidase, chondroitinase, and keratinase have shown valuable AR properties on polysaccharides-rich tissues and on glycosylated proteins [32]. The pH and temperature of the solution are adjusted to the optimal activity of the enzyme, and time of digestion and the concentration of the enzyme have to be carefully set to avoid overdigestion of the tissue sections which will lead to a loss of tissue structure and organization.

Success of immunohistostaining mainly relies upon the quality of the antibody. Compared to soluble proteins, only few antibodies against ECM molecules are commercially available [32]. The ECM proteins are highly conserved in mammals making difficult the immunization of animals to generate an efficient antibody. Some antibodies are raised from synthetic peptides (5–20 amino acids) chosen from the primary amino-acid sequence of the target protein, but the epitope generated could be irrelevant to the secondary and tertiary structures of the native proteins [33]. Polysaccharides are either not or very slightly immunogenic, making very difficult to generate antibodies against the sugar part of proteoglycans. Moreover, ECM proteins are organized into dense fibers structures or meshwork or bear high polysaccharides chains that hinder the access of the antibody to the epitopes. In conclusion, IHC of the ECM is a delicate technique but remains the best option to obtain a picture of individual ECM components distribution within the different compartments of a tissue. As an example among several ones, by the means of monoclonal antibodies raised against laminin chains alpha-4 and -5, beta-1 and -2, and gamma-1, it has been possible to elucidate the particular composition and organization of the basement membrane surrounding islets of Langerhans in human pancreas [34]. The identification of a duplex BM surrounding intra-islets vessels with a specific laminin composition for each of the two BM has led to the proposition of a double-basement membrane model of human islets of Langerhans clearly distinct from the organization of basement membrane surrounding islets in mouse [35].

2.3. Identification of ECM composition by proteomics with mass spectrometry

The proteomic strategy is based on the isolation of a complex mixture of proteins from cells, tissues, or a whole organism and their identification by mass spectrometry and genomic database. Mass spectrometers commonly used for protein identification are MALDI–TOF (for Matrix–Assisted Lazer Desorption Ionization—Time of Flight) and ESI-Q-TOF (for Electro Spray Ionization—Quadripole—Time of Flight) that have their own characteristics and performances but do not change the general flow-chart of the sample preparation and identification [36,37]. After extraction and purification, proteins are separated by 1D or 2D sodium dodecyl sulphate–polyacrylamide gel electrophoresis (SDS-PAGE), respectively to their molecular weight (1D) or by both their isoelectric point (pH which net charge of protein is neutral) and molecular weight (2D). Mass spectrometers only detect charged molecules with an accuracy and sensibility that depends on the ratio of mass over charge (m/z), so the proteins samples have to be hydrolyzed into peptides before mass spectrometry analysis to obtain spectra at atomic resolution. Protein bands (1D) or spots (2D) are excised from the gel, hydrolysed into peptides by a proteolytic enzyme (frequently Trypsin), and loaded in the mass spectrometer to measure the exact mass of the peptides. Each protein from the original mixture is identified by matching the measured masses of their peptides with the expected masses of peptides calculated in silico from genomic database [37,38]. This technique allows a large-scale identification of components without the bias of predetermined molecular candidates as with antibody detection. It is thus possible in theory to have the exact protein composition of a tissue and follow its modification with time or diseases [39].

The total or relative amount of identified proteins can also be addressed. The SDS-PAGE migration pattern and intensity of protein band (1D) or spot (2D) give a "map" of the protein content of the target tissue or organ and can be used to identify particular band/spot that are modified in specific conditions, enabling discovery of new therapeutic targets [40]. Labeling methods exist to generate quantitative data with mass spectrometry. Samples of the control conditions are modified with nuclear isotopes 13C, 15N, or 18O, whereas the treated sample is left unmodified, and the relative abundance of both isotopic pics is compared [41]. A direct semiquantitative approach is also possible, with the mathematical integration of ion counts of the peptides identifying each protein to describe its relative abundance [42]. In the ECM analysis, relative quantification is a remarkable tool to identify the specific isoform of some multimeric ECM proteins, such as collagens and laminins, as the relative amount of each monomer will indicate under which form the ECM molecule is present in the tissue. To be more specific, collagen type V exists in the common heterotrimeric isoform $[\alpha1(V)]_2\alpha2(V)$ and a more scarce homotrimeric isoform $[\alpha1(V)]_3$. The relative amount of ion counts for the $\alpha1(V)$ chain over $\alpha2(V)$ chain will indicate if the $\alpha1(V)$ chain is associated only with $\alpha2(V)$ ($\alpha1$ chain signal twofold of $\alpha2$ chain signal) or if the investigated tissue contains both heterotrimeric $[\alpha1(V)]_2\alpha2(V)$ and homotrimeric $[\alpha1(V)]_3$ isoforms ($\alpha1$ chain signal >> twofold of $\alpha2$ chain signal). However, quantification by mass spectrometry can be restricted by the ionization properties of some proteins that will make them less detected and consequently under-represented in the final analysis. Nevertheless, this highlights the potentials of proteomics and mass spectrometry in the study of ECM proteins, as such characterization of ECM proteins isoform will require several antibodies (i.e., one per protein chain) to identify one isoform by western blot or IHC [43].

The most critical steps of a proteomic analysis are the purification of the protein mixture and their identification from database. ECM proteins have a high molecular weight and are tightly associated with each other by covalent cross links that make them mostly insoluble. An important point in the analysis of ECM by mass spectrometry proteomic will be the proper solubilization of the ECM [44]. The tissue has to be first carefully decellularized to purify the ECM and eliminate the remaining intracellular proteins. This step requires the use of a detergent like SDS and will eliminate from the ECM part of the loosely bounded proteins like remodeling enzymes or growth factors [40]. The purified ECM can be solubilized by a combination of physical, chemical, and enzymatic methods. A physical method is the mechanical breaking with a French press or grinding with mortar and pestle in liquid nitrogen. This step is important to homogenize correctly the purified ECM and make the following solubilization treatment effective. Ultrasound can also be used, but this process yields heat that can denaturate and break the proteins creating smears instead of protein bands or spots during SDS-PAGE separation if temperature is not carefully controlled. Homogenized ECM can be solubilized with a chaotropic agent like concentrated urea or guanidium chloride [45]. These molecules are efficient for solubilization, but a too high concentration is not compatible with SDS-PAGE separation and can impair the trypsin digestion. Highly cross-linked collagen fibrils or elastin microfibrils can remain insoluble after chaotropic extraction. Partial digestion with proteolytic enzymes such as pepsin is also used to favor ECM solubilization, but again, it has to be done carefully to not hydrolyze the ECM sample before SDS-PAGE separation. Deglycosylation with glucosidase such as PNGase or chondroitinase can unravel parts of the dense polysaccharide network of proteoglycan and unleash trapped ECM proteins [45]. Moreover, deglycosylation is also favorable for further trypsin digestion and peptide identification from database. ECM proteins undergo several posttranslational modifications, such as hydroxylation, disulfide bridges, and glycosylation that can block digestive sites for trypsin, leading to inefficient peptide generation, or resulted in peptides of different masses than the expected masses from the genomic data base, leading to inappropriate identification of the protein. All these biochemical specificities of ECM proteins make proteomic discovery challenging and explain why only a few proteome of ECMs have been published so far. Nevertheless, this approach has a huge potential and consequently efficient solubilization and identification protocols are under development to make this technique more routinely usable in ECM and biomaterials characterization.

2.4. Three-dimensional organization of ECMs studied by electron microscopy

Electron microscopy gives higher spatial resolution than light microscopy with the use a shorter wave-length from an electron beam. With resolution at the nanometres scale, and below with high resolution microscopes, it gives access to the molecular structure of ECM proteins and can image their supramolecular organization (i.e., fibrils and fibers assemblies) that are hardly distinguishable with optical microscopes. Transmission electronic microscopes (TEM) are built on the same scheme as optical/visible-light microscopes and so, equivalent techniques and processing of samples are required for both type of microscopy. In TEM, the electron beam pass through the samples to give rise to a projected image on an electron-sensitive surface like a phosphorescent screen, on a silver-film plate to record the image or nowadays on CCD cameras. The electron beam requires a vacuum pressure and cannot pass through thick samples of several micrometers which both represent a challenge for biological samples that are mainly wet, thick, and soft materials [46]. Biological tissues have to be fixed, dehydrated, and embedded in hard material (epoxy resin) and sliced with a diamond knife ultra-microtome into hundreds nanometres slices to be investigated by TEM. The electron beam interacts poorly with low atomic numbers atoms, such as carbon, oxygen, and nitrogen found in biological samples, so sections are stained with heavy metal solutions (commonly tungsten in phosphotungstic acid, uranium in uranyl acetate, and lead in lead citrate) to give contrast [47]. Compared to histological staining, negative staining is more commonly used to prepare TEM sample to improve the contrast of organic materials: heavy metals dyes are absorbed by the background that creates contrast to the slightly stained specimen. The observation of ECM by

TEM is nearly concomitant of its apparition in the late 1930s. Native collagen fibrils extracted from tissues and stained negatively with phosphotungtsic acid present a typical cross-striated pattern with a series of dark and light bands, spaced with a regular period of 67 nm. These observations have allowed the establishment of the assembly model of collagen molecules into collagen fibrils, known as the quarter-stagger model from Hodge and Petruska (1963). This model proposes a lateral stacking of collagen molecules, creating overlaps that exclude phosphotungstic dye and appear light, and a longitudinal collinear succession of collagen molecule spaced with a constant gap filled by Tungsten dye and appears dark under electron beam [48]. The cross-striated pattern is characteristic of fibrillary collagen, that are collagen type I, type II, and type III. On the other hand, network forming collagen type IV do not present any bands on TEM but is seen as a meshwork of hexagonal structures [49]. The resolution (roughly 1–5 nm) of TEM allows analyzing single macromolecules deposited on carbon film and stained by rotary-shadowing, creating a 3D electron sensitive replica of the specimen [50]. This method has revealed the semiflexible rod structure of collagen molecules terminated by a globular C-term pro-peptide and the cross-shaped triple chain structure of laminin molecules. TEM is particularly accurate to measure length of ECM molecules and diameters of fibrils and fibers assembly. These last parameters are important when analyzing a tissue because ECM fibrils diameters appear to be tissue-specific and modification of their size can be induced by pathologies such as diabetes, fibrosis, cancer, or aging and consequently impair tissue organization and function [51]. Compact bone tissue which supports most of the load of the body and muscles anchorage has to resist strong mechanical solicitations, but it is surprisingly light in weight structure if compared to human-engineered buildings. Bone tissue is made of an abundant organic ECM, strengthened with a mineral phase, and has highly hierarchical structure with length scales ranging from meters to nanometers that give its overall mechanical properties [52]. The shaft of long bones is organized in cylindrical osteons formed by successive concentric lamellae, themselves constituted by compact assemblies of collagen fibrils. An oblique transverse section of successive concentric lamellae made with ultra-microtome and observed with TEM revealed coexistence of two patterns of organization for collagen fibrils [53]. One is an alternation of parallel and orthogonal fibrils, with a regular 90° shift of fibrils orientation from one lamellae to another. The second is seen under TEM as arced structures, as if collagen fibrils were bent in between two series of longitudinal fibrils. The arced pattern is the consequence of the oblique sections into succession of collagen fibrils rotating with a tiny and constant angle from one lamellae to another, creating the illusion of bend structures [54]. From these TEM observations, a twisted plywood model of collagen fibril organization in bone has been proposed. This particular constant angle twist recall the organization observed in some liquid crystal phase, and it has been suggested that collagen molecules could have a liquid crystal behavior and autoassemble in higher-scales structures [21,55]. This finally underlines the potentials of transmission electron microscopy (TEM) to address ECM architectures in tissues. As mentioned above, similar techniques and processing of samples used in optical microscopy are also applied with TEM. The different components of the ECM can be identified by immunolabeling with the same limitations for the necessity to retrieve antigens from the fixation and embedding processes. The antibodies are covalently linked to a gold particle to be seen by TEM and multiple labeling is possible with the use of a specific size of gold particle for each antibody [30].

TEM, however, needs a very thin specimen and cannot directly image a 3D structure. Unlike TEM, scanning electron microscope (SEM) uses electrons reflected from the surface of the sample as signals for image generation and provides information on surface topography, fibrillar organization, porosity, and also atomic composition of a bulk sample [56]. Samples have to be dehydrated to enter the low-vacuum chamber of the microscope and coated with an electron conducting layer (commonly gold) to ensure an adequate contrast and avoid charging phenomenon on the sample which are deleterious for the quality of the image. To keep their native 3D structure intact, biological samples are usually dehydrated by ethanol treatment and to a critical point drying. This procedure has enabled the evaluation of collagen fibrils diameters and spatial organization in reconstituted collagen hydrogels [57]. In biological tissues, the higher proportion of cells compared to ECM can minimize the access to the ECM fibrillar network. By a gentle decellularization method, the cellular counterpart of the tissue is removed and the native ECM frame remains [58]. This process mainly keeps in their original shape the reticular fibers of collagen and elastin but degrade most part of the laminins and GAGs network.

3. Extracellular matrix-inspired biomaterials

The deep exploration of ECMs composition, organization, and biological functions associated with the development of methods to produce new biocompatible materials has enabled material scientists to recreate *ex vivo* some of the key characteristics of ECM [59]. This section focuses on how the structural and functional characteristics derived from the knowledge of the native cell microenvironment have been applied to design biologically relevant biomaterials. Different strategies currently exist to build 3D models of the ECM: tissue-derived ECM, use of natural or synthetic polymers, and formulation into hydrogel or porous 3D materials. Some biomaterials are designed to recreate the composition of the ECM and thus offering the right environment for studying cell adhesion and anchorage-associated cell phenotypes. Other materials are developed to recreate the 3D architectures of ECM, proposing fibrillary structures with similar organization and mechanical properties of native tissues. These examples represent preliminary attempts of *ex vivo* models of ECM that will most likely be improved and increase with an overcoming of technical hurdle faced by material scientists and with rising interest of cell biologists for 3D models that will ask for more refined and specific materials to answer fundamental questions on cell biology.

3.1. Strategies to engineer 3D models of ECMs

3.1.1. Tissue-derived ECM: the gold standard Matrigel®

A basement membrane-derived tissue isolated from Engelberth–Holm–Swarm (EHS) mouse sarcoma is commercially available under the brand name Matrigel® (BD Biosciences) and has become widely used to evaluate cell migration, cancer cells behavior, and to create organoids *in vitro*. EHS-sarcoma produces a large amount of ECM rich in collagen type I, laminin-111, heparin sulfate proteoglycans (Perlecan), and Nidogen that are the main constituents of BM [60]. This basement membrane extract is liquid at 4°C and turns into a gel at 37°C under physiological pH and ionic strength. Matrigel® is currently the gold standard in most of the 3D assays performed in cell biology. Indeed, there is not yet any material in the market that is able to better reproduce the composition and partially the organization of BM, in particular because Matrigel® is obtained directly from animals and is not an *ex vivo* engineered ECM. It is a ready-to-use solution that allows user-defined utilization, even if manipulating Matrigel® requires skills and experience. The success of Matrigel® is also due to its biological activity that allows under normal conditions of culture the differentiation of several cell types and the formation of complex structures like vessels or mammary glands acinar structures [61,62]. Besides structural ECM molecules, Matrigel® contains soluble growth factors such as fibroblast growth factors (FGF), epidermal growth factor (EGF), and transforming growth factor-β 19 (TGF-β) and matrix metalloproteinases (MMPs) including MMP-2 and -9. The unique coexistence in an *ex vivo* substrate for cell culture of native and organized fibrous ECM proteins associated with soluble factors explains its genuine bioactivity. Concomitantly, this represents its main drawback to serve as an ECM model for 3D cell cultures. Because Matrigel® is produced and purified from an animal, there is a lack of control on its exact composition and a batch-to-batch variability of its content [63]. Moreover, the presence of growth factors in an unknown and uncontrollable amount can interfere (positively or negatively) with the intended parameters to be studied, like with the evaluation signaling molecule or a drug, and invalidate the use of Matrigel® in any experiments where the role of a growth factors in a cellular process would like to be addressed. The work of Edna Cukierman has demonstrated the dramatic changes in cell morphology in-between 2D and 3D cell culture systems with a massive reorganization of cell cytoskeleton and a modification of integrins expression [64]. This has increasingly recognized the importance of studying cells, in particular their adhesion and migration, within a 3D environment. Thus, Matrigel® should be considered with caution and with the full awareness of its limitations when it is chosen for a 3D model and so not creating a "black-box" during the switch of cell biology from 2D to 3D.

3.1.2. Biological polymers

Biological polymers were first used as a coating of tissue culture dishes to favor cell adhesion and spreading, and then incorporated into 3D materials under different forms (hydrogels, freeze-dried materials, and surface coating of bulk inorganic materials). Most of the biological polymers used in biomaterials are structural molecules derived from mammalian ECM such as collagens (type I–IV), elastin, fibronectin, laminin (mainly laminin-111), fibrin, and glycoaminoglycans (Hyaluronan, Chondroitin sulfate, and Heparan sulfate) [65]. They are classically purified from animals' ECM-rich tissues such as dermis and tendons (collagen type I, elastin), cartilages (collagen type II and GAGs), tumors (laminins and collagen IV) or directly from blood (fibrin and fibronectin). Nowadays, it is possible to obtain some of these molecules from DNA-recombinant sources. This allows to work with human ECM molecules, produced with a high degree of purity and free of many pathogens [66]. However, DNA-recombinant production of multimeric high molecular weight molecules is still a challenge and it minimizes the number and the amount of molecules available from this source and raises dramatically their prices. Nonmammalian ECM molecules are also widely used in the design of biomaterials, mainly for their ability to self-assemble in 3D structures. Numerous materials are engineered in the form of hydrogel from chitin/chitosan (polysaccharide purified from fungi or arthropods exoskeleton), agarose or alginate (both polysaccharides purified from algae). Silk fibroin, a protein found in silkworm cocoon and spiders, is also used to produce fibrous materials [67]. Because of their origin, nonmammalian biological polymers lack many of the cell adhesion cues on their structures and should be most often supplemented with adhesive molecules or peptides to obtain a biologically active material. Their inherent abilities to form 3D materials remain their best interests. Last type of biological polymers usable in the design of 3D materials are growth factors (FGF, EGF, platelet derived growth factor [PDGF], and vascular endothelial growth factor [VEGF]) that give important biological signals to the cells and enzymes (MMPs, proteolytic enzymes) that can favor a remodeling and a progression with time of the material, like that is observed in the healing processes and during tumor invasion. However, as mentioned in the first section of this chapter, these molecules are not structural proteins and thus no 3D material can be raised from them. Nevertheless, both growth factors and enzymes represent essential molecules to reinforce the basic 3D scaffold in a relevant ECM model. For that purpose, there are strategies to incorporate these soluble molecules into a material and trigger their appropriate release at specific time points or location in the material [59].

3.1.3. Synthetic polymers

Synthetic organic polymers offer a large panel of creativity to produce 3D materials. They inherently lack basic biological activity but possess a great processing flexibility. They are easier to produce as well as purify in large quantities and finally, are free from animal contaminations. Synthetic polymers are suitable for many types of chemical modifications such as chemical grafting of adhesive peptides or incorporation of bioactive molecules and can be processed into 3D materials with many types of techniques (electrospinning, foaming, hydrogel, and sheets), some of them not bearable by biological polymers. The diversity of synthetic polymers used in biomaterials is large, including polyacrylamides, polyacrylates, polyethers (e.g., polyethylene glycol), polyesters (e.g., polycaprolactone), polyhydroxy acids (e.g., poly lactic acid, poly glycolic acid, and copolymers poly lactic-co-glycolic acid), polyfumarates, and polyphosphazenes [68]. Polyesters and polyhydroxy acids have both biodegradable properties with presence of hydrolysable bonds in their backbone, whereas polyacrylamides and polyacrylates are almost unbreakable under cell culture conditions. Chemically or genetically engineered peptides or protein-like polymers with amphiphilic or autoassembling properties offer a direct incorporation of bioactive cues into fibrillar materials. Inorganic materials such as metal (e.g., titanium, stainless, and cobalt) and mineral alloys (e.g., ceramics) are widely used as bone and dental substitutes, but their bulk and stiff structures associated with poor possibilities of chemical modifications make them rarely used as *ex vivo* engineered ECM [69]. Synthetic polymers are usually functionalized by chemical grafting of peptides which are recognized for cell adhesion such as the well-known Arg-Gly-Asp (RGD) motif found in numerous ECM molecules (fibronectin, collagen, and vitronectin) and bind by several integrins. More specific ECM-derived peptides can also be grafted, such as DGEA and GFPGER sequences found in collagen I and IV, respectively, or IKLLI and YIGSR sequences from laminin alpha-1 and beta-1 chain, respectively. Functionalization strategies use the inherent properties of the synthetic polymers to form a 3D scaffold and to contain active chemical groups (amine, acid, or alcohol functions) sensitive to chemical reactions and so providing an ECM model with well-defined characteristics in order to answer a specific question. Synthetic polymers can also be functionalized with whole proteins such as growth factors or enzymes. Through the mediation of a peptide spacer incorporating a cleavage sequence, the release of the bioactive compounds may be triggered at the required time or progressively by the proper proteolytic action of the cells [70]. In conclusion, synthetic polymers represent an engineering solution to rationalized parameters tested in 3D model of ECM and can introduce a dynamic aspect into the system [71,72].

3.1.4. *Hydrogels*

Hydrogels are produced from polymers in solutions which are gelated into 3D materials that possess high water content. To some extent, hydrogels behave in the same way than proteoglycans in ECM and represent a valuable mimicry of soft-tissues ECM structures and physical properties due to its hydrated state. They are most often produced by mild, physiological conditions and, as such, are mainly compatible with cell encapsulation and culture, as water is the natural solvent for living organisms and biochemical reactions. The water content of the gels also facilitates the diffusion of low molecular weight nutriments, oxygen, and metabolic waste. Hydrogel can be produced from a large variety of polymers, including both natural and synthetic substrates, offering a large diversity of biochemical, physical, and mechanical properties. Depending on polymers characteristics, the gelation can be reversible or irreversible and triggered by different factors such as multivalent ions, chemical covalent cross-linking (including aldehyde fixatives), and physical phase-transition induced by temperature, pH, or concentration [73]. Covalent cross-linking can form hydrogel with almost all types of polymers in solutions, but the toxicity of cross-linking molecules must be considered, and furthermore, most of the cross-links are nondegradable by cells or in physiological conditions. Moreover, the gelating process can be deleterious for cells, and thus it will dictate the ways that cells are associated with the scaffold (i.e., before or after gelation). Mixing the cells with the polymers prior gelation allows a homogenous distribution of the cells inside the material, which could be more difficult to achieve when seeding the cells on an already formed hydrogel [74].

Natural polymer hydrogels can be made with collagen, fibrin, hyaluronan, alginate, and agarose. Collagen hydrogels are formed by pH neutralization of acid solutions that trigger the assembly of the collagen molecules into fibrils and fibers which will stabilize the structure of the gel. Mechanical properties of collagen hydrogels can be finely modulated by adjusting their concentrations [75]. Fibrin gels are formed following the same reaction that occurs in the body after an injury or an inflammatory response: fibrinogen is mixed in solution with thrombin, a serine protease, which hydrolyses the N-terminal peptide of fibrinogen to create fibrin monomers that assembles into fibrils that will produce an interconnected fibrous hydrogel. Mechanical properties of fibrin gel are tuned by adjusting fibrinogen and thrombin concentrations. However, fibrin gels are highly sensitive to several proteases, which are secreted by cells, and a gel degradation regularly occurs in long-term cell cultures if no protease inhibitors are added in the culture media [76]. Hyaluronan hydrogels are formed by covalent cross-linking with hydrazide derivatives that react with the carboxylic acid groups of the polymer. They bear inherent biological properties and are susceptible for further chemical modifications which modulate these properties, such as the grafting of fibrous proteins or adhesive peptides. However, hyaluronan hydrogels are mechanically poor, and hyaluronan is more often incorporated into other materials than being the main frame of a hydrogel. Several different types of materials are produced from alginate hydrogels. Alginate polymerization is triggered by addition of divalent cations Ca2+, Mg2+, Ba2+, or Sr2+ that create bridges in-between negative charges of alginate monomers. Processing of alginate hydrogels enables an easy cell encapsulation, but a progressive diffusion and solution-exchange of divalent cation can undermine its structure. Agarose gels are formed by heating (near boiling temperature) of the solution that gelate with cooling. Different porous architectures and mechanical properties are obtained with modulation of agarose concentration.

Hydrogels can be formed with synthetic polymers including polyethylene oxide, polyvinyl alcohol or in a blended solution or copolymer association with poly lactic acid. Again, the interest of synthetic polymers to form hydrogels is the diversity and the reproducibility of materials that they are able to produce, with versatile biophysical, mechanical and biological properties. Poly ethylene glycol (PEG) polymers can be modified to allow *in situ* gelation by visible UV-light induction [77]. Photopolymerization is based on the interaction of light with photo initiators that create free radicals which react with the polymer and initiate cross-links. Compared to other type of chemical cross-linkers, photopolymerization is fast (second or minutes) and allows a spatial control of the polymerization. It works under physiological conditions and the radical species, even harsh for living species, are quickly removed by the polymerization process, making these materials more free of adverse chemical compounds as it may occur with aldehyde or hydrazide derivatives. Photopolymerization of PEG-based solutions was used to encapsulate chondrocytes that were cultured for 14 days with progressive deposition of a cartilaginous native ECM showing the biocompatibility of the process [77]. Synthetic peptides containing cell adhesion cues can also be designed for autoassembling into supramolecular structures able to form hydrogels. Engineered proteins produced by recombinant DNA can, in the same way, present adhesive domains and reactive groups such as thiols and amines for being the specific targets of polymerization reactants [72]. Finally, hydrogel-forming properties of synthetic polymers can be used to include biological polymers unable to form otherwise materials on their own and then offering a mixed inert 3D structures with biological functions.

3.1.5. Porous material

Hydrogels show a good biocompatibility due to their water content but present most often low mechanical properties, high degradation rates, and a compromised deep diffusion of large molecules like proteins. To overcome these issues, porous materials with interconnected pore networks and surfaces or fibers to sustain cell adhesion as well as cell phenotype have been designed. These materials are discriminated in microporous scaffolds where embedded cells will mainly attach on the surface of pores of >100μm diameter and will more represent curved 2D surfaces, and in nanoporous scaffolds where pore structures are in the range of the cell diameter (roughly 10 μm) and represent more the native 3D environment of ECMs [78]. Nevertheless, microporous structures allow a more effective cell penetration and migration into the material while smaller-sized pores can represent a barrier for cell colonization of the material, and limit the cell interactions at the edges of the material [79]. This particular point of cell colonization represents one of the challenges with nanoporous materials, and definitely a limitation when directly compared with both the microporous and the hydrogel materials. Similar to hydrogels, porous materials can be prepared with both natural and synthetic polymers with a large panel of techniques including, but not restricted to electrospinning, phase separation, templating, and vapor-phase polymerization [79]. Electrospinning forms fibrillar materials with control on the fibers diameters, fibers alignment, and fibers interdistances that dictate the overall porosity of the material. A polymeric solution is injected through a capillary tube into a high electric field that forms a Taylor cone with application of the electrostatic forces. On the opposite side of the capillary, the collector of the electrospun fibers is charged or grounded. The electric field in-between capillary and collector accelerates the flow of the polymer solution that evaporates the solvent and generate coalescence of the polymers that finally form solid fibers arrays. This technique is mainly used with synthetic polymers (poly lactic/glycolic acid and poly caprolactone) but is also suitable for natural polymers such as collagen and silk fibroin [80]. Electrospinning produces thin materials not exceeding millimeters and with a high-packing density of fibers that can impair a deep colonization by cells, but intensive researches are presently conducted to overcome these issues [81]. Collagen porous materials can be formed by thermally induced phase separation leading to the production of the so-called "collagen sponges" due to their high porous structures. Acidic collagen solutions are frozen which induces a phase separation of collagen molecules from the water-based solvent that is then eliminated by lyophilization. Materials are then made of dense collagen walls with unstriated microfibrils [57] or with native cross-striated fibrils [82]. Size and interconnectivity of the pores can be modified by modulation of the length and temperature of the phase separation, by modulation of the collagen concentration or by mixing the collagen solutions with other natural polymers such as GAGs [82] or with synthetic polymers such as poly lactic acids. Freeze-dried collagen materials have weak mechanical properties and are often strengthened by the addition of GAGs, or cross-linked by chemical species (aldehydes) or dehydrothermal processes [83].

The various polymers and processes presented above enable the engineering of different types of 3D materials designed to study *ex vivo* a large variety of cellular phenomenon such as cell differentiation, drug response, tumor formation, cell migration, cell morphology and cytoskeleton organization, cell death and proliferation, tissue architecture, and coculture behavior of cells [84]. For that purpose, 3D materials are built either to mimic ECM composition, ECM organization, or ECM mechanical properties. We present here examples of such engineered

3D models of ECM with description of the material characteristics and cellular outcomes observed.

3.2. Three-dimensional models mimicking ECM composition

Recreating *ex vivo* the ECM composition in 3D materials offers the possibility to address in a relative more biologically relevant environment the specific function of one or several of its components on several cell mechanisms. It can be used to recreate cell adhesion and cell migration or to offer adhesive cues that are specially organized.

A minimal system has been designed with a gelatin hydrogel mixed with PEG and functionalized with RGD and PHSRN adhesion peptides to evaluate the inflammatory response of adhering immune cells [85]. Monocytes are circulating immune cells which are among the first cells to react at the site of an injury. Once they have migrated and attached to the site of aggression, they start to secrete cytokines and ECM remodeling enzymes. The hydrogel was built with PEG-diacrylate (PEG-da) photopolymerizable polymers mixed with gelatin monomers covalently grafted with PEG-RGD and PEG-PHSRN, two amino acid sequences found in fibronectin and recognized by several types of integrins. The hydrogel mixture was made with different ratios of Gelatin-PEG-peptides compared to the PEG-da polymers for the formation of hydrogels with different adhesive peptide densities (from 30 to 50% of gelatin-modified polymers). The study showed that monocyte adhesion and cytokine secretion reached a plateau when the gelatin-modified polymers concentration was above 40%, indicating that rising adhesive-peptides densities did not improve their accessibility for monocytes. The nature of the ligands has also an influence on monocytes behavior, with a higher level of MMP-9 expression when hydrogels are incorporating PHSRN peptides, while more interlekin-1β was secreted in RGD hydrogels. These data suggest a specific monocyte inflammatory response for each of these adhesive cues. This model can be developed and used, for instance, to study the different parameters that initiates the inflammatory response at the site of injury.

Metastatic cancerous cells escape the original tumor site to attach and invade a distant organ. The switch of the cellular microenvironment from primary tumor site to the metastatic organ can affect the metastatic cells response to therapeutics which was originally designed to act on the original cancerous cells in the specific microenvironment of the primary tumor site. Blehm and coworkers have engineered a 3D model of the ECM of the perivascular niche of the brain, a metastatic target of cutaneous melanomas, to address the effect of anti-MAPK therapeutics (ERK inhibitor and Darafemib) on metastatic cutaneous melanomas cells lines [86]. Cells were seeded in a huyaluronic acid-PEG-diacrylate hydrogel supplemented with purified full-length proteins, laminin-111 and fibronectin, or with RGD-cyclic peptides. Diminution of Darafemib drug efficacy is reported for one melanoma cell line in hydrogels supplemented with fibronectin, whereas laminin-supplemented hydrogels protected against ERK inhibition. With another cell line, drug efficacy was weakened for both therapeutics only on laminin-supplemented hydrogels, reinforcing the importance of the cellular microenvironment for drug testing.

Natural polymers, including ECM proteins, GAGs, and growth factors, with a high degree of purity were successively assembled to form molecularly defined materials evaluated by subcutaneous implantation as well as in a cartilage reconstruction model [87]. Collagen type I solution purified from bovine Achilles tendon was mixed with insoluble elastin, purified from equine ligamentum nuchae and homogenized in a grinder before molding and freeze-drying of the solution. This process formed a porous scaffold (collagen sponge-based material) with both collagen and elastin fibrils. Porosity of the material was controlled by the freezing temperatures, as higher rates of freezing provided materials with smaller pore diameter. Complexity and ECM mimicry of the material is extended by incorporation of chondroitin sulfate or heparan sulfate. Collagen-elastin material is soaked with either chondroitin or heparan sulfate solution, and GAGs are covalently bounded to the collagen-elastin fibers by carbodiimide cross-linking. The amount of GAGs effectively grafted to the materials is a function of GAGs structures and their respective reactiveness to the cross-linker, with a final grafting of 10% of the original heparan sulfate solution and 6% of the chondroitin sulfate solution. Finally, growth factors such as bFGF and VEGF were loaded into the Collagen-Elastin-GAG material to form a valuable model of the different classes of natural polymers found in ECMs. Incubation of growth factor solution with the material only allowed a partial adsorption as growth factors are rapidly lost after a few hours of cultures. However, heparan sulfate bears native biding sites for bFGF, enabling a more efficient loading of bFGF into the material and a progressive and extended release during cell culture. Two-weeks after subcutaneous implantation in rat, collagen-heparan sulfate materials showed a higher vascularization than collagen alone. After 10 weeks, collagen materials loaded with bFGF exhibited a lower cell invasion and vascularization than the collagen-heparan sulfate-bFGF material. Even evaluated *in vivo*, where cells are in a more positive environment than cultured *in vitro*, this study showed the possibility to engineer a complex material reproducing the different class of natural polymers found in ECMs. A critical parameter for the use of such porous material as an *ex vivo* ECM model is the possibility to seed evenly the cells within the material as compared to hydrogel-embedding. However, this can be overcome by creation and control of an open and interconnected porosity within the material.

3.3. Mimicking ECM organization and mechanics

Two-dimensional cell cultures on TCPS have been coated with ECM molecules to give more biologically relevant ligands for cell adhesion, spreading, migration, and expression of phenotypes. This has partly raised the idea that instead of coating, ECM-derived adhesive cues should be displayed in 3D to be more close to the cell environment, and we have presented above 3D materials that recreate ECM composition. However, this approach does not generally consider the reconstruction of the spatial organization of ECM that is of equal importance as the ECM composition [88].

The study of collagen-rich connective tissues such as skin, tendons, and bones by polarized-light microscopy and TEM revealed that type I collagen is highly structured, with a spatial geometry specific for each tissue. This spatial organization of collagen I in tissues can be reproduced *in vitro*, using the autoassembling properties of collagen molecules in acidic

http://dx.doi.org/10.5772/62979

solutions which will self-organize in different liquid-crystal organizations dependent on the collagen concentration [89]. Regulated evaporation of solvent raise progressively the collagen concentration and modify collagen molecules organization. These organized molecular textures are stabilized by neutralization of the pH of the solution and further triggering the assembly of soluble collagen molecules into insoluble cross-striated native collagen fibrils [54,90]. The supramolecular assembly of collagen molecules into collagen fibrils also induce the transition from the original collagen solution to a solid hydrogel structure (i.e., so-called "sol/gel transition") that retains the molecular tissue-like organization of collagen molecules. This leads to materials with collagen fibrils architecture which may mimic those observed in tendon, skin, and bones [21,91] but that also reproduce the mechanical features of collagen-rich tissues from dermis to bone, depending on the collagen concentration of the material [75]. Ordered collagen hydrogels have been used to study myofibroblasts migration as a model of wound-healing [92], behavior of human osteoblasts in long-term *in vitro* cultures [57,93] and also as a model to study the kinetics of noncollagenous ECM proteins secretion in relation to the biomineralization processes [94]. However, tissue-like organization of collagen fibrils resulted in mild (20–40 mg/mL) or high (>80 mg/mL) collagen concentrations, which mainly restricted cell seeding to the surface of the materials.

In the depth of a tissue, different ECM organization could be found, like in the skin with different ECM compositions, organizations, and properties of epidermis and dermis, or in articular cartilage from the GAGs-rich upper surface to the deeper interface with the underlying bone tissue. A layer-by-layer approach, based on successive freeze-drying of different natural polymer solutions to modulate material organization, was used to recreate the discrepancies in the ECM organization observed in the cartilage tissue [95]. The first layer intends to recreate main features of the deeper osteochondral tissue architecture. Collagen type I in acidic solution is mixed with hydroxyapatite, the mineral phase found in bones, and freeze-dried at a constant rate to obtain a dense porous structure. The subsequent porous material was cross-linked with carbodiimide to strengthen its structure. The second layer mimics the bone–cartilage interface and is made with an equal amount of mixture of the collagen type I, main collagen of bone tissue, and of the collagen type II that is the main collagen of cartilage tissue. Collagen I and II mixture is supplemented with hydroxyapatite at a final concentration five-times lower than on the first layer. This mixture is poured on the surface of the first layer and freeze-dried to finally form a two-layer material, with two distinct compositions and porous structures. The last layer mimics the articular cartilage tissue. This layer is made from a mixture solution of 25%/75% of collagen type I and II, respectively, and then further supplemented with hyaluronic acid to recreate the GAGs content of the articular cartilage. The last layer is again freeze-dried with time-prolonged freezing and drying steps to ensure the proper porous structure. The overall material is strengthened by a dehydrothermal process that creates amine-based cross-links. This process forms a material with three specific porous structures as well as a molecular composition and with the opened and interconnected porosity that allows an effective cell colonization of each of the three layers.

The specific composition and spatial organization of the ECM dictate its overall mechanical properties, that cells are able to sense through their integrin receptors and their cytoskeletons.

On 2D surfaces, it has been shown that cells are responsive to surface rigidity and that it influences the commitment of mesenchymal stem cells toward differentiation in a specific lineage [96]. Again, surface stiffness applies forces which are unevenly distributed on the cells (i.e., only located at the cell-surface interface). Developing a 3D material with tuned and controllable mechanical properties will generate a more biologically relevant environment to evaluate the role of ECM mechanobiology on cells functions and differentiation processes. To study the influence of mechanical stiffness on mesenchymal stem cells differentiation, a series of alginate gels with elastic modulus ranging from 2.5 to 110 kPa has been developed [97]. Mechanical properties of alginate gels are modulated by the percentage of alginate polymers in the final hydrogel. Because alginate is not sensitive to the degradation of hydrolytic enzymes of mammalian cells, the elastic modulus of the mechanical properties of the material are expected to remain constant all along the study (7 days of cell culture). RGD-peptides are covalently grafted to alginate polymers prior to hydrogel formation to give to the cells adhesive cues. The more rigid materials trigger mesenchymal stem cells differentiation toward bone lineage with an expression of the bone-related molecular markers such as alkaline phosphatase and osteopontin after 7 days of culture. On the other hand, alginate gels with the lower elastic modulus (softer material) triggers an accumulation of oil-droplets into stem cells, indicating adipose tissue differentiation. The density of RGD-peptides incorporated into these materials did not modify the cell fate related to the elastic modulus of the material, but induce a higher level of expression of the lineage markers for both bone and adipose-committed cells.

3.4. Conclusion: toward a gold-standard of 3D model of ECMs?

As mentioned at the beginning of this section, the ECM-derived Matrigel® represents currently the most often used material for 3D experiments in cell biology. Despite that Matrigel®-related drawbacks are of importance, Matrigel®is a widely and available model to investigate many fundamental questions in cell biology, from cell adhesion and tumor formation, to drug testing. We have presented in this chapter a large panel of techniques, methodologies, and engineering processes that allow the exploration of ECM organization and permit to recreate *ex vivo* some of their key features. At the conclusion of this chapter and after the review of several studies investigating various 3D materials, it appears that no material can represent the unique and ideal answer for all cell investigations in 3D [63]. A modular approach should be taken by rationalizing the biological question to be studied and the parameter of ECM intended to be recreated. Nevertheless, more and more complex materials are engineered that will finally be able to mimic simultaneously several key factors of ECM composition, architecture, or mechanical properties, and so enabling investigation of multiple parameters for cell biology experiments. An important drawback with engineered 3D materials is to create a "black-box" where undefined and uncontrollable parameters may influence the cellular outcomes to be investigated. To avoid part of this problem, rigorous attention should be paid on the purity of polymers used to build the material, in particular with biological polymers. The development of DNA-recombinant production of ECM proteins can overcome this problem, even if this will raise ultimately the cost of the final material. The structural characteristics of the final material (porosity, polymers distribution, and fibrils diameter) should be consistently reproducible and addressed. To do so, an "easy," meaning straightforward, process of the material should be

http://dx.doi.org/10.5772/62979

sought and preferred rather than a more complex multistep fabrication process. Biomaterial scientists propose continuously new design and approaches to engineer *ex vivo* ECMs. The production of a gold-standard material may become possible for a specific biological question. It might happen with the existence of a deep and intelligible dialog in-between material scientists, whom brings engineering strategies, and the cell biologists, that implement the material design to mimic the biological process that has to be investigated *ex vivo*. This collaboration may result in major advances for science and medicine.

Author details

Sylvain Vigier[1*] and Tamas Fülöp[2*]

*Address all correspondence to: sylvain.vigier@gmail.com and Tamas.fulop@usherbrooke.ca

1 Department of Chemical and Biotechnological Engineering, Université de Sherbrooke, Sherbrooke, Québec, Canada

2 Department of Medicine, Geriatric Division Research Center on Aging, Université de Sherbrooke, Sherbrooke, Qc, Canada

References

[1] Shaw TJ, Martin P. Wound repair at a glance. J Cell Sci. 2009;122(Pt 18):3209–13.

[2] Frantz C, Stewart KM, Weaver VM. The extracellular matrix at a glance. J Cell Sci. 2010;123(24):4195–200.

[3] Kadler KE, Baldock C, Bella J, Boot-Handford RP. Collagens at a glance. J Cell Sci. 2007;120(Pt 12):1955–8.

[4] Ottani V, Raspanti M, Ruggeri A. Collagen structure and functional implications. Micron. 2001 avril;32(3):251–60.

[5] LeBleu VS, Macdonald B, Kalluri R. Structure and function of basement membranes. Exp Biol Med Maywood NJ. 2007;232(9):1121–9.

[6] Ayad S, Boot-Handford R, Humphries MJ, Kadler KE, Shuttleworth A. 15—Collagen type IV. In: Ayad S, Boot-Handford R, Humphries MJ, Kadler KE, Shuttleworth A, editors. The Extracellular Matrix Facts Book (Second Edition) [Internet]. San Diego, California: Academic Press; 1998 [cited 2016 Jan 7]. pp. 54–62.

[7] Aumailley M. The laminin family. Cell Adhes Migr. 2013;7(1):48–55.

[8] Ushiki T. Collagen fibers, reticular fibers and elastic fibers. A comprehensive understanding from a morphological viewpoint. Arch Histol Cytol. 2002;65(2):109–26.

[9] Wierzbicka-Patynowski I, Schwarzbauer JE. The ins and outs of fibronectin matrix assembly. J Cell Sci. 2003;116(Pt 16):3269–76.

[10] Ayad S, Boot-Handford R, Humphries MJ, Kadler KE, Shuttleworth A. 39 - Fibronectin. In: Ayad S, Boot-Handford R, Humphries MJ, Kadler KE, Shuttleworth A, editors. The Extracellular Matrix FactsBook (Second Edition) [Internet]. San Diego, California: Academic Press; 1998 [cited 2016 Jan 7]. pp. 149–52.

[11] Prydz K. Determinants of Glycosaminoglycan (GAG) Structure. Biomolecules. 2015;5(3):2003–22.

[12] Vigetti D, Karousou E, Viola M, Deleonibus S, De Luca G, Passi A. Hyaluronan: biosynthesis and signaling. Biochim Biophys Acta BBA—Gen Subj. 2014 août;1840(8): 2452–9.

[13] Iozzo RV, Schaefer L. Proteoglycan form and function: a comprehensive nomenclature of proteoglycans. Matrix Biol J Int Soc Matrix Biol. 2015;42:11–55.

[14] Milev P, Monnerie H, Popp S, Margolis RK, Margolis RU. The core protein of the chondroitin sulfate proteoglycan phosphacan is a high-affinity ligand of fibroblast growth factor-2 and potentiates its mitogenic activity. J Biol Chem. 1998;273(34):21439–42.

[15] Horobin RW. Biological staining: mechanisms and theory. Biotech Histochem off Publ Biol Stain Comm. 2002;77(1):3–13.

[16] Monte-Alto-Costa A, Porto LC. Special stains for extracellular matrix. Methods Mol Biol Clifton NJ. 2010;611:131–40.

[17] Ramos-Vara JA. Technical aspects of immunohistochemistry. Vet Pathol Online. 2005;42(4):405–26.

[18] Foster BL. Methods for studying tooth root cementum by light microscopy. Int J Oral Sci. 2012;4(3):119–28.

[19] Levingstone TJ, Thompson E, Matsiko A, Schepens A, Gleeson JP, O'Brien FJ. Multi-layered collagen-based scaffolds for osteochondral defect repair in rabbits. Acta Biomater. 2016;32:149–60.

[20] Lattouf R, Younes R, Lutomski D, Naaman N, Godeau G, Senni K, et al. Picrosirius red staining: a useful tool to appraise collagen networks in normal and pathological tissues. J Histochem Cytochem off J Histochem Soc. 2014;62(10):751–8.

[21] Giraud-Guille M-M, Besseau L, Martin R. Liquid crystalline assemblies of collagen in bone and in vitro systems. J Biomech. 2003;36(10):1571–9.

[22] Howie AJ, Brewer DB, Howell D, Jones AP. Physical basis of colors seen in Congo red-stained amyloid in polarized light. Lab Investig J Tech Methods Pathol. 2008;88(3):232–42.

[23] Montes GS. Structural biology of the fibres of the collagenous and elastic systems. Cell Biol Int. 1996;20(1):15–27.

[24] Hyllested JL, Veje K, Ostergaard K. Histochemical studies of the extracellular matrix of human articular cartilage—a review. Osteoarthr Cartil OARS Osteoarthr Res Soc. 2002;10(5):333–43.

[25] Leong TY-M, Cooper K, Leong AS-Y. Immunohistology—past, present, and future. Adv Anat Pathol. 2010;17(6):404–18.

[26] Schacht V, Kern JS. Basics of immunohistochemistry. J Invest Dermatol. 2015;135(3):e30.

[27] O'Leary TJ, Fowler CB, Evers DL, Mason JT. Protein fixation and antigen retrieval: chemical studies. Biotech Histochem Off Publ Biol Stain Comm. 2009;84(5):217–21.

[28] Shi SR, Key ME, Kalra KL. Antigen retrieval in formalin-fixed, paraffin-embedded tissues: an enhancement method for immunohistochemical staining based on microwave oven heating of tissue sections. J Histochem Cytochem Off J Histochem Soc. 1991;39(6):741–8.

[29] Shi S-R, Shi Y, Taylor CR. Antigen retrieval immunohistochemistry: review and future prospects in research and diagnosis over two decades. J Histochem Cytochem off J Histochem Soc. 2011;59(1):13–32.

[30] Griffiths G, Lucocq JM. Antibodies for immunolabeling by light and electron microscopy: not for the faint hearted. Histochem Cell Biol. 2014;142(4):347–60.

[31] D'Amico F, Skarmoutsou E, Stivala F. State of the art in antigen retrieval for immunohistochemistry. J Immunol Methods. 2009;341(1-2):1–18.

[32] Hayes AJ, Hughes CE, Caterson B. Antibodies and immunohistochemistry in extracellular matrix research. Methods San Diego Calif. 2008;45(1):10–21.

[33] Saper CB. A guide to the perplexed on the specificity of antibodies. J Histochem Cytochem Off J Histochem Soc. 2009;57(1):1–5.

[34] Virtanen I, Banerjee M, Palgi J, Korsgren O, Lukinius A, Thornell L-E, et al. Blood vessels of human islets of Langerhans are surrounded by a double basement membrane. Diabetologia. 2008;51(7):1181–91.

[35] Otonkoski T, Banerjee M, Korsgren O, Thornell L-E, Virtanen I. Unique basement membrane structure of human pancreatic islets: implications for beta-cell growth and differentiation. Diabetes Obes Metab. 2008;10 Suppl 4:119–27.

[36] Holčapek M, Jirásko R, Lísa M. Recent developments in liquid chromatography-mass spectrometry and related techniques. J Chromatogr A. 2012;1259:3–15.

[37] Roepstorff P. Mass spectrometry based proteomics, background, status and future needs. Protein Cell. 2012;3(9):641–7.

[38] Eng JK, Searle BC, Clauser KR, Tabb DL. A face in the crowd: recognizing peptides through database search. Mol Cell Proteomics MCP. 2011;10(11):R111.009522.

[39] Naba A, Clauser KR, Ding H, Whittaker CA, Carr SA, Hynes RO. The extracellular matrix: tools and insights for the "omics" era. Matrix Biol J Int Soc Matrix Biol. 2016 Jan;49:10–24.

[40] Wilson R, Diseberg AF, Gordon L, Zivkovic S, Tatarczuch L, Mackie EJ, et al. Comprehensive profiling of cartilage extracellular matrix formation and maturation using sequential extraction and label-free quantitative proteomics. Mol Cell Proteomics MCP. 2010;9(6):1296–313.

[41] Rotilio D, Della Corte A, D'Imperio M, Coletta W, Marcone S, Silvestri C, et al. Proteomics: bases for protein complexity understanding. Thromb Res. 2012;129(3):257–62.

[42] Ishihama Y, Oda Y, Tabata T, Sato T, Nagasu T, Rappsilber J, et al. Exponentially modified protein abundance index (emPAI) for estimation of absolute protein amount in proteomics by the number of sequenced peptides per protein. Mol Cell Proteomics. 2005;4(9):1265–72.

[43] Wilson R. The extracellular matrix: an underexplored but important proteome. Expert Rev Proteomics. 2010;7(6):803–6.

[44] Byron A, Humphries JD, Humphries MJ. Defining the extracellular matrix using proteomics. Int J Exp Pathol. 2013 Apr;94(2):75–92.

[45] de Castro Brás LE, Ramirez TA, DeLeon-Pennell KY, Chiao YA, Ma Y, Dai Q, et al. Texas 3-step decellularization protocol: looking at the cardiac extracellular matrix. J Proteomics. 2013;86:43–52.

[46] Egerton RF. Physical Principles of Electron Microscopy [Internet]. Boston, MA: Springer US; 2005 [cited 2016 Mar 2]. Available from: http://link.springer.com/10.1007/b136495

[47] Quantock AJ, Winkler M, Parfitt GJ, Young RD, Brown DJ, Boote C, et al. From nano to macro: studying the hierarchical structure of the corneal extracellular matrix. Exp Eye Res. 2015;133:81–99.

[48] Holmes DF, Graham HK, Trotter JA, Kadler KE. STEM/TEM studies of collagen fibril assembly. Micron Oxf Engl 1993. 2001;32(3):273–85.

[49] Khoshnoodi J, Pedchenko V, Hudson BG. Mammalian collagen IV. Microsc Res Tech. 2008;71(5):357–70.

[50] Hendricks GM. Metal shadowing for electron microscopy. Methods Mol Biol Clifton NJ. 2014;1117:73–93.

[51] Eyden B, Tzaphlidou M. Structural variations of collagen in normal and pathological tissues: role of electron microscopy. Micron. 2001;32(3):287–300.

[52] Dalle Carbonare L, Valenti MT, Bertoldo F, Zanatta M, Zenari S, Realdi G, et al. Bone microarchitecture evaluated by histomorphometry. Micron. 2005;36(7–8):609–16.

[53] Giraud-Guille MM. Twisted plywood architecture of collagen fibrils in human compact bone osteons. Calcif Tissue Int. 1988;42(3):167–80.

[54] Mosser G, Anglo A, Helary C, Bouligand Y, Giraud-Guille M-M. Dense tissue-like collagen matrices formed in cell-free conditions. Matrix Biol J Int Soc Matrix Biol. 2006;25(1):3–13.

[55] Giraud-Guille MM. Twisted liquid crystalline supramolecular arrangements in morphogenesis. Int Rev Cytol. 1996;166:59–101.

[56] Mammadov R, Tekinay AB, Dana A, Guler MO. Microscopic characterization of peptide nanostructures. Micron. 2012 février;43(2–3):69–84.

[57] Vigier S, Helary C, Fromigue O, Marie P, Giraud-Guille M-M. Collagen supramolecular and suprafibrillar organizations on osteoblasts long-term behavior: benefits for bone healing materials. J Biomed Mater Res A. 2010;94(2):556–67.

[58] Ohtani O. Three-dimensional organization of the connective tissue fibers of the human pancreas: a scanning electron microscopic study of NaOH treated-tissues. Arch Histol Jpn Nihon Soshikigaku Kiroku. 1987;50(5):557–66.

[59] Hubbell JA. Materials as morphogenetic guides in tissue engineering. Curr Opin Biotechnol. 2003;14(5):551–8.

[60] Benton G, Kleinman HK, George J, Arnaoutova I. Multiple uses of basement membrane-like matrix (BME/Matrigel) in vitro and in vivo with cancer cells. Int J Cancer J Int Cancer. 2011;128(8):1751–7.

[61] Arnaoutova I, George J, Kleinman HK, Benton G. The endothelial cell tube formation assay on basement membrane turns 20: state of the science and the art. Angiogenesis. 2009;12(3):267–74.

[62] Ghajar CM, Bissell MJ. Extracellular matrix control of mammary gland morphogenesis and tumorigenesis: insights from imaging. Histochem Cell Biol. 2008;130(6):1105–18.

[63] Serban MA, Prestwich GD. Modular extracellular matrices: solutions for the puzzle. Methods San Diego Calif. 2008;45(1):93–8.

[64] Cukierman E, Pankov R, Stevens DR, Yamada KM. Taking cell-matrix adhesions to the third dimension. Science. 2001;294(5547):1708–12.

[65] Jafari M, Paknejad Z, Rad MR, Motamedian SR, Eghbal MJ, Nadjmi N, et al. Polymeric scaffolds in tissue engineering: a literature review. J Biomed Mater Res B Appl Biomater. 2015.

[66] Nagaoka M, Jiang H-L, Hoshiba T, Akaike T, Cho C-S. Application of recombinant fusion proteins for tissue engineering. Ann Biomed Eng. 2010;38(3):683–93.

[67] Kapoor S, Kundu SC. Silk protein-based hydrogels: promising advanced materials for biomedical applications. Acta Biomater. 2016 février;31:17–32.

[68] Seal BL, Otero TC, Panitch A. Polymeric biomaterials for tissue and organ regeneration. Mater Sci Eng R Rep. 2001;34(4–5):147–230.

[69] Saha K, Pollock JF, Schaffer DV, Healy KE. Designing synthetic materials to control stem cell phenotype. Curr Opin Chem Biol. 2007;11(4):381–7.

[70] Zisch AH, Lutolf MP, Ehrbar M, Raeber GP, Rizzi SC, Davies N, et al. Cell-demanded release of VEGF from synthetic, biointeractive cell ingrowth matrices for vascularized tissue growth. FASEB J Off Publ Fed Am Soc Exp Biol. 2003;17(15):2260–2.

[71] Liu J, He X, Corbett SA, Lowry SF, Graham AM, Fässler R, et al. Integrins are required for the differentiation of visceral endoderm. J Cell Sci. 2009;122(2):233–42.

[72] Lutolf MP, Hubbell JA. Synthetic biomaterials as instructive extracellular microenvironments for morphogenesis in tissue engineering. Nat Biotechnol. 2005;23(1):47–55.

[73] Lee KY, Mooney DJ. Hydrogels for tissue engineering. Chem Rev. 2001;101(7):1869–79.

[74] Drury JL, Mooney DJ. Hydrogels for tissue engineering: scaffold design variables and applications. Biomaterials. 2003;24(24):4337–51.

[75] Ramtani S, Takahashi-Iñiguez Y, Helary C, Geiger D, Guille MMG. Mechanical behavior under unconfined compression loadings of dense fibrillar collagen matrices mimetic of living tissues. J Mech Med Biol. 2010;10(01):35–55.

[76] DeVolder R, Kong H-J. Hydrogels for in vivo-like three-dimensional cellular studies. Wiley Interdiscip Rev Syst Biol Med. 2012;4(4):351–65.

[77] Nguyen KT, West JL. Photopolymerizable hydrogels for tissue engineering applications. Biomaterials. 2002;23(22):4307–14.

[78] Tibbitt MW, Anseth KS. Hydrogels as extracellular matrix mimics for 3D cell culture. Biotechnol Bioeng. 2009;103(4):655–63.

[79] Puppi D, Zhang X, Yang L, Chiellini F, Sun X, Chiellini E. Nano/microfibrous polymeric constructs loaded with bioactive agents and designed for tissue engineering applications: a review. J Biomed Mater Res B Appl Biomater. 2014;102(7):1562–79.

[80] Kundu B, Rajkhowa R, Kundu SC, Wang X. Silk fibroin biomaterials for tissue regenerations. Adv Drug Deliv Rev. 2013 avril;65(4):457–70.

[81] Guimarães A, Martins A, Pinho ED, Faria S, Reis RL, Neves NM. Solving cell infiltration limitations of electrospun nanofiber meshes for tissue engineering applications. Nanomed. 2010;5(4):539–54.

[82] O'Brien FJ, Harley BA, Yannas IV, Gibson LJ. The effect of pore size on cell adhesion in collagen-GAG scaffolds. Biomaterials. 2005 février;26(4):433–41.

[83] Tierney CM, Haugh MG, Liedl J, Mulcahy F, Hayes B, O'Brien FJ. The effects of collagen concentration and crosslink density on the biological, structural and mechanical properties of collagen-GAG scaffolds for bone tissue engineering. J Mech Behav Biomed Mater. 2009;2(2):202–9.

[84] Ravi M, Paramesh V, Kaviya SR, Anuradha E, Solomon FDP. 3D cell culture systems: advantages and applications. J Cell Physiol. 2015;230(1):16–26.

[85] Chung AS, Waldeck H, Schmidt DR, Kao WJ. Monocyte inflammatory and matrix remodeling response modulated by grafted ECM-derived ligand concentration. J Biomed Mater Res A. 2009;91(3):742–52.

[86] Blehm BH, Jiang N, Kotobuki Y, Tanner K. Deconstructing the role of the ECM microenvironment on drug efficacy targeting MAPK signaling in a pre-clinical platform for cutaneous melanoma. Biomaterials. 2015 juillet;56:129–39.

[87] Geutjes PJ, Daamen WF, Buma P, Feitz WF, Faraj KA, van Kuppevelt TH. From molecules to matrix: construction and evaluation of molecularly defined bioscaffolds. Adv Exp Med Biol. 2006;585:279–95.

[88] Nelson CM, Tien J. Microstructured extracellular matrices in tissue engineering and development. Curr Opin Biotechnol. 2006;17(5):518–23.

[89] Martin R, Farjanel J, Eichenberger D, Colige A, Kessler E, Hulmes DJ, et al. Liquid crystalline ordering of procollagen as a determinant of three-dimensional extracellular matrix architecture. J Mol Biol. 2000;301(1):11–7.

[90] Besseau L, Giraud-Guille MM. Stabilization of fluid cholesteric phases of collagen to ordered gelated matrices. J Mol Biol. 1995;251(2):197–202.

[91] Giraud-Guille MM, Mosser G, Helary C, Eglin D. Bone matrix like assemblies of collagen: from liquid crystals to gels and biomimetic materials. Micron Oxf Engl 2005;36(7–8):602–8.

[92] Helary C, Ovtracht L, Coulomb B, Godeau G, Giraud-Guille MM. Dense fibrillar collagen matrices: a model to study myofibroblast behaviour during wound healing. Biomaterials. 2006;27(25):4443–52.

[93] Vigier S, Catania C, Baroukh B, Saffar J-L, Giraud-Guille M-M, Colombier M-L. Dense fibrillar collagen matrices sustain osteoblast phenotype in vitro and promote bone formation in rat calvaria defect. Tissue Eng Part A. 2011;17(7-8):889–98.

[94] Silvent J, Nassif N, Helary C, Azaïs T, Sire J-Y, Guille MMG. Collagen osteoid-like model allows kinetic gene expression studies of non-collagenous proteins in relation with mineral development to understand bone biomineralization. PloS One. 2013;8(2):e57344.

[95] Levingstone TJ, Matsiko A, Dickson GR, O'Brien FJ, Gleeson JP. A biomimetic multi-layered collagen-based scaffold for osteochondral repair. Acta Biomater. 2014;10(5):1996–2004.

[96] Engler AJ, Sen S, Sweeney HL, Discher DE. Matrix elasticity directs stem cell lineage specification. Cell. 2006;126(4):677–89.

[97] Huebsch N, Arany PR, Mao AS, Shvartsman D, Ali OA, Bencherif SA, et al. Harnessing traction-mediated manipulation of the cell/matrix interface to control stem-cell fate. Nat Mater. 2010;9(6):518–26.

Chapter 14

Extracellular Matrix Enhances Therapeutic Effects of Stem Cells in Regenerative Medicine

Yan Nie, Shuaiqiang Zhang, Na Liu and Zongjin Li

Additional information is available at the end of the chapter

http://dx.doi.org/10.5772/62229

Abstract

Stem cell therapy is a promising option for regenerative of injured or diseased tissues. However, the extremely low survival and engraftment of transplanted cells and the obviously inadequate recruitment and activation of the endogenous resident stem cells are the major challenges for stem cell therapy. Fortunately, recent progresses show that extracellular matrix (ECM) could not only act as a spatial and mechanical scaffold to enhance cell viability but also provide a supportive niche for engraftment or accelerating stem cell differentiation. These findings provide a new approach for increasing the efficiency of stem cell therapy and may lead to substantial changes in cell administration. In order to take a giant stride forward in stem cell therapy, we need to know much more about how the ECM affects cell behaviours. In this chapter, we provide an overview of the influence of ECM on regulating stem cell maintenance and differentiation. Moreover, the enhancement of supportive microenvironment function of natural or synthetic ECMs in stem cell therapy is discussed.

Keywords: extracellular matrix (ECM), stem cell therapy, microenvironment, growth factor, regenerative medicine

1. Introduction

Stem cells reside within a specific extracellular microenvironment, which consists of a complex mixture of soluble and insoluble, short- and long-range signals [1]. Extracellular matrix (ECM) to which stem cells adheres is one of the microenvironmental parameters regulated stem-cell fates [2–4]. Once moving outside of their niche, stem cells will quickly lose their developmental potential, which limits the application of stem cell therapy [5]. Besides, mounting evidence on stem cells and their niche indicates that transplanted stem cells are unable to survive and adapt at the site of administration where there is lack of functional vascular network to transport blood, supply oxygen and nutrients, and remove metabolites [6].

Through enhancing cell retention and engraftment after transplantation, modulating stem cell fate, and promoting functional vasculature formation, co-transplantation stem cells with natural or synthetic ECM that mimic natural extracellular milieu could be a potentially powerful tool to break the current bottleneck and maximize the effectiveness of stem cell therapy [7–9]. These strategies provide considerable hope for the development of stem cell therapy in degenerative diseases. This chapter will provide the insights into the interaction between stem cells and ECM, as well as current knowledge and involvement of stem cell therapy. Moreover, we will discuss the strategy of co-transplantation stem cells and ECM for tissue regeneration with enhanced therapeutic efficacy.

2. Why extracellular matrix is necessary for stem cell therapy

With the capacity of self-renewal and differentiation, stem cells have shown promising potential in regenerative medicine and tissue engineering. So far, stem cell transplantation have been proposed as future therapies for degenerative diseases or injury, including Alzheimer's disease [10], type 1 diabetes [11], Parkinson's disease [12], cardiac disease [13], muscle damage [14] and many others [15–17]. However, some studies showed that stem cell therapy only had modest improvement, which could be attributed to the fact that transplanted cells were unable to survive and adapt in the diseased area. For instance, low cell retention and engraftment and remarkably cell death after transplantation have been observed by using bioluminescence imaging (BLI) [18].

Though it is not clear what signals and underlying pathways cause the acute donor cell death following transplantation, increasing evidence suggests that that a supportive microenvironment is of crucial importance for stem cell survival, proliferation and differentiation [5,19]. For this reason, the strategy to seed stem cells on biomaterials that mimic the biochemical and biophysical properties of native niche could be a viable solution to the above mentioned problems [20] and optimize functional recovery of injured tissue (**Figure 1**). For instance, Matrigel, a product derived from the Engelbroth-Holm-Swarm (EHS) mouse sarcoma, is one of the most commonly used plate-coating materials for stem cell culture in vitro and effectively applied vehicles for transplanted stem cells [21]. Mounting evidence has demonstrated that Matrigel could affect cell fate in a variety of dimensions [21,22]. However, Matrigel is a complex with unknown variable matrices and numerous mixed growth factors, which makes it impossible for us to get further insights into the interplay of stem cells and ECM. Besides, another reason for safety concern is that Matrigel has been reported contaminated with Lactate Dehydrogenase Elevating Virus [23]. To avoid these problems, artificially synthetic ECM with both high purity and defined components in qualitative and quantitative measures for safe application is strongly demanded [24,25]. Recently, developments in engineered ECM-based microenvironments have gradually exhibited their ability for directing stem cell behaviours, such as adhesion, proliferation, and differentiation [26].

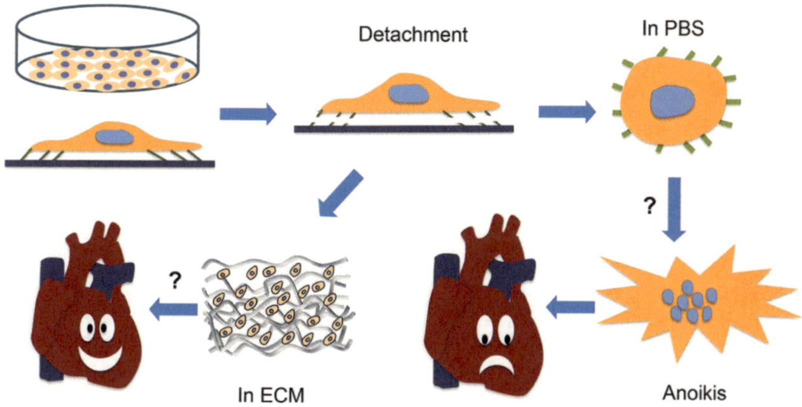

Figure 1. Extracellular matrix(ECM) is necessary for stem cell therapy. A form of apoptosis, called "anoikis", will be initiated once interactions between stem cells and ECM are cut off. Re-establish the connection between ECM and detached cells could increase cell viability and promote function recovery of injured tissue [4]. Reprinted by permission of the publisher.

3. Influence of the extracellular matrix on stem cell behaviour

Extracellular matrix (ECM), acting in conjunction with the biophysical properties and biochemical extracellular stimuli, is critical to regulate stem cell maintenance and differentiation [27,28]. It has been reported that a form of apoptosis, called "anoikis", would be initiated when interactions between stem cells and ECM were cut off [19]. Great effort has been made in an attempt to detail the mechanisms, which provides some key information for cell–ECM interplay. For example, recent study investigated changes of genes' expression after cell detachment by using PCR Array [2]. In that study, researchers found that adhesion molecules expression had no significant difference between cultured human embryonic stem cell-derived

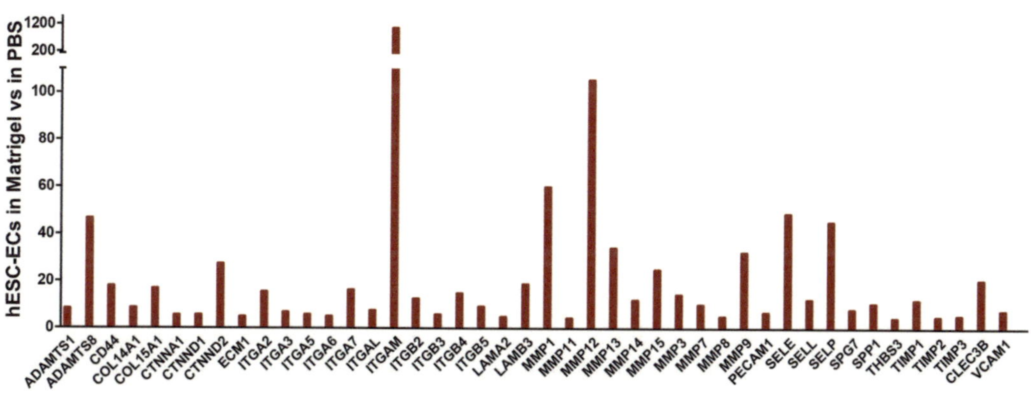

Figure 2. ECM and cell adhesion related gene-expression patterns of hESC-EC at different conditions. Expression of more than 4-fold changes genes of hESC-ECs in Matrigel compared with hESC-ECs in PBS [4]. Reprinted by permission of the publisher.

endothelial cells (hESC-ECs) and enzymatically dispersed hESC-ECs suspended in Matrigel. However, a series of ECM and adhesion molecules-specific genes was considerably down-regulated in hESC-EC suspended in PBS (**Figure 2**). These gene-expression data indicated that adding ECM to detached cells could reverse genes down-regulation of ECM pathway, cell adhesion molecules pathway, ECM and adhesion signalling.

3.1. Biochemical stimulus

The assignment of cell fate results from a response to sophisticated extracellular signals [29,30]. There is mounting evidence suggesting that ECM could deliver numerous soluble and immobilized factors that play vital roles in making the fate choice between self-renewal and lineage commitment [31]. Further insights and exquisite control of signals transported by ECM could provide opportunities for enhancing the regenerative efficacy in both in vitro and in vivo and further accelerating the translation of basic science to the clinical setting.

3.1.1. Release of soluble factors

The propagation of soluble signalling molecules controls a great variety of cellular responses, including proliferation [32], polarity [33], migration [34], and differentiation [35]. It has been

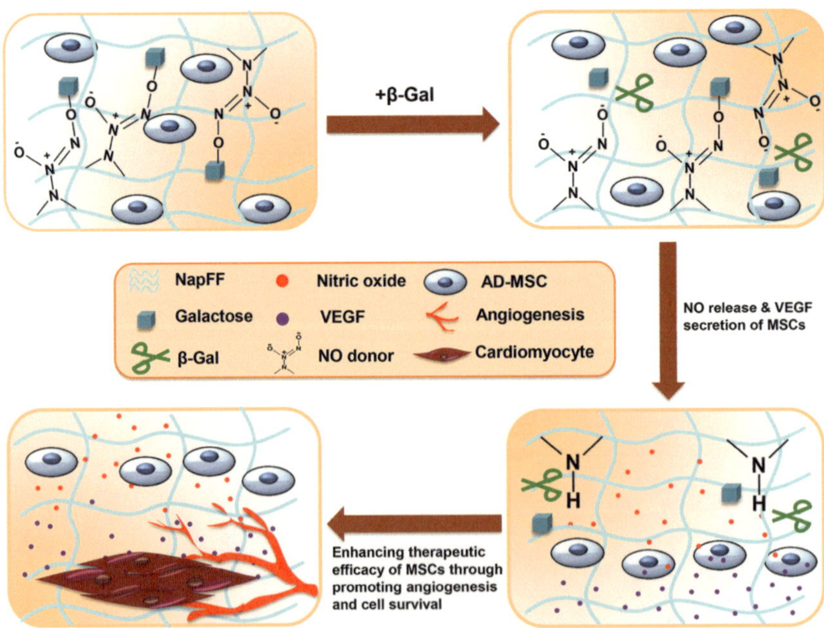

Figure 3. Putative model outlining the controlled nitric oxide (NO)-releasing hydrogel enhances the therapeutic effect of adipose derived-mesenchymal stem cells (ADSCs) for myocardial infarction. Encapsulation of ADSCs by NO-releasing hydrogel prevented transplanted cells effusing from injection positions. NO molecule released from the hydrogel catalyzed by β-galactosidase can facilitate angiogenic cytokines secretion of ADSCs, resulting in promoting angiogenesis, ADSCs survival and cardiac function. β-gal, β-galactosidase [9]. Reprinted by permission of the publisher.

demonstrated that a specific interaction between cells and ECM is required for the ultimate biological response of soluble molecules [29,36,37]. Interactions with ECM can affect the responses of cell toward signalling messengers. For instance, insulin-triggered activation of insulin receptor substrate (IRS) was intensely enhanced in cells cultured on basement membrane than on collagen I, whereas higher levels of tyrosine phosphorylation of the EGF receptor and Erk was triggered by EGF in cell adhesion to collagen I [36].

In living systems, the coordinated effort among cells, growth factors and ECM is required for the successful tissue regeneration. The ability of manipulating biological signals transduced by ECM in a controlled and spatiotemporal manner that mimic the natural regenerative process could provide specific control over the stem cell-based regenerative therapy [38]. The potential therapeutic effect of a peptide-based ECM with the capacity of controlled release nitric oxide (NO), NapFF-NO, was tested in a mouse model of myocardial infarction [9]. The therapeutic effect of adipose-derived-mesenchymal stem cells (ADSCs) was elevated through co-transplantation with NapFF-NO and on-demand NO release (**Figure 3**). Additionally, the administration of growth factors within the context of the ECM niche could accentuate their therapeutic effects for tissue repair [39]. It was reported that a recombinant fragment of fibronectin (FN) could significantly enhance the regenerative effects of growth factors in models of chronic wounds and bone defects [40].

3.1.2. Immobilized factors

Sustained release and improved local retention of regenerative factors, such as growth factors and extracellular substances, are required during tissue regeneration [41]. However, these molecules are suffering from rapid degradation, and therefore they will quickly lose their functionality and clinical efficacy [42]. Additionally, there is evidence that cellular processes are also affected by the interactions between cells and non-soluble constituents of the ECM [43]. For these reasons, immobilization of signalling molecules or functional components to ECM could be suitable for stabilizing these highly reactive molecules, increasing local concentration of biochemical stimuli, and increasing the bioactivity of engineered ECM.

A growing number of studies have utilized short synthetic peptides to mimic the biological properties of full-length growth factors and to substitute parent proteins [44,45]. For example, insulin-like growth factor 1 (IGF-1) is considered as an essential biochemical stimulus in tissue regeneration. The C domain of IGF-1 (IGF-1C), a 12 amino acids sequence, had already been proved as the active region of IGF-1 [46]. IGF-1C has been used as substitute for IGF-1 and applied into hydrogel biomaterials as biomimetic material for tissue engineering and regenerative medicine. The proliferation, apoptosis resistance, and paracrine effects of ADSCs were significantly enhanced after they were seeded on chitosan (CS) hydrogel with immobilization of IGF-1C [47]. When co-transplanted ADSCs with CS-IGF-1C hydrogel into ischemic organ, this biomimetic matrix could create a favourable microenvironment for the survival and adaptation of transplanted cells and further promote functional and structural recovery of injured organ (**Figure 4**).

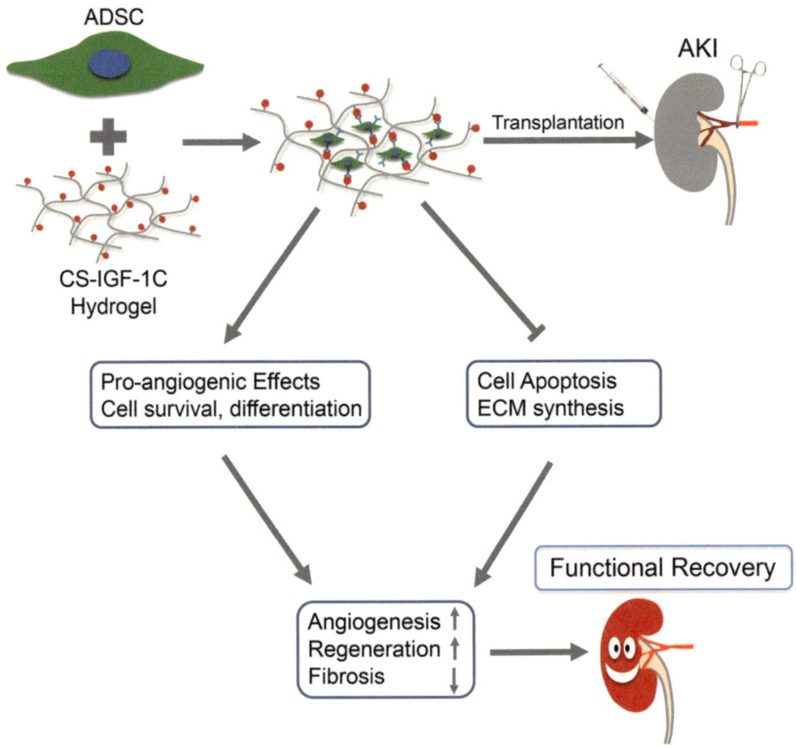

Figure 4. Schema of renoprotective effects of ADSCs and synthetic ECM (CS-IGF-1C hydrogel). When co-transplanted into AKI model, CS-IGF-1C hydrogel could protect delivered ADSCs, facilitated their paracrine and anti-inflammatory effects, and inhibit ECM synthesis in kidney, which result in enhanced angiogenesis, regeneration and alleviated fibrosis after kidney injuries. Consequently, CS-IGF-1C hydrogel therapy leads to improved functional and structural recovery of kidney [47]. Reprinted by permission of the publisher.

3.2. Physical interaction

Although it has commonly acknowledged that signals transduced by ECM could direct stem cell fate, there is increasing evidence that physical properties of ECM could also make a great impact on cell behaviours [48–50]. Some of these factors are proven to be of great influence, but we still have a long way to go and a lot of work to do to establish a complete theory. For example, in response to injury, the accumulation of ECM is excess and abnormal, which would cause significant changes to the stiffness of ECM and ultimately lead to tissue fibrosis [51,52].

3.2.1. Stiffness and elasticity

To test the effect of different stiffness (EY) on cell behaviors, substrates with EY ranging from <1 kPa to 30 kPa were synthesized [53]. The results showed that ECM stiffness has influence on cell proliferation as well as cell differentiation. For instance, neural stem/progenitor cells (NSPCs) could proliferated on substrates with EY <10 kPa. On soft substrates (<1 kPa), neuronal differentiation was promoted; whereas, on relatively stiff substrates (>7 kPa), oligodendrocyte differentiation was favoured. This consequence indicated that matrix stiffness had effect on lineage choice and differentiation. In light of previous data that stiffness was a regulator of differentiation, Shih et al. further explore how matrix affects the osteogenic phenotype of MSCs [54]. They found that the matrix rigidity promoted osteogenic commitment through a α2-integrin-ROCK-FAK-ERK1/2 axis.

Fascinatingly, it was observed that mesenchymal stem cells all underwent osteogenic differentiation on both stiff and soft polydimethylsiloxane (PDMS) substrates; whereas, the osteoblast differentiation of the same cells was promoted on stiff polyacrylamide (PAAm) hydrogels, and more cell differentiated into adipocytes on soft PAAm hydrogels [55]. The different cellular responses to different substrates indicated that stiffness was not an independent stimulus for differentiation. Further data provided in this study suggested that the differentiation of human mesenchymal stem cells on PAAm was also regulated by the elastic modulus. Consistent with the previous study, Xue et al. reported that matrix elasticity was the main physical parameter directing stem cell differentiation at low cell density; with increased cell density, the cell–cell contact force and interactions took priority over the matrix elasticity [56]. Most notably, although cell differentiation was influenced by elastic modulus, recent discovery found that matrix-promoted adipogenic or osteogenic differentiation could not maintain when the cells were re-seeded into tissue culture plastic (TCP) [57]. Furthermore, global gene expression profiles and DNA-methylation profiles revealed no significant impact caused by matrix with different elasticity. These results indicated that matrix elasticity only exerted a transient influence on stem cell lineage commitment.

3.2.2. Ligands and ligand densities

When cells were seeded on ECM with different ligand densities, changes in stem cell viability, size, and shape provided the direct evidence that ligand immobilized to ECM could not be easily separated from the biophysical effects of matrix [58,59]. The spatial arrangement of ligands had a significant influence on MSC behavior [44]. Through manipulating of the ratio of polystyrene-block-poly (ethylene oxide) copolymers (PS-PEO-Ma) in mixtures of block copolymer and polystyrene homopolymer, the lateral spacing of RGD (arginine-glycine-aspartic acid) peptides was under control. With increased lateral spacing, osteogenesis of MSC was reduced while adipogenic differentiation was increased, which was consistent with the results of gene expression levels and alkaline phosphatase activities.

Moreover, the type of ligands covalently linked to ECM could also influence stem cell fate determination. The differentiation of MSCs on different composition of adhesion ligands with the same concentration was various. MSCs cultured on fibronectin or laminin matrices tended to undergo adipogenic differentiation; whereas, MSCs cultured on ECM containing collagen preferred to adopt a neurogenic outcome [60].

3.2.3. Macro/nano-scale topography

Recent development also demonstrated that macro/nano-scale of ECM was another important physical parameter that could not only change the shape of stem cells but also influence the behaviour of stem cells. A preliminary study demonstrated Zyxin played an important role in nanotopographical feature-facilitated changes in stem cells [48]. On 350 nm grating, expression of Zyxin was down-regulated, which was associated with the accelerated speed of migration and the decreased intracellular tension. Likewise, McMurray et al. revealed that modification in surface nanotopography of thermoplastic polycaprolactone (PCL) would influence intracellular tension, which could maintain the multipotency of stem cells and diminish spontaneous differentiation of MSC [61]. Moreover, his current study further illustrated that nanoscale spatial organization of cell-adhesive ligands bound to ECM could affect lineage commitment of MSCs [62]. By using nanopatterning techniques, arginine-glycine-aspartate (RGD) was covalently linked to the surface of poly (ethyleneglycol) (PEG) hydrogels with different nanospacing. It was interesting to identify that large RGD nanospacing was beneficial for osteogenesis; small RGD nanospacing was conducive to adipogenesis.

4. ECM augments therapeutic effects of stem cell therapy

Many attempts at cell therapy have employed ECM to improve efficacy for the following reasons. First, the major obstacle to the application of stem cells, which is known as the extremely poor survival and engraftment of transplanted stem cells, could be minimized by co-transplantation stem cells with ECM [51,63,64]. Second, engineered ECM mimicking the natural stem-cell microenvironments could provide plenty of subtle and instructive cues to control the fate of both transplanted and endogenous cells, including stem cell self-renewal, differentiation, and migration [7,65,66]. Taken together, the development of engineered matrices is promising for the application of stem cells in regenerative medicine.

4.1. Enhance efficacy of transplanted cells

For both experimental studies and clinical applications, transplanted stem cells are commonly prepared for transplantation as single cells. During this process, interactions between cells and ECM are lost and adhesion-related survival signals are down-regulated, which could cause a decrease in cell viability and initiated apoptosis [67]. Fortunately, recent research discovered that the down-regulated molecules of detached cells could be regained in the presence of Matrigel [4], which provided a theoretical rationale for using ECM as a protective scaffold to enhance viability and to stimulate self-renewal of the transplanted cells.

In support of this finding, recent study demonstrated that biomimetic scaffold could protect the transplanted stem cells, and further promote functional and structural recovery from acute kidney injury (AKI). Through immobilization of the C-domain of insulin-like growth factor 1 (IGF-1C) to chitosan (CS) hydrogel (CS-IGF-1C), an artificial microenvironment for supporting growth of stem cells was synthesized. The pro-proliferative, anti-apoptotic, and pro-angiogenic effects of CS-IGF-1C were demonstrably beneficial for enhancing survival of transplanted stem cells, which could ameliorate renal function [47] (**Figure 5**).

http://dx.doi.org/10.5772/62229

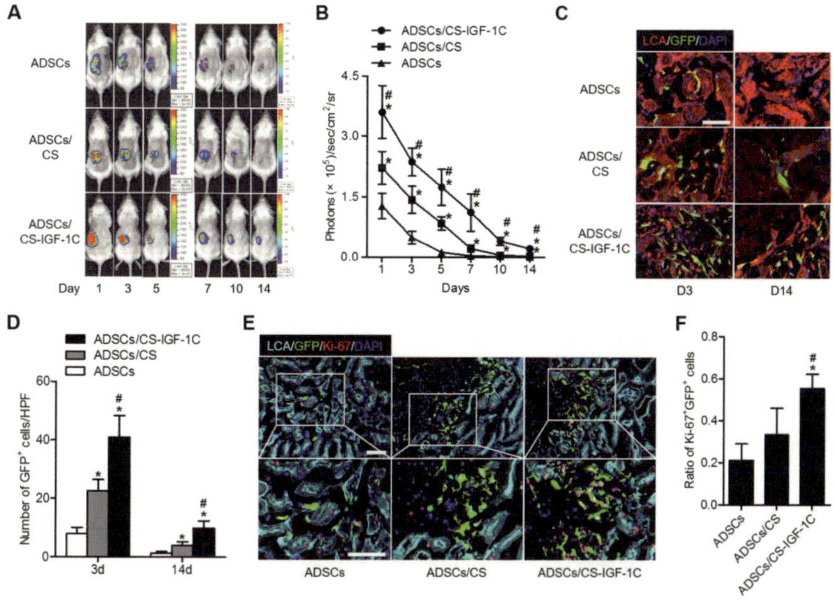

Figure 5. CS-IGF-1C hydrogel increases ADSCs viability in vivo. (A) The fate of ADSCs after transplantation was tracked by molecular imaging. Images are from representative animals receiving $1{\times}10^6$ ADSCs alone, with chitosan hydrogel or CS-IGF-1C hydrogel. (B) Quantitative analysis of BLI signals demonstrated that cell survival was improved by chitosan hydrogel and CS-IGF-1C hydrogel application at all time-points. CS-IGF-1C hydrogel group showed significantly better cell survival. Data are expressed as mean ± SEM. (C) Representative photomicrographs displayed the engraftment of ADSCs (GFP, green) within kidneys at day 3 and 14. Proximal tubular epithelial cells were stained by rhodamine-labeled lens culinaris agglutinin (LCA, red). (D) Quantitative analysis data revealed that chitosan hydrogel improved cell engraftment and CS-IGF-1C hydrogel further increased this effect. Data are expressed as mean ± SEM. *P<0.05 vs. ADSCs, #P<0.05 vs. ADSCs/CS. (E) Representative images showing the proliferation (Ki-67, red) of transplanted ADSCs (GFP, green) in the border regions 3 days after AKI. DyLight 649-labeled LCA staining (cyan) was performed to reveal renal structure. (F) Quantification of the proliferation index of ADSCs. Data are expressed as mean ± SEM. *P<0.05 vs. ADSCs, #P<0.05 vs. ADSCs/CS. (47). Reprinted by permission of the publisher.

Besides, we could attribute the efficacy of stem cell therapy partly to the pluripotency of stem cells [68,69]. A morphological study of MSCs in collagen type I (Col I) hydrogel and in interfacial polyelectrolyte complexation (IPC) based hydrogels containing Col I discovered that cells were neatly arranged and closely packed in IPC- Col I hydrogel [70]. This uniform arrangement results in notably enhanced commitment to the chondrogenic lineage of MSC, which could be an attractive source of cartilage equivalents for tissue engineering.

Recently, a variety of studies have demonstrated that decellularized matrix could provide tissue specific cues for cell growth and lineage commitment [71–73]. Decellularized myocardial matrix hydrogel, which keeps the original structure and natural heart ECM, is the most compelling example among these biomaterials. One of the most inspiring finding is that a mouse heart could contract and beat again after removing its own cells and repopulating the decellularized whole-heart ECMs (DC-ECMs) with multipotential cells that could differentiate in response to the signals from the DC-ECMs. Through repopulating decellularized mouse hearts with induced pluripotent stem cell (iPSC)-derived earliest heart progenitors, the recellularized DC-ECMs exhibited myocardium, vessel-like structures, intracellular Ca^{2+} transients (CaiT), spontaneous heart contractions and significant response to numerous drug interventions [74].

4.2. Support the function of endogenous cells

MSCs could mobilize into circulating blood and be recruited to the injury site, which was consistent with the evidence that numbers of MSCs were increased considerably in peripheral blood [75]. Several approaches were used in an attempt to investigate this cell recruitment event. It was unexpected to find that ECM was indispensable for the homing of MSCs toward sites of injury [76]. The homing effect could be inhibited through adding inhibitor of serine proteases and leupeptin to ECM, which illustrated the key role of matrix remodelling in MSC migration. In addition, evidence also indicated that exposing MSCs to injury-associated ECM prior to transplantation could augment the efficiency of MSCs' intrinsic tropism for injury [77].

As resident stem cells and progenitor cells could be activated to participate tissue regeneration after injury [78,79], ECM designed for cell seeding should also benefit the growth of host cells and support the function of endogenous cells. Encouragingly, evidence suggested that host cells could respond to ECM in the site of injury in vivo. Firstly, immune responses were elicited in hosts, which was identified by the quickly infiltrated $CD68^+$ cells throughout the entire ECM within 3 days after implantation. Then there were indications of myogenesis in the muscle injury area, which was confirmed by morphology and myosin heavy chain positive staining [80].

Furthermore, accumulating data suggested that human mesenchymal stem cells (hMSCs) could modulate immune system response through their paracrine effect and then create a pro-regenerative environment in situ. Their paracrine effects could be optimized through encapsulating hMSCs into protective ECM [81]. The recruitment of endogenous macrophages and the M1/M2 polarization were modulated by the trophic factors secreted by hMSCs, which was possibly capable of counteracting the hostile environment and sustaining tissue regeneration. This cell-friendly microenvironment could also be established by administration of ECM alone. Increased stem cell tropism, revascularization, and improved cardiac function induced by chitosan-based ECM were observed in ischemic myocardium [82], which may be attributing to the mechanical support provided by ECM and the therapeutic biomolecules enriched by ECM [83–85].

5. Future perspectives

The ECM is not only a simple scaffold that provides physical supports for stem cells but also a dynamic and complex environment that is capable of regulating cell behaviours. Although the application of natural or synthetic ECM with the aim to enhance therapeutic effects of stem cells is highly appealing for promoting regenerative processes, issues related to efficiency and safety limit their translational use as regenerative medicine. Further identifying specific biochemical and biophysical properties of ECM and understanding the interplay between stem cells and ECM will provide knowledge of stem cell biology and fuel the development of regenerative therapies based on stem cells.

Acknowledgements

This work was partially supported by grants from the National Natural Science Foundation of China (81371620, 81320108014), Tianjin Natural Science Foundation (14JCZDJC35200) and the Program for Changjiang Scholars and Innovative Research Team in University (IRT13023).

Author details

Yan Nie[1], Shuaiqiang Zhang[1], Na Liu[2] and Zongjin Li[1,2*]

*Address all correspondence to: zongjinli@nankai.edu.cn

1 Nankai University School of Medicine, Tianjin, P.R. China

2 The Key Laboratory of Bioactive Materials, Ministry of Education, College of Life Sciences, Nankai University, Tianjin, P.R. China

References

[1] Gattazzo F, Urciuolo A, Bonaldo P. Extracellular matrix: A dynamic microenvironment for stem cell niche. Biochim Biophys Acta. 2014;1840(8):2506-2519. DOI:10.1016/j.bbagen.2014.01.010.

[2] Higuchi A, Ling Q-D, Hsu S-T, Umezawa A. Biomimetic cell culture proteins as extracellular matrices for stem cell differentiation. Chemical Reviews. 2012;112(8): 4507-4540. DOI:10.1021/cr3000169.

[3] Murphy WL, Mc Devitt TC, Engler AJ. Materials as stem cell regulators. Nature Materials. 2014;13(6):547-557. DOI:10.1038/nmat3937.

[4] He N, Xu Y, Du W, Qi X, Liang L, Wang Y, et al. Extracellular matrix can recover the downregulation of adhesion molecules after cell detachment and enhance endothelial cell engraftment. Sci Rep. 2015;5:10902. DOI:10.1038/srep10902.

[5] Taddei ML, Giannoni E, Fiaschi T, Chiarugi P. Anoikis: An emerging hallmark in health and diseases. The Journal of Pathology. 2012;226(2):380-393. DOI:10.1002/path.3000.

[6] Prestwich GD, Healy KE. Why regenerative medicine needs an extracellular matrix. Expert Opin Biol Ther. 2015;15(1):3-7. DOI:10.1517/14712598.2015.975200.

[7] Zhang J, Klos M, Wilson GF, Herman AM, Lian X, Raval KK, et al. Extracellular matrix promotes highly efficient cardiac differentiation of human pluripotent stem cells: The matrix sandwich method. Circulation Research. 2012;111(9):1125-1136. DOI:10.1161/CIRCRESAHA.112.273144.

[8] Khetan S, Guvendiren M, Legant WR, Cohen DM, Chen CS, Burdick JA. Degradation-mediated cellular traction directs stem cell fate in covalently crosslinked three-dimensional hydrogels. Nat Mater. 2013;12(5):458-465. DOI:10.1038/nmat3586.

[9] Yao X, Liu Y, Gao J, Yang L, Mao D, Stefanitsch C, et al. Nitric oxide releasing hydrogel enhances the therapeutic efficacy of mesenchymal stem cells for myocardial infarction. Biomaterials. 2015;60:130-140. DOI:10.1016/j.biomaterials.2015.04.046.

[10] Israel MA, Yuan SH, Bardy C, Reyna SM, Mu Y, Herrera C, et al. Probing sporadic and familial alzheimer's disease using induced pluripotent stem cells. Nature. 2012;482(7384):216-220. DOI:10.1038/nature10821.

[11] Jurewicz M, Yang S, Augello A, Godwin JG, Moore RF, Azzi J, et al. Congenic mesenchymal stem cell therapy reverses hyperglycemia in experimental type 1 diabetes. Diabetes. 2010;59(12):3139-3147. DOI:10.2337/db10-0542.

[12] Lindvall O, Kokaia Z. Prospects of stem cell therapy for replacing dopamine neurons in parkinson's disease. Trends Pharmacol Sci. 2009;30(5):260-267. DOI:10.1016/j.tips.2009.03.001.

[13] Karantalis V, Hare JM. Use of mesenchymal stem cells for therapy of cardiac disease. Circ Res. 2015;116(8):1413-1430. DOI:10.1161/circresaha.116.303614.

[14] Quintero AJ, Wright VJ, Fu FH, Huard J. Stem cells for the treatment of skeletal muscle injury. Clin Sports Med. 2009;28(1):1-11. DOI:10.1016/j.csm.2008.08.009.

[15] Fadini GP, Agostini C, Avogaro A. Autologous stem cell therapy for peripheral arterial disease meta-analysis and systematic review of the literature. Atherosclerosis. 2010;209(1):10-17. DOI:10.1016/j.atherosclerosis.2009.08.033.

[16] Ichim TE, Alexandrescu DT, Solano F, Lara F, Campion Rde N, Paris E, et al. Mesenchymal stem cells as anti-inflammatories: Implications for treatment of duchenne muscular dystrophy. Cell Immunol. 2010;260(2):75-82. DOI:10.1016/j.cellimm.2009.10.006.

[17] Trounson A, Thakar RG, Lomax G, Gibbons D. Clinical trials for stem cell therapies. BMC Med. 2011;9:52. DOI:10.1186/1741-7015-9-52.

[18] van der Bogt KE, Hellingman AA, Lijkwan MA, Bos EJ, de Vries MR, van Rappard JR, et al. Molecular imaging of bone marrow mononuclear cell survival and homing in

murine peripheral artery disease. JACC Cardiovascular imaging. 2012;5(1):46-55. DOI: 10.1016/j.jcmg.2011.07.011.

[19] Gilmore AP. Anoikis. Cell Death Differ. 2005;12 Suppl 2:1473-1477. DOI:10.1038/sj.cdd.4401723.

[20] Jhala D, Vasita R. A review on extracellular matrix mimicking strategies for an artificial stem cell niche. Polymer Reviews. 2015;55(4):561-595. DOI: 10.1080/15583724.2015.1040552.

[21] Ou L, Li W, Zhang Y, Wang W, Liu J, Sorg H, et al. Intracardiac injection of matrigel induces stem cell recruitment and improves cardiac functions in a rat myocardial infarction model. J Cell Mol Med. 2011;15(6):1310-1318. DOI:10.1111/j.1582-4934.2010.01086.x.

[22] Uemura M, Refaat MM, Shinoyama M, Hayashi H, Hashimoto N, Takahashi J. Matrigel supports survival and neuronal differentiation of grafted embryonic stem cell-derived neural precursor cells. J Neurosci Res. 2010;88(3):542-551. DOI:10.1002/jnr.22223.

[23] Carlson Scholz JA, Garg R, Compton SR, Allore HG, Zeiss CJ, Uchio EM. Poliomyelitis in mulv-infected icr-scid mice after injection of basement membrane matrix contaminated with lactate dehydrogenase-elevating virus. Comp Med. 2011;61(5):404-411.

[24] Furue MK, Na J, Jackson JP, Okamoto T, Jones M, Baker D, et al. Heparin promotes the growth of human embryonic stem cells in a defined serum-free medium. Proc Natl Acad Sci U S A. 2008;105(36):13409-13414. DOI:10.1073/pnas.0806136105.

[25] Nagaoka M, Si-Tayeb K, Akaike T, Duncan SA. Culture of human pluripotent stem cells using completely defined conditions on a recombinant e-cadherin substratum. BMC Dev Biol. 2010;10:60. DOI:10.1186/1471-213x-10-60.

[26] Rowland TJ, Miller LM, Blaschke AJ, Doss EL, Bonham AJ, Hikita ST, et al. Roles of integrins in human induced pluripotent stem cell growth on matrigel and vitronectin. Stem Cells Dev. 2010;19(8):1231-1240. DOI:10.1089/scd.2009.0328.

[27] Du J, Chen XF, Liang XD, Zhang GY, Xu J, He LR, et al. Integrin activation and internalization on soft ecm as a mechanism of induction of stem cell differentiation by ecm elasticity. Proc Natl Acad Sci U S A. 2011;108(23):9466-9471. DOI:10.1073/pnas.1106467108.

[28] Swift J, Ivanovska IL, Buxboim A, Harada T, Dingal PC, Pinter J, et al. Nuclear lamin-a scales with tissue stiffness and enhances matrix-directed differentiation. Science. 2013;341(6149):1240104. DOI:10.1126/science.1240104.

[29] Kim SH, Turnbull J, Guimond S. Extracellular matrix and cell signalling: The dynamic cooperation of integrin, proteoglycan and growth factor receptor. J Endocrinol. 2011;209(2):139-151. DOI:10.1530/joe-10-0377.

[30] Zouani OF, Kalisky J, Ibarboure E, Durrieu MC. Effect of bmp-2 from matrices of different stiffnesses for the modulation of stem cell fate. Biomaterials. 2013;34(9): 2157-2166. DOI:10.1016/j.biomaterials.2012.12.007.

[31] Seif-Naraghi SB, Horn D, Schup-Magoffin PJ, Christman KL. Injectable extracellular matrix derived hydrogel provides a platform for enhanced retention and delivery of a heparin-binding growth factor. Acta Biomater. 2012;8(10):3695-3703. DOI:10.1016/ j.actbio.2012.06.030.

[32] Somaiah C, Kumar A, Mawrie D, Sharma A, Patil SD, Bhattacharyya J, et al. Collagen promotes higher adhesion, survival and proliferation of mesenchymal stem cells. PLo S One. 2015;10(12):e0145068. DOI:10.1371/journal.pone.0145068.

[33] Venero Galanternik M, Kramer KL, Piotrowski T. Heparan sulfate proteoglycans regulate fgf signaling and cell polarity during collective cell migration. Cell Rep. 2015 DOI:10.1016/j.celrep.2014.12.043.

[34] Bowen CJ, Zhou J, Sung DC, Butcher JT. Cadherin-11 coordinates cellular migration and extracellular matrix remodeling during aortic valve maturation. Dev Biol. 2015;407(1):145-157. DOI:10.1016/j.ydbio.2015.07.012.

[35] Lu S, Lam J, Trachtenberg JE, Lee EJ, Seyednejad H, van den Beucken JJ, et al. Dual growth factor delivery from bilayered, biodegradable hydrogel composites for spatially-guided osteochondral tissue repair. Biomaterials. 2014;35(31):8829-8839. DOI: 10.1016/j.biomaterials.2014.07.006.

[36] Lee Y-J, Streuli CH. Extracellular matrix selectively modulates the response of mammary epithelial cells to different soluble signaling ligands. Journal of Biological Chemistry. 1999;274(32):22401-22408. DOI:10.1074/jbc.274.32.22401.

[37] Jeanes AI, Wang P, Moreno-Layseca P, Paul N, Cheung J, Tsang R, et al. Specific β-containing integrins exert differential control on proliferation and two-dimensional collective cell migration in mammary epithelial cells. Journal of Biological Chemistry. 2012;287(29):24103-24112. DOI:10.1074/jbc.M112.360834.

[38] Kim PH, Yim HG, Choi YJ, Kang BJ, Kim J, Kwon SM, et al. Injectable multifunctional microgel encapsulating outgrowth endothelial cells and growth factors for enhanced neovascularization. J Control Release. 2014;187:1-13. DOI:10.1016/j.jconrel.2014.05.010.

[39] Martino MM, Briquez PS, GüçE, Tortelli F, Kilarski WW, Metzger S, et al. Growth factors engineered for super-affinity to the extracellular matrix enhance tissue healing. Science. 2014;343(6173):885-888. DOI:10.1126/science.1247663.

[40] Martino MM, Tortelli F, Mochizuki M, Traub S, Ben-David D, Kuhn GA, et al. Engineering the growth factor microenvironment with fibronectin domains to promote wound and bone tissue healing. Sci Transl Med. 2011;3(100):100ra189. DOI:10.1126/ scitranslmed.3002614.

[41] Kolambkar YM, Dupont KM, Boerckel JD, Huebsch N, Mooney DJ, Hutmacher DW, et al. An alginate-based hybrid system for growth factor delivery in the functional repair

of large bone defects. Biomaterials. 2011;32(1):65-74. DOI:10.1016/j.biomaterials.2010.08.074.

[42] Grassian AR, Schafer ZT, Brugge JS. Erbb2 stabilizes epidermal growth factor receptor (egfr) expression via erk and sprouty2 in extracellular matrix-detached cells. Journal of Biological Chemistry. 2011;286(1):79-90. DOI:10.1074/jbc.M110.169821.

[43] Walters NJ, Gentleman E. Evolving insights in cell-matrix interactions: Elucidating how non-soluble properties of the extracellular niche direct stem cell fate. Acta Biomater. 2015;11:3-16. DOI:10.1016/j.actbio.2014.09.038.

[44] Frith JE, Mills RJ, Cooper-White JJ. Lateral spacing of adhesion peptides influences human mesenchymal stem cell behaviour. Journal of Cell Science. 2012;125(2):317-327. DOI:10.1242/jcs.087916.

[45] Meng Q, Man Z, Dai L, Huang H, Zhang X, Hu X, et al. A composite scaffold of msc affinity peptide-modified demineralized bone matrix particles and chitosan hydrogel for cartilage regeneration. Sci Rep. 2015;5:17802. DOI:10.1038/srep17802.

[46] Jansen M, van Schaik FM, Ricker AT, Bullock B, Woods DE, Gabbay KH, et al. Sequence of cdna encoding human insulin-like growth factor i precursor. Nature. 1983;306(5943): 609-611.

[47] Feng G, Zhang J, Li Y, Nie Y, Zhu D, Wang R, et al. Igf-1c modified hydrogel enhances cell therapy for acute kidney injury. Journal of the American Society of Nephrology. 2015. in press. doi: 10.1681/ASN.2015050578.

[48] Kulangara K, Yang Y, Yang J, Leong KW. Nanotopography as modulator of human mesenchymal stem cell function. Biomaterials. 2012;33(20):4998-5003. DOI:10.1016/j.biomaterials.2012.03.053.

[49] Wen JH, Vincent LG, Fuhrmann A, Choi YS, Hribar KC, Taylor-Weiner H, et al. Interplay of matrix stiffness and protein tethering in stem cell differentiation. Nat Mater. 2014;13(10):979-987. DOI:10.1038/nmat4051.

[50] Huebsch N, Lippens E, Lee K, Mehta M, Koshy ST, Darnell MC, et al. Matrix elasticity of void-forming hydrogels controls transplanted-stem-cell-mediated bone formation. Nat Mater. 2015;14(12):1269-1277. DOI:10.1038/nmat4407.

[51] Georges PC, Hui JJ, Gombos Z, Mc Cormick ME, Wang AY, Uemura M, et al. Increased stiffness of the rat liver precedes matrix deposition: Implications for fibrosis. Am J Physiol Gastrointest Liver Physiol. 2007;293(6):G1147-1154. DOI:10.1152/ajpgi.00032.2007.

[52] Chaudhuri O, Koshy ST, Branco da Cunha C, Shin JW, Verbeke CS, Allison KH, et al. Extracellular matrix stiffness and composition jointly regulate the induction of malignant phenotypes in mammary epithelium. Nat Mater. 2014;13(10):970-978. DOI: 10.1038/nmat4009.

[53] Leipzig ND, Shoichet MS. The effect of substrate stiffness on adult neural stem cell behavior. Biomaterials. 2009;30(36):6867-6878. DOI:10.1016/j.biomaterials.2009.09.002.

[54] Shih YR, Tseng KF, Lai HY, Lin CH, Lee OK. Matrix stiffness regulation of integrin-mediated mechanotransduction during osteogenic differentiation of human mesenchymal stem cells. J Bone Miner Res. 2011;26(4):730-738. DOI:10.1002/jbmr.278.

[55] Trappmann B, Gautrot JE, Connelly JT, Strange DG, Li Y, Oyen ML, et al. Extracellular-matrix tethering regulates stem-cell fate. Nat Mater. 2012;11(7):642-649. DOI:10.1038/nmat3339.

[56] Xue R, Li JY, Yeh Y, Yang L, Chien S. Effects of matrix elasticity and cell density on human mesenchymal stem cells differentiation. J Orthop Res. 2013;31(9):1360-1365. DOI:10.1002/jor.22374.

[57] Schellenberg A, Joussen S, Moser K, Hampe N, Hersch N, Hemeda H, et al. Matrix elasticity, replicative senescence and DNA methylation patterns of mesenchymal stem cells. Biomaterials. 2014;35(24):6351-6358. DOI:10.1016/j.biomaterials.2014.04.079.

[58] Choi JS, Harley BA. The combined influence of substrate elasticity and ligand density on the viability and biophysical properties of hematopoietic stem and progenitor cells. Biomaterials. 2012;33(18):4460-4468. DOI:10.1016/j.biomaterials.2012.03.010.

[59] Kilian KA, Mrksich M. Directing stem cell fate by controlling the affinity and density of ligand-receptor interactions at the biomaterials interface. Angew Chem Int Ed Engl. 2012;51(20):4891-4895. DOI:10.1002/anie.201108746.

[60] Lee J, Abdeen AA, Zhang D, Kilian KA. Directing stem cell fate on hydrogel substrates by controlling cell geometry, matrix mechanics and adhesion ligand composition. Biomaterials. 2013;34(33):8140-8148. DOI:10.1016/j.biomaterials.2013.07.074.

[61] Mc Murray RJ, Gadegaard N, Tsimbouri PM, Burgess KV, Mc Namara LE, Tare R, et al. Nanoscale surfaces for the long-term maintenance of mesenchymal stem cell phenotype and multipotency. Nat Mater. 2011;10(8):637-644. DOI:10.1038/nmat3058.

[62] Ye K, Wang X, Cao L, Li S, Li Z, Yu L, et al. Matrix stiffness and nanoscale spatial organization of cell-adhesive ligands direct stem cell fate. Nano Letters. 2015;15(7): 4720-4729. DOI:10.1021/acs.nanolett.5b01619.

[63] Mathieu E, Lamirault G, Toquet C, Lhommet P, Rederstorff E, Sourice S, et al. Intramyocardial delivery of mesenchymal stem cell-seeded hydrogel preserves cardiac function and attenuates ventricular remodeling after myocardial infarction. PLo S One. 2012;7(12):e51991. DOI:10.1371/journal.pone.0051991.

[64] Tzouanas SN, Ekenseair AK, Kasper FK, Mikos AG. Mesenchymal stem cell and gelatin microparticle encapsulation in thermally and chemically gelling injectable hydrogels for tissue engineering. Journal of Biomedical Materials Research Part A. 2014;102(5): 1222-1230. DOI:10.1002/jbm.a.35093.

[65] Lei Y, Gojgini S, Lam J, Segura T. The spreading, migration and proliferation of mouse mesenchymal stem cells cultured inside hyaluronic acid hydrogels. Biomaterials. 2011;32(1):39-47. DOI:10.1016/j.biomaterials.2010.08.103.

[66] Singelyn JM, Sundaramurthy P, Johnson TD, Schup-Magoffin PJ, Hu DP, Faulk DM, et al. Catheter-deliverable hydrogel derived from decellularized ventricular extracellular matrix increases endogenous cardiomyocytes and preserves cardiac function post-myocardial infarction. Journal of the American College of Cardiology. 2012;59(8): 751-763. DOI:10.1016/j.jacc.2011.10.888.

[67] Livshits G, Kobielak A, Fuchs E. Governing epidermal homeostasis by coupling cell-cell adhesion to integrin and growth factor signaling, proliferation, and apoptosis. Proc Natl Acad Sci U S A. 2012;109(13):4886-4891. DOI:10.1073/pnas.1202120109.

[68] Toh WS, Lim TC, Kurisawa M, Spector M. Modulation of mesenchymal stem cell chondrogenesis in a tunable hyaluronic acid hydrogel microenvironment. Biomaterials. 2012;33(15):3835-3845. DOI:10.1016/j.biomaterials.2012.01.065.

[69] Ansboro S, Hayes JS, Barron V, Browne S, Howard L, Greiser U, et al. A chondromimetic microsphere for in situ spatially controlled chondrogenic differentiation of human mesenchymal stem cells. J Control Release. 2014;179:42-51. DOI:10.1016/j.jconrel. 2014.01.023.

[70] Raghothaman D, Leong MF, Lim TC, Toh JK, Wan AC, Yang Z, et al. Engineering cell matrix interactions in assembled polyelectrolyte fiber hydrogels for mesenchymal stem cell chondrogenesis. Biomaterials. 2014;35(9):2607-2616. DOI:10.1016/j.biomaterials. 2013.12.008.

[71] Yin Z, Chen X, Zhu T, Hu JJ, Song HX, Shen WL, et al. The effect of decellularized matrices on human tendon stem/progenitor cell differentiation and tendon repair. Acta Biomater. 2013;9(12):9317-9329. DOI:10.1016/j.actbio.2013.07.022.

[72] Cheung HK, Han TT, Marecak DM, Watkins JF, Amsden BG, Flynn LE. Composite hydrogel scaffolds incorporating decellularized adipose tissue for soft tissue engineering with adipose-derived stem cells. Biomaterials. 2014;35(6):1914-1923. DOI:10.1016/ j.biomaterials.2013.11.067.

[73] Wang RM, Christman KL. Decellularized myocardial matrix hydrogels: In basic research and preclinical studies. Adv Drug Deliv Rev. 2015;96:77–82. DOI:10.1016/ j.addr.2015.06.002.

[74] Lu TY, Lin B, Kim J, Sullivan M, Tobita K, Salama G, et al. Repopulation of decellularized mouse heart with human induced pluripotent stem cell-derived cardiovascular progenitor cells. Nat Commun. 2013;4:2307. DOI:10.1038/ncomms3307.

[75] Chen G, Tian F, Li C, Zhang Y, Weng Z, Zhang Y, et al. In vivo real-time visualization of mesenchymal stem cells tropism for cutaneous regeneration using nir-ii fluorescence imaging. Biomaterials. 2015;53:265-273. DOI:10.1016/j.biomaterials.2015.02.090.

[76] Mauney J, Olsen BR, Volloch V. Matrix remodeling as stem cell recruitment event: A novel in vitro model for homing of human bone marrow stromal cells to the site of injury shows crucial role of extracellular collagen matrix. Matrix Biol. 2010;29(8): 657-663. DOI:10.1016/j.matbio.2010.08.008.

[77] Smith CL, Chaichana KL, Lee YM, Lin B, Stanko KM, O'Donnell T, et al. Pre-exposure of human adipose mesenchymal stem cells to soluble factors enhances their homing to brain cancer. Stem Cells Transl Med. 2015;4(3):239-251. DOI:10.5966/sctm.2014-0149.

[78] Segers VF, Lee RT. Stem-cell therapy for cardiac disease. Nature. 2008;451(7181): 937-942. DOI:10.1038/nature06800.

[79] Kim JY, Xin XJ, Moioli EK, Chung J, Lee CH, Chen M, et al. Regeneration of dental-pulp-like tissue by chemotaxis-induced cell homing. Tissue Engineering Part A. 2010;16(10):3023-3031. DOI:10.1089/ten.tea.2010.0181.

[80] Wolf MT, Daly KA, Brennan-Pierce EP, Johnson SA, Carruthers CA, D'Amore A, et al. A hydrogel derived from decellularized dermal extracellular matrix. Biomaterials. 2012;33(29):7028-7038. DOI:10.1016/j.biomaterials.2012.06.051.

[81] Caron I, Rossi F, Papa S, Aloe R, Sculco M, Mauri E, et al. A new three dimensional biomimetic hydrogel to deliver factors secreted by human mesenchymal stem cells in spinal cord injury. Biomaterials. 2016;75:135-147. DOI:10.1016/j.biomaterials. 2015.10.024.

[82] Liu Z, Wang H, Wang Y, Lin Q, Yao A, Cao F, et al. The influence of chitosan hydrogel on stem cell engraftment, survival and homing in the ischemic myocardial microenvironment. Biomaterials. 2012;33(11):3093-3106. DOI:10.1016/j.biomaterials. 2011.12.044.

[83] Lord MS, Cheng B, McCarthy SJ, Jung M, Whitelock JM. The modulation of platelet adhesion and activation by chitosan through plasma and extracellular matrix proteins. Biomaterials. 2011;32(28):6655-6662. DOI:10.1016/j.biomaterials.2011.05.062.

[84] Gu Y, Zhu J, Xue C, Li Z, Ding F, Yang Y, et al. Chitosan/silk fibroin-based, schwann cell-derived extracellular matrix-modified scaffolds for bridging rat sciatic nerve gaps. Biomaterials. 2014;35(7):2253-2263. DOI:10.1016/j.biomaterials.2013.11.087.

[85] Lin CY, Li LT, Su WT. Three dimensional chitosan scaffolds influence the extra cellular matrix expression in schwann cells. Mater Sci Eng C Mater Biol Appl. 2014;42:474-478. DOI:10.1016/j.msec.2014.05.063.

Chapter 15

New and Improved Tissue Engineering Techniques: Production of Exogenous Material-Free Stroma by the Self-Assembly Technique

Ingrid Saba, Weronika Jakubowska,
Stéphane Chabaud and Stéphane Bolduc

Additional information is available at the end of the chapter

http://dx.doi.org/10.5772/62588

Abstract

Tissue engineering results from the use of cells and scaffolds to reproduce structural and spatial organization or function of a tissue. The Production of an ideal engineered tissue depends on its designed purpose. For clinical applications, the main concerns are biocompatibility and the generation of a tissue able to mimic most of its original biological functions. Moreover, the viability of an implanted tissue is associated with its stability to support vascular networks. This chapter summarizes the theory of the self-assembly approach for tissue engineering. Adjustments and modifications in stromal thickness and extracellular matrix composition for various self-assembled tissues are discussed. Methods developed to generate tissue closely mimicking the native morphology and structure, to incorporate capillary-like networks, and to reduce production time and costs are also reviewed. The self-assembly technique leads to the production of a stroma free of exogenous material and can be adapted to generate fastest, inexpensive, and near-to-native tissue bioengineering for medical and fundamental research applications.

Keywords: Tissue engineering, Self-assembly, Stroma, Epithelial cells, Endothelial cells

Abbreviations

AGE	Advanced glycation end-products
Arg	Arginine

ASC	Adipose tissue-derived stem cells
EC	Endothelial cells
ECM	Extracellular matrix
HA	Hyaluronic acid
HMVEC	Human microvascular endothelial cell
HUVEC	Human umbilical vein endothelial cell
L–Arg	L–arginine
LOX	Lysyl oxidase enzymes
LOXL	Lysyl oxidase homologues
LPA	Lysophosphatidic acid
MMP	Matrix metalloproteinases
MSC	Mesenchymal stem/stromal cells
NG2	Neuron glial-2
PLGA	Poly-lactic-co-glycolic acid
SVF	Adipose stromal vascular fraction
TGF-β	Transforming growth factor-beta
TIMP	Tissue inhibitors of metalloproteinases
VEGF	Vascular endothelial growth factor
2D	Two-dimensional
3D	Three-dimensional

1. Introduction

The extracellular matrix (ECM) is present within all tissues and organs. It constitutes the noncellular microenvironment around the cells that plays an important role in modulating their behavior and functions [1]. This elaborated milieu is very dynamic and extremely adaptable [2–4]. ECM is composed of several components that include proteoglycans, as well as collagen proteins and noncollagenous glycoproteins. Each component has several subcategories of molecules that influence the ECM physical and biochemical properties [5].

1.1. ECM deposition and assembly

The homeostasis of epithelial tissues depends on a dynamic interaction of the stroma components, such as fibroblasts, adipocytes, and nonactivated immune players [6]. In fact, fibroblasts were reported to secrete and organize type I and type III collagens, elastin, fibronectin, tenascin, and a repertoire of proteoglycans (hyaluronic acid (HA) and decorin), that maintains interstitial ECM integrity [7]. The ECM is constantly remodeled to allow the healthy tissue to resist to a wide range of tensile pressures [8, 9]. This remodeling occurs through the synthesis of elastin, which originates secreted tropoelastin, the precursor of elastin, that assembles into fibers and becomes cross-linked on lysine residues by members of the lysyl oxidase (LOX) enzymes and lysyl oxidase homologues (LOXL) molecules [10]. LOX and LOXL catalyze the first step in the formation of collagens and elastins, a very conserved process that plays an important role in cell growth, chemotaxis, or sprouting of new blood vessels [11, 12]. On the one hand, coordinated secretion of matrix metalloproteinases (MMPs) by fibroblasts mediates ECM remodeling [13]. On the other hand, the mesh networks are counterbalanced by tissue inhibitors of metalloproteinases (TIMPs) [14] or by other enzymes such as LOX molecules and transglutaminases that stiffen the ECM [10].

2. Cellular players and ECM production

2.1. Mesenchymal stem/stromal cells

Adult mesenchymal stem/stromal cells (MSCs) are found in all postnatal organs and tissues, and they play important functions in tissue injury repair and general homeostasis [15]. These cells are one of the principal adult stem cells and the most promising tool for regenerative medicine because of their sustained proliferative capacity and their multipotent differentiation potential [15–17].

2.1.1. Fibroblasts

Fibroblasts are nonterminally differentiated mesenchymal cells derived from the embryonic mesoderm [18]. They are found in the connective tissue, a tissue that supports the whole body. Fibroblasts are spread in the ECM containing fibrous proteins and gel-like substances. In fact, fibroblasts produce the ECM proteins, such as fibrous collagen and elastin, as well as adhesive proteins such as laminin and fibronectin. Fibroblasts are also the major source of glycosaminoglycans (hyaluronan and glycoproteins) [19]. Interconnecting meshworks of extracellular protein fibers and connector proteins provide the architectural tissue structure. Moreover, this milieu forms the connections needed for cellular migration of fibroblasts, immune cells, and endothelial cells (ECs) during angiogenesis [19].

Most tissues are composed of a simple or multiple layers of epithelial cells that exhibit an apical–basal polarization. The basal part is in contact with the basement membrane, whereas the apical side is oriented toward the fluid-filled lumen [20]. Fibroblasts form a basement membrane, composed of a layer of basal lamina and a layer of reticular lamina. This basement membrane serves essentially as a structural scaffold that maintains the dynamics of a three-dimensional (3D) engineered tissue. It is also critical for tissue regeneration in wound healing and acts as a cell barrier. The basement membrane acts as a cell barrier by segregating epithelial cells from endothelial cells (ECs), thus preventing tumor invasion or metastasis.

2.1.2. *Myofibroblasts: functions and origins*

Myofibroblasts, or activated fibroblasts, are contractile, resistant to apoptosis and have an upregulated rate of matrix deposition. They also express different cytokine and chemokine receptors that enable fiber regulation and wound contraction at injury sites [21–24]. In this sense, activated fibroblasts not only favor wound healing, but can also cause injuries when their activation is uncontrolled, producing a pathological fibrotic response [25, 26]. The precursor of myofibroblast is not precisely known, but many cells can differentiate into myofibroblasts through different signaling pathways or gene regulation. Some of these precursors include epithelial cells, ECs, pericytes, multipotent monocytes, and fibroblasts.

2.2. Adipose tissue–derived stem cells

Fat is an abundant and accessible source of stem cells. Adipose tissue–derived stem cells (ASCs) include preadipocytes, and a subpopulation of stromal cells able to differentiate into multilineages, including neuronal cells, chondrocytes, and osteoblasts [27, 28]. Moreover, these MSCs are able to secrete cytokines and growth factors promoting regenerative processes because they can influence cell recruitment, proliferation rates, or inhibit apoptosis [29, 30]. ASCs extracted from liposuctions can be expanded in culture and used as building blocks for tissue engineering. Both connective and adipose tissues were engineered in vitro using ASCs [31–33]. Both allogeneic and xenogeneic ASCs can be transplanted in patients regardless of their immunocompatibility and without the need of immunosuppression therapy, making them an unlimited source for regenerative medicine applications [34].

3. Tissue engineering history and techniques

As medical treatments and expanded lifespan expectancies in both males and females have improved, the number of individuals waiting for organ transplants or blood vessel bypasses is constantly increasing but the availability of organs does not often match the demand. To circumvent this shortage in organ and tissue supplies, many efforts in cell culture methods were deployed to engineer tissues that could be used as an alternative therapeutical option.

3.1. Self-assembly technique

The self-assembly technique is based on the ability of MSCs to secrete and organize their own ECM to produce sheets. This tissue engineering method allows the production of autologous living tissues, free of exogenous biomaterials [35–37]. The self-assembly technique has exploited the inherent characteristics of MSCs to produce ECM. For instance, it was well documented that ascorbic acid, a vitamin C derivate, promotes collagen protein synthesis and deposition of sulfated glycosaminoglycans in human skin substitutes [38]. Once fibroblasts supplemented with ascorbic acid are cultured for 21–35 days, they form sheets of matrix where stromal cells are embedded within [35]. These sheets can be peeled from the culture dish and superimposed. The superimposed layers are maintained for an additional week for further cell-matrix reorganization and layer fusion (**Figure 1A**). Holes can be made in the multilayer dermal equivalent and hair follicles can be added to mimic the presence of native skin component. An additional culture time is required before seeding keratinocytes. Thereafter, the skin equivalent is maintained for 21 days at an air–liquid interface to induce the cornification of the epidermis [36, 39].

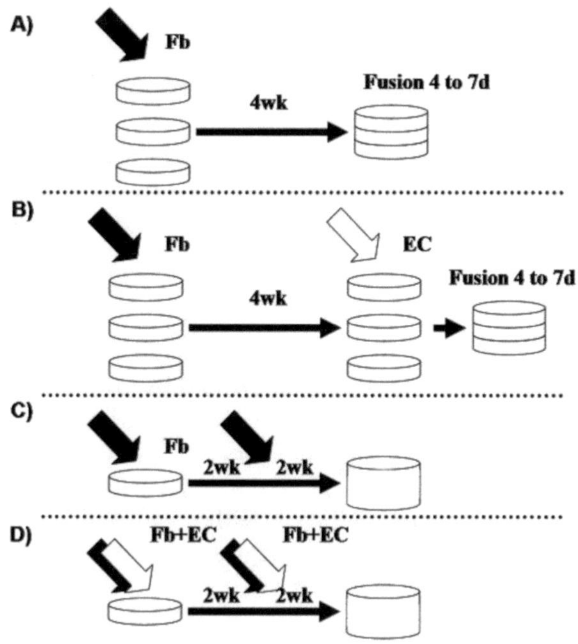

Figure 1. Schematic representation of self-assembled tissues. **A**) The classical self-assembly technique requires stacking of three fibroblast (Fb) cell sheets that are cultured for 4 weeks in the presence of ascorbic acid. Sheet fusion lasts 4 to 7 days. **B**) After 4 weeks of Fb cultures, the stroma is seeded with endothelial cells (ECs) and then a fusion set is carried on to generate endothelialized tissues. **C**) Modification to the self-assembly technique consists of an initial Fb culture for 2 weeks followed by an additional Fb reseeding. The culture is maintained for an additional 2 weeks before using the stroma without stacking. **D**) Fb and EC are co-seeded in order to generate endothelialized tissues with a 3D capillary-like network.

3.1.1. *Engineered human skin substitutes*

One of the great breakthroughs in medicine was achieved by engineering human skin substitutes for grafting purposes. Self-assembled skin substitutes were generated by extracting the patient's own cells, thus avoiding immunological incompatibility problems upon grafting. These skin substitutes were characterized and showed a fully differentiated epidermis, structural and morphological resemblance to native human skin. Moreover, these in vitro engineered tissues were able to deliver cytokines, chemokines, and growth factors at the grafted site, improving the wound closure [36, 40–43]. Because of all these characteristics, self-assembled human skin is clinically used for wound healing and burn treatments [44, 45]. Self-assembled skin substitutes possess a near-to-native architecture and maintain their cell growth potential and matrix deposition. Therefore, these equivalents are free of exogenous material, cytotoxicity, and have clinically reduced morbidity in burnt patients (reviewed in Refs. [46, 47].

Over the years, self-assembled skin substitutes were also produced from extracted cells of patients having psoriasis [48]. The reconstructed tissues from psoriatic donors represent an ideal model to study one of the most common human skin diseases. In fact, this model outlines the excessive growth and aberrant differentiation of keratinocytes. It offers a reliable in vitro mean to measure the efficacy of appropriate treatments, perform tests directly on human primary cells, and avoid animal use [48–52]. Furthermore, self-assembled skin substitutes were used to extensively characterize cellular and molecular players involved in the pathogenesis of hypertrophic scars and scleroderma [53–55]. More recently, a skin substitute derived from patients diagnosed with amyotrophic lateral sclerosis (ALS) was similarly generated. As one of the early perturbations in ALS patients is skin alterations that often precede the neurological symptoms, this human skin model is designed to better identify disease-specific biomarkers and early diagnostic tools to monitor disease progression [56].

3.1.2. Cardiovascular tissues

The self-assembly technique was customized to engineer human blood vessel that displayed excellent physiological and mechanical properties without the need for any exogenous scaffold [35, 57]. Blood vessels are constituted of a functional endothelium seeded onto an internal membrane of human skin fibroblasts. In order to mimic the shape of a blood vessel, a smooth muscle cell (SMC) sheet is first rolled, followed by the fibroblast sheet around a cylindrical

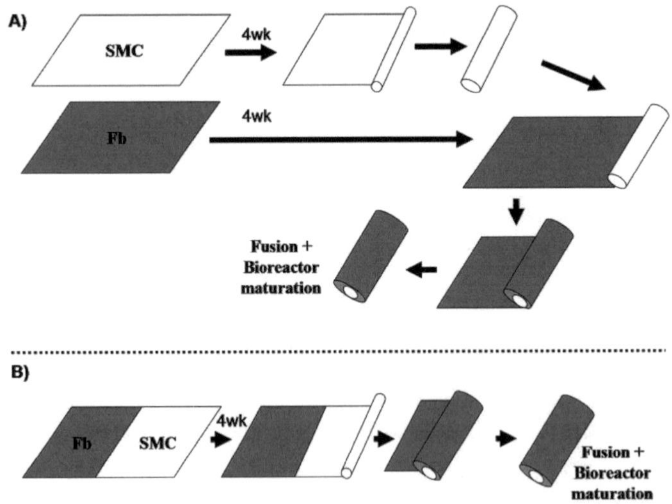

Figure 2. Vascular self-assembled tissue. A) A sheet of smooth muscle cells (SMC) (first) and a sheet of fibroblasts (Fb) (second) are rolled around a cylindrical support after 4 weeks of culture. The construct is allowed to fuse and mature in a bioreactor. **B**) Modification to the technique shown in (**A**) consists of co-seeding SMCc and Fb each at an extremity of the same sheet, which is rolled around a cylindrical support after 4 weeks of culture and then allowed to fuse and mature in a bioreactor.

support, and cultured until fusion (**Figure 2A**). Analyses of these in vitro engineered vessels confirmed the presence of numerous ECM proteins (collagen types I, III, IV, laminin, fibronectin, and chondroitin sulfates) and a functional endothelium [35, 58, 59].

Progress in developing self-assembled valves was reported over the years [60, 61]. Valve leaflets made of self-assembled tissue sheets can organize into a characteristic three-layer structure featuring appropriate dynamic fluidics [60]. This tissue remains to be grafted into living recipients in order to assess in vivo survival and behavior of the transplanted valve. Nevertheless, this stentless bioprosthetic offers a great alternative to artificial valves for cardiovascular surgeries [60, 61].

3.1.3. Cornea

Corneal tissue engineering was developed in an attempt to cure corneal opacity by replacing the damaged area with a clear substitute. Proulx *et al.* [62] generated a self-assembled three-layer equivalent of human cornea. These substitutes presented near-to-native stromal, endothelial, and epithelial morphology with an intact basement membrane filled with laminin V and collagen VII proteins. The differentiated epithelial layer had defined basal and wing cells that expressed Na^+/K^+ ATPase $\alpha 1$ protein, keratin 3/12, and basic keratins. This human cornea model was also used to study the pattern of MMP genes expression during corneal wound healing [63].

3.1.4. ASCs for the production of adipose tissues and other connective tissues

Adipose cell sheets can be generated in vitro using the self-assembly technique supplemented with ascorbic acid and adipogenic differentiation factors. These sheets share many adipocyte features [64]. ASCs have the ability to respond to media composition and motion allowing them to be an optimal cell type for tissue engineering. Using the self-assembly technique, fully autologous vascular tissues were also engineered from ASCs in vitro, with an organized structure and matrix components [65]. Other studies showed that ASCs could be used to bioengineer near-to-native skin [31] and bladder mucosa equivalents [66] in vitro.

3.1.5. Urogenital tissues

In vitro reconstruction of a bladder substitute using the self-assembly technique was first documented by Magnan *et al.* [67], where a single porcine biopsy was processed and allowed the generation of an endothelialized bladder equivalent. Subsequently, seeding urothelial cells on a dermal fibroblast stroma generated tubular urethral grafts for in vivo replacement. To mimic in vivo tissue architecture, the engineered construct was placed under perfusion in a bioreactor [68]. ECs were also added to the model [69]. The human tissue–engineered bladder model can be used to screen common prescribed medicine. For instance, this model was used to study ketamine, an anesthetic agent and a drug used in chronic pain management, which is excreted in the urine. The drug application on the 3D bladder model showed that ketamine directly damages the urothelium, especially the structure and the interconnections that characterize the intermediate epithelial layers, by inducing apoptosis [70].

4. Improvements of stromal thickness and mechanical properties of self-assembled tissue

The self-assembly technique has great therapeutic potential because it uses autologous cells that produce their own ECM, thus reducing allogeneic graft rejection. Although the self-assembly approach is suitable for clinical applications, the time required for tissue reconstruction and the costs are important drawbacks hindering its wider use. Hence, many strategies to reduce tissue reconstruction time and the cost associated with cell culture were investigated. Efforts to stimulate collagen deposition and matrix reorganization are detailed in this section.

4.1. Mechanical stimulation

Mechanical stimuli induce major biological modifications in the organization of the cells cytoskeleton and their ECM composition [71, 72]. For instance, the mechanical stimulation of blood flow induces the realignment of collagen fibers and strengthening of the tissue [73, 74]. At the molecular level, these changes are triggered by the activation of mechanoreceptors such as the ones containing the Arginyl-Glycyl-Aspartic acid attachment site that bind to integrins [75]. This mechanical stimulation often results in activation of extracellular signal–regulated kinase, ERK, and the c-Jun N-terminal kinase, JNK, signaling pathways that will induce cellular responses in order to adapt to new environments [75]. In some studies G-proteins seem to be also involved in the molecular signaling [76]. In response to mechanical stimuli, cells can also secrete growth factors such as transforming growth factor-beta (TGF-β) [77] that will exert paracrine or autocrine functions. Furthermore, cells can secrete and/or activate latent MMPs and other proteases [78, 79], which affect the balance between synthesis of ECM elements and their degradation. Accordingly, fibrosis was observed in mechanically overstimulated settings emphasizing that increased collagen deposition rates need to be controlled in order to remain reversible [80].

4.1.1. Improved blood vessels generation

The quality of engineered vascular tissues can be improved in a bioreactor by applying the appropriate laminar/cyclic flow. Other modification to the self-assembled blood vessel generation, such as co-seeding fibroblasts and SMCs, each at their respective half of the same sheet, before rolling around a cylindrical support [81] (**Figure 2B**) was reported. The fully autologous vascular substitutes possess high-grade mechanical strength to sustain engraftment and are readily available when needed without any immunosuppressive treatments [81–83].

4.1.2. Specific culture surface can influence fiber alignment in engineered tissues

Tissue functions can be improved using microstructured surfaces that control the interactions between cells and the ECM. With the use of a specific surface topography on an elastomeric material, it was observed that the first cell layer followed the same patterns and orientation as the material. Subsequently, this orientation influenced the second cell layer to follow a physiologically similar alignment mimicking the structure of the native tissue. Furthermore, secreted ECM followed cell orientation in every layer, resulting in very well-structured self-assembled sheets for cornea, vascular, and dermis. A micropatterned surface on which cells are seeded have the capacity to generate multiple layers, in which cells and the ECM spontaneously organize in patterns consistent with the original tissue [84].

4.1.3. ASCs cultures in dynamic conditions

In order to reduce culture time required for tissue production, human ASCs were used to replace dermal fibroblasts in some self-assembled tissues. Self-assembled stromas generated with dermal fibroblast or ASCs can be subjected to static or dynamic conditions [85], as they can be mechanically stimulated on a 3D shaker platform. Dynamic culture conditions increased (1.5- to 2-fold) the thickness of tissues derived from ASCs compared to static conditions. Moreover, culture time could be reduced in dynamic conditions. Yet, mechanical properties of these tissues were not measured.

4.2. Enzymatic reactions and chemical stimulation

Although ascorbic acid is an essential element that contributes to collagen deposition, an increase in its concentration does not lead to enhanced collagen deposition. Ascorbic acid is an enzymatic cofactor of prolyl- and lysyl-hydroxylase [86], and its action reaches a plateau when these enzymes achieve their peak of activity. Independently of its role as a cofactor, ascorbic acid is responsible for a certain level of collagen secretion in fibroblast cultures, until it reaches its biological limits [87], albeit it can be toxic for fibroblasts if present in high dose [23].

Chemical inhibitors of MMP could also increase ECM production by restricting the extent of protease activity. Among them, galardin was used to produce self-assembled tissues and it significantly increased the thickness of treated tissues [54]. Currently, the cost associated with the use of galardin is too expensive to be a promising solution.

L-arginine (L-Arg) is converted in ornithine followed by glutamine semialdehyde and finally proline, an important amino acid that is metabolized during collagen synthesis. L-Arg supplementation to culture media was evaluated, when the stroma was produced using the self-assembly method. Although an increase in collagen synthesis and secretion (20% more collagen type-I) was observed, collagen deposition remained unchanged when compared to controls [88]. A plausible explanation would be that enzymes involved in collagen maturation were not sufficient to process the surplus of this amino acid in vitro.

4.3. Biological stimulation

Biological stimulation of ECM deposition in the field of tissue engineering is a challenge. This complexity is due to pleiotropic roles of multiple bioactive agents and their subtle effects, which could appear after a long period of time, for instance after tissue implantation. In contrary to monolayer culture studies in which experiments rarely exceed days, tissue engineering methods can be carried out for months, especially if it involves in vivo implantation. Many proteins, peptides, and lipids can be used to stimulate collagen synthesis and deposition. Most of them are involved in fibrosis and need to be carefully handled to avoid production of pathological-like tissue.

4.3.1. *Polysaccharides*

Beta-glucans constitute a family of carbohydrates that stimulates fibroblasts to produce collagen [89, 90]. For instance, laminaran, a glucan from *Saccharina longicruris* seaweed, increased collagen secretion when added to dermal fibroblast culture [91]. An increase in collagen synthesis and secretion was observed in self-assembled tissues. Moreover, thicker stroma could be obtained without significant increase in cell proliferation and alpha-smooth muscle actin content, a hallmark of fibrosis [92]. The authors argued that the aggregation properties of laminaran triggered a net increase of collagen secretion without inducing a fibrotic phenotype [92, 93].

Tissue engineering often relies on glucose-rich media because glucose is the primary source of energy that allows MSCs to produce ECM compounds. Advanced glycation end-products (AGE) result from glucose metabolism and are found in elderly tissues or in diabetic patient tissues [94]. Unfortunately, the glucose concentration used during the production of most self-assembled tissues is too high. Consequently, it was reported that AGE are involved in the process of skin aging, which has an impact on mechanical and biological parameters [95]. New approaches to circumvent this issue are currently being developed and should generate promising alternatives.

4.3.2. *Insulin and hypoxia*

In addition to mediating glucose entry in cells, insulin also plays an active role in collagen synthesis and deposition [96]. Insulin has a long history of safety use for human therapies and microencapsulated insulin-secreting cells in hydrogels can improve collagen fiber density in diabetic mouse models [97]. Poly-lactic-co-glycolic acid (PLGA) alginate structure that releases insulin in rats was also found to increase collagen deposition and maturation [98]. In a clinical setting, wound healing is problematic for diabetic patients because their insulin metabolism is altered. Also, their tissues are less irrigated because of microvascular network changes caused by the loss of ECs. When capillary networks are altered, the surrounding tissues undergo hypoxia. In such an environment, fibroblasts change to a fibrotic phenotype. Fibrosis is induced by factors that are released by damaged ECs [99], as well as by other unknown mechanisms [96, 100]. Insulin and hypoxia exert a synergic effect on self-assembled tissues. They increase collagen deposition as demonstrated by tests on human and animal cell cultures [101] (unpublished data). Nevertheless, long-term effect of hypoxia exposure (more than 2 weeks) induced acidification of the cell culture medium and a thinning of the engineered tissues [102]. Hence, cyclic hypoxia seems a better alternative than constitutive hypoxia because it produces thicker tissues in vitro.

4.3.3. *Adenosine*

Adenosine and other derivatives have been used to enhance the rate of wound healing [103]. Their receptors were also found to be involved in fibrosis. Activation of A_{2B}-adenosine receptors resulted in an increase of collagen synthesis and a decrease in MMP-9 activity [104, 105]. This molecule was successfully tested to produce rabbit tissues by the self-assembly approach [101]. Effects of adenosine on human cultures remain to be evaluated.

4.3.4. *Lysophosphatidic acid*

Lysophosphatidic acid (LPA) is a bioactive lipid found in blood. LPA binds to its receptors at the surface of many cells and activates pathways leading to proliferation, migration, and secretion of cytokines. LPA expression is upregulated in disease conditions such as in fibrosis and cancer or cancer [106, 107]. As LPA is naturally present in human blood, it was used in vitro and approved by regulatory agencies. LPA-cultured fibroblasts showed increased collagen type-I and fibronectin deposition in a dose-dependent manner that could be completely reversible. No adverse effects were noted: alpha-smooth muscle actin was not overexpressed and cell proliferation rates remained normal [108]. Thicker stroma and enhanced collagen deposition kinetics suggested that the production time could be reduced by 25% when LPA was added to the cell culture medium.

5. Bioengineering substitutes that resemble native tissues

Classical self-assembly technique involves sheet stacking in order to generate a tissue with sufficient mechanical strength. The superimposition of sheets Influences cell distribution. Although fusion of all sheets occurs following sheet stacking, a pattern at the site of each sheet fusion remains visible. Epithelial cell seeding has been noted to reduce sheet demarcations after sheet stacking. Nevertheless, different layers are visible in the 3D self-assembled tissue, Which does not correspond to native stroma architecture and weakens tissue mechanical strength [31, 66, 85, 88]. To outwit this issue, a newly reseeding self-assembly protocol was elaborated and allowed a more uniform distribution of cells throughout the tissue without delineation marks [88].

5.1. Reseeding self-assembly technique

Ascorbic acid triggers collagen deposition that reaches a plateau level after 2 weeks of fibroblast culture [88]. This time period also correlates with the thickness reached by self-assembled tissue [85]. When fibroblasts reach confluence, the cells begin to secrete and deposit collagen to form the ECM, a step that lasts 2 weeks before collagen synthesis rate decreases. These observations led to the generation of engineered tissues by reseeding of cells instead of sheet stacking (**Figure 1C**). The new reseeding approach is based on the fact that a second layer of fibroblasts seeded onto the first sheet will concomitantly induce a transitory peak of MMP activity and a boost of collagen secretion. The fibroblasts in the first sheet play a role in this remodeling, and after an additional 2 weeks of culture, the reseeding process results in the generation of a stroma with the same thickness as the one obtained by classical stacking of three sheets produced without reseeding. This dense stroma supported the development and maturation of the epithelium [88]. The reseeding technique offers a remarkable alternative to the classical self-assembly protocol because it is faster and it reduces costs associated with extensive culture medium consumption as well as material [88] (**Figure 3**).

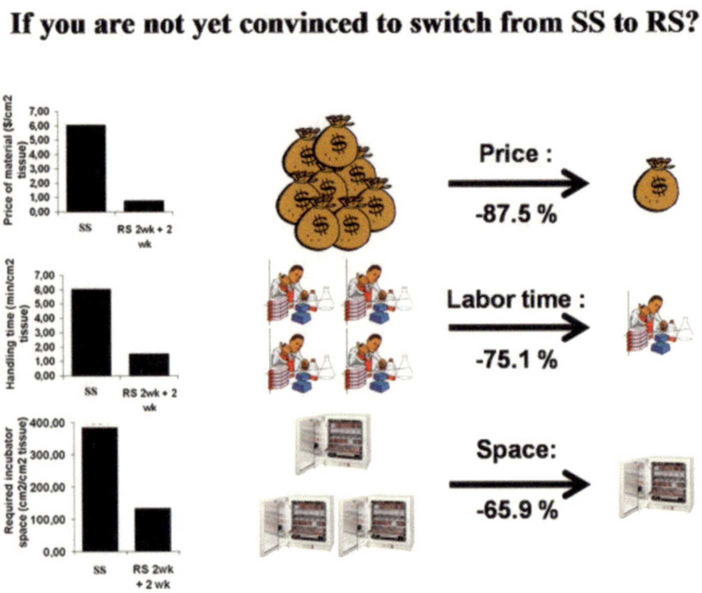

Figure 3. Graphical illustration of the improvements to the self-assembly technique (SS) when reseeding of cells (RS) is applied. Reseeding allows the reduction of costs, handling time, and incubator space requirements throughout the steps of cell culture.

5.2. Organ-specific stroma

Over the years, it has been shown that the origin of mesenchymal cells has a direct impact on the quality of bioengineered tissues. Carrier *et al.* [109] showed that reconstructed human cornea substitutes had great macroscopic and histological differences; especially in the corneal epithelium thickness and differentiation whether dermal fibroblasts or keratocytes (corneal fibroblasts) were used to produce the ECM. Constructs made with dermal fibroblasts were less transparent and lacked ultraviolet absorption characteristics compared to corneal tissues which were produced using autologous keratocytes [109]. The importance of this cross talk was further evaluated by mass spectrometry analyses performed on human stromal and epithelial layers of corneal substitute. Not only is the origin of mesenchymal cell important, but a fully differentiated and stratified epithelium is required for appropriate ECM synthesis and organization. This reciprocal regulation between the ECM and epithelial components is initiated as epithelial cells adhere to the stroma, they emit a continuous signal that mediates ECM remodeling accordingly [63].

Modifications to the original protocol have generated near-to-native self-assembled bladder and urethral human tissues. Many improvements include mechanical stimulation [68], the use of autologous human stromal cells, urothelial cells, urine[110, 111], and a new proposed reseeding technique of stromal cells [88]. In particular, the absence of an air/liquid interface and the presence of urine allowed the new bladder mucosa model [110] to be continuously cultured in submerged conditions. Consequently, these modifications generated a bladder model that preserved the best urothelial cell properties and uroplakin distribution [110, 111].

5.3. Importance of blood vessels for tissue homoeostasis

Blood vessel formation can occur through two distinct mechanisms: angiogenesis and vasculogenesis. Vasculogenesis involves the recruitment of progenitors of ECs from the bone marrow, which leads to the formation of a vascular plexus *de novo*, whereas angiogenesis occurs when pre-existing vessels form new branches or sprouts [112, 113]. Finally, inosculation of a tissue is the anastomosis of the pre-established capillary-like network within the graft or engineered tissue with the host's vasculature [114].

ECs are in a stable quiescent state, however, they can become activated upon angiogenic stimuli in engineered tissues [115]. For instance, in response to conditions, such as tissue ischemia or chronic hypoxia, new collateral vessels can grow. Endothelium proliferation is stimulated by growth factors such as vascular endothelial growth factor (VEGF) that induce sprouting of new blood vessels. Additionally, proangiogenic signals increase MMP activity that prompts ECs to break apart their basement membrane allowing sprouting [116].

5.3.1. Vascularization strategies in tissue engineering

ECs are a promising angiogenic cell source for therapeutic vasculogenesis because they have the potential to proliferate and rearrange themselves into functional capillary-like networks. Human umbilical vein endothelial cells (HUVEC) are an important source of ECs widely used in vasculogenesis [117]. Advantages of this cell type use are the noninvasive cell source, the profusion of medical wastes, and the impressive source of ECs in umbilical cords and placental tissues. Although the therapeutic use of HUVECs is limited because of their allogeneic nature, they remain a valuable EC source for basic and applied research needs [117].

To overcome this issue, the potential of ASCs for their differentiation into ECs was explored. Isolated adipose stromal vascular fraction (SVF) from white human adipose tissue is rich in adult stem cell populations, including EC progenitors. Freshly harvested SVF containing mixed white adipose stromal cells and white adipose ECs was cultured in 3D collagen hydrogels. Within the first week, the culture showed a formation of capillary network with continuous lumen [118], and after 3 weeks it gave rise to a functional 3D vascularized skin substitute that responded well to implantation in mice. This experiment demonstrates the synergy of vascular and stromal cells in blood network formation de novo. Hence, white ASCs Demonstrate promising results with minimal cell handling. Finally, human microvascular endothelial cells (HMVEC), which originate from small superficial capillaries, also represent a promising avenue for tissue endothelialization. These cells have been incorporated in in vitro models using the self-assembly method and formed vascular networks with lumen [115]. HMVECs are an easily accessible source because they can be derived from a skin biopsy or any other tissue. The use of HMVECs could be particularly suitable for therapeutic application because it is best adapted for organ specific reconstructs.

5.3.1.1. Vascularization of self-assembled tissues

The co-culture of dermal fibroblasts and keratinocytes with HUVEC on a chitosan/collagen sponge showed the establishment of a capillary-like network similar to the microvasculature found in vivo [119]. Prevascularization of tissues prior to implantation has yield impressive improvements in regenerative medicine. In 2005, human endothelialized reconstructed skin models revealed an important reduction in the delay of functional vascularization after implantation in mice. Early signs of vascularization were observed in the endothelialized human skin grafts within 4 days following tissue implantation, as opposed to 14 days in the nonendothelialized reconstructed skin. Mouse blood vessels were only detected after 14 days in both models demonstrating that neovascularization is a latter process. The uniform distribution of ECs across the reconstruct ensures adequate perfusion of the entire graft. The colocalization of human and host mouse ECs inside a human capillary within the graft suggests the formation of chimeric microvessels and confirms inosculation between both microvascular networks [114] (later confirmed in Gibot *et al.* [120] generated self-assembled tissue).

The progress of endothelialized tissue–engineered dermal substitutes lead to the introduction of a new in vitro model of capillary-like network formation in self-assembled skin substitutes without the use of an exogenous scaffold. In this approach, stromal sheets, formed by culturing dermal fibroblast during 4 weeks, were seeded with ECs. To generate the 3D skin, two endothelialized stromal sheets were stacked and allowed to fuse [121] (**Figure 1B**). Although a capillary network was observed, the fact that ECs were seeded in a single plane orientation, on top of the stromal sheets, resulted in a vascularized skin model with mainly a 2D vascular network rather than a 3D network. In order to provide the reconstructed skin with the optimal 3D capillary network, ECs were co-seeded with fibroblasts (**Figure 1D**). Incorporation of ECs in the reconstructed model using the reseeding technique produced a capillary-like network with increased tissue elasticity and mechanical strength [88]. Moreover, because fibroblasts were seeded at high density, ECM was readily generated and allowed the dermal stroma to be rapidly embedded with ECs [88].This vascularized stroma had pericyte-like cells that expressed the neuron-glial 2 (NG2) marker, which characterizes the surrounding of capillary-like structures.

6. Conclusions

The self-assembly approach is used to generate several tissues for fundamental and clinical research applications. Over the years, adjustments to the stroma elaboration protocols and especially the ECM generation were proposed to improve the quality of the bioengineered substitutes. As one of the main objectives is to reduce the production time and costs, mechanical, biological, and chemical modifications were also introduced. Organ-specific ECM was associated with a better epithelial differentiation and an overall tissue architecture that closely mimics native tissues. To improve clinical applications, endothelialized tissues were generated and grafted with better survival and functions compared to nonvascularized substitutes.

Author details

Ingrid Saba[1], Weronika Jakubowska[1], Stéphane Chabaud[1] and Stéphane Bolduc[1,2*]

*Address all correspondence to: stephane.bolduc@fmed.ulaval.ca

1 Experimental organogenesis research center of Laval University /LOEX, University Hospital (CHU) of Quebec - Laval University, Enfant-Jésus Hospital, Quebec City, QC, Canada

2 Department of Surgery, Faculty of Medicine, Laval University, Quebec City, QC, Canada

References

[1] Hynes RO. The extracellular matrix: not just pretty fibrils. Science. 2009;326(5957):1216–9.

[2] Har-el R, Tanzer ML. Extracellular matrix. 3: evolution of the extracellular matrix in invertebrates. FASEB J. 1993;7(12):1115–23.

[3] Engler AJ, Humbert PO, Wehrle-Haller B, Weaver VM. Multiscale modeling of form and function. Science. 2009;324(5924):208–12.

[4] Ozbek S, Balasubramanian PG, Chiquet-Ehrismann R, Tucker RP, Adams JC. The evolution of extracellular matrix. Mol Biol Cell. 2010;21(24):4300–5.

[5] Lu P, Takai K, Weaver VM, Werb Z. Extracellular matrix degradation and remodeling in development and disease. Cold Spring Harb Perspect Biol. 2011;3(12).

[6] Ronnov-Jessen L, Petersen OW, Bissell MJ. Cellular changes involved in conversion of normal to malignant breast: importance of the stromal reaction. Physiol Rev. 1996;76(1): 69–125.

[7] Paszek MJ, Zahir N, Johnson KR, Lakins JN, Rozenberg GI, Gefen A, et al. Tensional homeostasis and the malignant phenotype. Cancer Cell. 2005;8(3):241–54.

[8] Egeblad M, Rasch MG, Weaver VM. Dynamic interplay between the collagen scaffold and tumor evolution. Curr Opin Cell Biol. 2010;22(5):697–706.

[9] Kass L, Erler JT, Dembo M, Weaver VM. Mammary epithelial cell: influence of extracellular matrix composition and organization during development and tumorigenesis. Int J Biochem Cell Biol. 2007;39(11):1987–94.

[10] Lucero HA, Kagan HM. Lysyl oxidase: an oxidative enzyme and effector of cell function. Cell Mol Life Sci. 2006;63(19–20):2304–16.

[11] Csiszar K. Lysyl oxidases: a novel multifunctional amine oxidase family. Prog Nucleic Acid Res Mol Biol. 2001;70:1–32.

[12] Molnar J, Fong KS, He QP, Hayashi K, Kim Y, Fong SF, et al. Structural and functional diversity of lysyl oxidase and the LOX-like proteins. Biochim Biophys Acta. 2003;1647(1–2):220–4.

[13] Mott JD, Werb Z. Regulation of matrix biology by matrix metalloproteinases. Curr Opin Cell Biol. 2004;16(5):558–64.

[14] Cruz-Munoz W, Khokha R. The role of tissue inhibitors of metalloproteinases in tumorigenesis and metastasis. Crit Rev Clin Lab Sci. 2008;45(3):291–338.

[15] Tuan RS, Boland G, Tuli R. Adult mesenchymal stem cells and cell-based tissue engineering. Arthritis Res Ther. 2003;5(1):32–45.

[16] Pittenger MF, Mackay AM, Beck SC, Jaiswal RK, Douglas R, Mosca JD, et al. Multilineage potential of adult human mesenchymal stem cells. Science. 1999;284(5411):143–7.

[17] da Silva Meirelles L, Chagastelles PC, Nardi NB. Mesenchymal stem cells reside in virtually all post-natal organs and tissues. J Cell Sci. 2006;119(Pt 11):2204–13.

[18] Dennis JE, Charbord P. Origin and differentiation of human and murine stroma. Stem Cells. 2002;20(3):205–14.

[19] Blaauboer ME, Boeijen FR, Emson CL, Turner SM, Zandieh-Doulabi B, Hanemaaijer R, et al. Extracellular matrix proteins: a positive feedback loop in lung fibrosis? Matrix Biol. 2014;34:170–8.

[20] Frantz C, Stewart KM, Weaver VM. The extracellular matrix at a glance. J Cell Sci. 2010;123(Pt 24):4195–200.

[21] Strehlow D, Korn JH. Biology of the scleroderma fibroblast. Curr Opin Rheumatol. 1998;10(6):572–8.

[22] Abraham DJ, Varga J. Scleroderma: from cell and molecular mechanisms to disease models. Trends Immunol. 2005;26(11):587–95.

[23] Larochelle S, Langlois C, Thibault I, Lopez-Valle CA, Roy M, Moulin V. Sensitivity of myofibroblasts to H2O2-mediated apoptosis and their antioxidant cell network. J Cell Physiol. 2004;200(2):263–71.

[24] Moulin V, Larochelle S, Langlois C, Thibault I, Lopez-Valle CA, Roy M. Normal skin wound and hypertrophic scar myofibroblasts have differential responses to apoptotic inductors. J Cell Physiol. 2004;198(3):350–8.

[25] Gabbiani G. The myofibroblast in wound healing and fibrocontractive diseases. J Pathol. 2003;200(4):500–3.

[26] Midwood KS, Williams LV, Schwarzbauer JE. Tissue repair and the dynamics of the extracellular matrix. Int J Biochem Cell Biol. 2004;36(6):1031–7.

[27] Zuk PA, Zhu M, Ashjian P, De Ugarte DA, Huang JI, Mizuno H, et al. Human adipose tissue is a source of multipotent stem cells. Mol Biol Cell. 2002;13(12):4279–95.

[28] Tsuji W, Rubin JP, Marra KG. Adipose-derived stem cells: implications in tissue regeneration. World J Stem Cells. 2014;6(3):312–21.

[29] Murphy MB, Moncivais K, Caplan AI. Mesenchymal stem cells: environmentally responsive therapeutics for regenerative medicine. Exp Mol Med. 2013;45:e54.

[30] Salgado AJ, Reis RL, Sousa NJ, Gimble JM. Adipose tissue derived stem cells secretome: soluble factors and their roles in regenerative medicine. Curr Stem Cell Res Ther. 2010;5(2):103–10.

[31] Trottier V, Marceau-Fortier G, Germain L, Vincent C, Fradette J. IFATS collection: using human adipose-derived stem/stromal cells for the production of new skin substitutes. Stem Cells. 2008;26(10):2713–23.

[32] Labbe B, Marceau-Fortier G, Fradette J. Cell sheet technology for tissue engineering: the self-assembly approach using adipose-derived stromal cells. Methods Mol Biol. 2011;702:429–41.

[33] Aubin K, Vincent C, Proulx M, Mayrand D, Fradette J. Creating capillary networks within human engineered tissues: impact of adipocytes and their secretory products. Acta Biomater. 2015;11:333–45.

[34] Lin CS, Lin G, Lue TF. Allogeneic and xenogeneic transplantation of adipose-derived stem cells in immunocompetent recipients without immunosuppressants. Stem Cells Dev. 2012;21(15):2770–8.

[35] L'Heureux N, Paquet S, Labbe R, Germain L, Auger FA. A completely biological tissue-engineered human blood vessel. FASEB J. 1998;12(1):47–56.

[36] Michel M, L'Heureux N, Pouliot R, Xu W, Auger FA, Germain L. Characterization of a new tissue-engineered human skin equivalent with hair. In Vitro Cell Dev Biol Anim. 1999;35(6):318–26.

[37] Germain L, Auger FA, Grandbois E, Guignard R, Giasson M, Boisjoly H, et al. Reconstructed human cornea produced in vitro by tissue engineering. Pathobiology. 1999;67(3):140–7.

[38] Hata R, Senoo H. L-ascorbic acid 2-phosphate stimulates collagen accumulation, cell proliferation, and formation of a three-dimensional tissuelike substance by skin fibroblasts. J Cell Physiol. 1989;138(1):8–16.

[39] Prunieras M, Regnier M, Woodley D. Methods for cultivation of keratinocytes with an air-liquid interface. J Invest Dermatol. 1983;81(1 Suppl):28s–33s.

[40] Cvetkovska B, Islam N, Goulet F, Germain L. Identification of functional markers in a self-assembled skin substitute in vitro. In Vitro Cell Dev Biol Anim. 2008;44(10):444–50.

[41] Pouliot R, Larouche D, Auger FA, Juhasz J, Xu W, Li H, et al. Reconstructed human skin produced in vitro and grafted on athymic mice. Transplantation. 2002;73(11):1751–7.

[42] Spiekstra SW, Breetveld M, Rustemeyer T, Scheper RJ, Gibbs S. Wound-healing factors secreted by epidermal keratinocytes and dermal fibroblasts in skin substitutes. Wound Repair Regen. 2007;15(5):708–17.

[43] Nowinski D, Lysheden AS, Gardner H, Rubin K, Gerdin B, Ivarsson M. Analysis of gene expression in fibroblasts in response to keratinocyte-derived factors in vitro: potential implications for the wound healing process. J Invest Dermatol. 2004;122(1): 216–21.

[44] Boa O, Cloutier CB, Genest H, Labbe R, Rodrigue B, Soucy J, et al. Prospective study on the treatment of lower-extremity chronic venous and mixed ulcers using tissue-engineered skin substitute made by the self-assembly approach. Adv Skin Wound Care. 2013;26(9):400–9.

[45] Beaudoin Cloutier C, Guignard R, Bernard G, Gauvin R, Larouche D, Lavoie A, et al. Production of a bilayered self-assembled skin substitute using a tissue-engineered acellular dermal matrix. Tissue Eng Part C Methods. 2015;21(12):1297–305.

[46] Pham C, Greenwood J, Cleland H, Woodruff P, Maddern G. Bioengineered skin substitutes for the management of burns: a systematic review. Burns. 2007;33(8):946–57.

[47] Auger FA, Berthod F, Moulin V, Pouliot R, Germain L. Tissue-engineered skin substitutes: from in vitro constructs to in vivo applications. Biotechnol Appl Biochem. 2004;39(Pt 3):263–75.

[48] Jean J, Lapointe M, Soucy J, Pouliot R. Development of an in vitro psoriatic skin model by tissue engineering. J Dermatol Sci. 2009;53(1):19–25.

[49] Jean J, Leroy M, Duque-Fernandez A, Bernard G, Soucy J, Pouliot R. Characterization of a psoriatic skin model produced with involved or uninvolved cells. J Tissue Eng Regen Med. 2015;9(7):789–98.

[50] Jean J, Soucy J, Pouliot R. Effects of retinoic acid on keratinocyte proliferation and differentiation in a psoriatic skin model. Tissue Eng Part A. 2011;17(13–14):1859–68.

[51] Ayata RE, Bouhout S, Auger M, Pouliot R. Study of in vitro capillary-like structures in psoriatic skin substitutes. Biores Open Access. 2014;3(5):197–205.

[52] Garcia-Perez ME, Royer M, Duque-Fernandez A, Diouf PN, Stevanovic T, Pouliot R. Antioxidant, toxicological and antiproliferative properties of Canadian polyphenolic extracts on normal and psoriatic keratinocytes. J Ethnopharmacol. 2010;132(1):251–8.

[53] Bellemare J, Roberge CJ, Bergeron D, Lopez-Valle CA, Roy M, Moulin VJ. Epidermis promotes dermal fibrosis: role in the pathogenesis of hypertrophic scars. J Pathol. 2005;206(1):1–8.

[54] Simon F, Bergeron D, Larochelle S, Lopez-Valle CA, Genest H, Armour A, et al. Enhanced secretion of TIMP-1 by human hypertrophic scar keratinocytes could contribute to fibrosis. Burns. 2012;38(3):421–7.

[55] Corriveau MP, Boufaied I, Lessard J, Chabaud S, Senecal JL, Grodzicky T, et al. The fibrotic phenotype of systemic sclerosis fibroblasts varies with disease duration and severity of skin involvement: reconstitution of skin fibrosis development using a tissue engineering approach. J Pathol. 2009;217(4):534–42.

[56] Pare B, Touzel-Deschenes L, Lamontagne R, Lamarre MS, Scott FD, Khuong HT, et al. Early detection of structural abnormalities and cytoplasmic accumulation of TDP-43 in tissue-engineered skins derived from ALS patients. Acta Neuropathol Commun. 2015;3:5.

[57] L'Heureux N, Germain L, Labbe R, Auger FA. In vitro construction of a human blood vessel from cultured vascular cells: a morphologic study. J Vasc Surg. 1993;17(3):499–509.

[58] Grenier G, Remy-Zolghadri M, Guignard R, Bergeron F, Labbe R, Auger FA, et al. Isolation and culture of the three vascular cell types from a small vein biopsy sample. In Vitro Cell Dev Biol Anim. 2003;39(3–4):131–9.

[59] L'Heureux N, Stoclet JC, Auger FA, Lagaud GJ, Germain L, Andriantsitohaina R. A human tissue-engineered vascular media: a new model for pharmacological studies of contractile responses. FASEB J. 2001;15(2):515–24.

[60] Tremblay C, Ruel J, Bourget JM, Laterreur V, Vallieres K, Tondreau MY, et al. A new construction technique for tissue-engineered heart valves using the self-assembly method. Tissue Eng Part C Methods. 2014;20(11):905–15.

[61] Dube J, Bourget JM, Gauvin R, Lafrance H, Roberge CJ, Auger FA, et al. Progress in developing a living human tissue-engineered tri-leaflet heart valve assembled from tissue produced by the self-assembly approach. Acta Biomater. 2014;10(8):3563–70.

[62] Proulx S, d'Arc Uwamaliya J, Carrier P, Deschambeault A, Audet C, Giasson CJ, et al. Reconstruction of a human cornea by the self-assembly approach of tissue engineering using the three native cell types. Mol Vis. 2010;16:2192–201.

[63] Couture C, Zaniolo K, Carrier P, Lake J, Patenaude J, Germain L, et al. The tissue-engineered human cornea as a model to study expression of matrix metalloproteinases during corneal wound healing. Biomaterials. 2016;78:86–101.

[64] Vermette M, Trottier V, Menard V, Saint-Pierre L, Roy A, Fradette J. Production of a new tissue-engineered adipose substitute from human adipose-derived stromal cells. Biomaterials. 2007;28(18):2850–60.

[65] Vallieres K, Laterreur V, Tondreau MY, Ruel J, Germain L, Fradette J, et al. Human adipose-derived stromal cells for the production of completely autologous self-assembled tissue-engineered vascular substitutes. Acta Biomater. 2015;24:209–19.

[66] Rousseau A, Fradette J, Bernard G, Gauvin R, Laterreur V, Bolduc S. Adipose-derived stromal cells for the reconstruction of a human vesical equivalent. J Tissue Eng Regen Med. 2015;9(11):E135-43.

[67] Magnan M, Berthod F, Champigny MF, Soucy F, Bolduc S. In vitro reconstruction of a tissue-engineered endothelialized bladder from a single porcine biopsy. J Pediatr Urol. 2006;2(4):261–70.

[68] Cattan V, Bernard G, Rousseau A, Bouhout S, Chabaud S, Auger FA, et al. Mechanical stimuli-induced urothelial differentiation in a human tissue-engineered tubular genitourinary graft. Eur Urol. 2011;60(6):1291–8.

[69] Imbeault A, Bernard G, Rousseau A, Morissette A, Chabaud S, Bouhout S, et al. An endothelialized urothelial cell-seeded tubular graft for urethral replacement. Can Urol Assoc J. 2013;7(1–2):E4-9.

[70] Bureau M, Pelletier J, Rousseau A, Bernard G, Chabaud S, Bolduc S. Demonstration of the direct impact of ketamine on urothelium using a tissue engineered bladder model. Can Urol Assoc J. 2015;9(9–10):E613-7.

[71] Grodzinsky AJ. Electromechanical and physicochemical properties of connective tissue. Crit Rev Biomed Eng. 1983;9(2):133–99.

[72] Wang JH, Thampatty BP. An introductory review of cell mechanobiology. Biomech Model Mechanobiol. 2006;5(1):1–16.

[73] Buck RC. Reorientation response of cells to repeated stretch and recoil of the substratum. Exp Cell Res. 1980;127(2):470–4.

[74] Dartsch PC, Hammerle H, Betz E. Orientation of cultured arterial smooth muscle cells growing on cyclically stretched substrates. Acta Anatomica. 1986;125(2):108–13.

[75] Mac Kenna DA, Dolfi F, Vuori K, Ruoslahti E. Extracellular signal-regulated kinase and c-Jun NH2-terminal kinase activation by mechanical stretch is integrin-dependent and matrix-specific in rat cardiac fibroblasts. J Clin Invest. 1998;101(2):301–10.

[76] Gudi SR, Lee AA, Clark CB, Frangos JA. Equibiaxial strain and strain rate stimulate early activation of G proteins in cardiac fibroblasts. Am J Physiol. 1998;274(5 Pt 1):C1424-8.

[77] Lee AA, Delhaas T, Mc Culloch AD, Villarreal FJ. Differential responses of adult cardiac fibroblasts to in vitro biaxial strain patterns. J Mol Cell Cardiol. 1999;31(10):1833–43.

[78] Adhikari AS, Chai J, Dunn AR. Mechanical load induces a 100-fold increase in the rate of collagen proteolysis by MMP-1. J Am Chem Soc. 2011;133(6):1686–9.

[79] Tyagi SC, Lewis K, Pikes D, Marcello A, Mujumdar VS, Smiley LM, et al. Stretch-induced membrane type matrix metalloproteinase and tissue plasminogen activator in cardiac fibroblast cells. J Cell Physiol. 1998;176(2):374–82.

[80] Carver W, Goldsmith EC. Regulation of tissue fibrosis by the biomechanical environment. Biomed Res Int. 2013;2013:101979.

[81] Gauvin R, Ahsan T, Larouche D, Levesque P, Dube J, Auger FA, et al. A novel single-step self-assembly approach for the fabrication of tissue-engineered vascular constructs. Tissue Eng Part A. 2010;16(5):1737–47.

[82] Gauvin R, Guillemette M, Galbraith T, Bourget JM, Larouche D, Marcoux H, et al. Mechanical properties of tissue-engineered vascular constructs produced using arterial or venous cells. Tissue Eng Part A. 2011;17(15–16):2049–59.

[83] Konig G, Mc Allister TN, Dusserre N, Garrido SA, Iyican C, Marini A, et al. Mechanical properties of completely autologous human tissue engineered blood vessels compared to human saphenous vein and mammary artery. Biomaterials. 2009;30(8): 1542–50.

[84] Guillemette MD, Cui B, Roy E, Gauvin R, Giasson CJ, Esch MB, et al. Surface topography induces 3D self-orientation of cells and extracellular matrix resulting in improved tissue function. Integr Biol (Camb). 2009;1(2):196–204.

[85] Fortier GM, Gauvin R, Proulx M, Vallee M, Fradette J. Dynamic culture induces a cell type-dependent response impacting on the thickness of engineered connective tissues. J Tissue Eng Regen Med. 2013;7(4):292–301.

[86] Cardinale GJ, Udenfriend S. Prolyl hydroxylase. Adv Enzymol Relat Areas Mol Biol. 1974;41(0):245–300.

[87] Murad S, Grove D, Lindberg KA, Reynolds G, Sivarajah A, Pinnell SR. Regulation of collagen synthesis by ascorbic acid. Proc Natl Acad Sci U S A. 1981;78(5):2879–82.

[88] Chabaud S, Rousseau A, Marcoux TL, Bolduc S. Inexpensive production of near-native engineered stromas. J Tissue Eng Regen Med. 2015.

[89] Alexakis C, Mestries P, Garcia S, Petit E, Barbier V, Papy-Garcia D, et al. Structurally different RGTAs modulate collagen-type expression by cultured aortic smooth muscle cells via different pathways involving fibroblast growth factor-2 or transforming growth factor-beta1. FASEB J. 2004;18(10):1147–9.

[90] Wei D, Zhang L, Williams DL, Browder IW. Glucan stimulates human dermal fibroblast collagen biosynthesis through a nuclear factor-1 dependent mechanism. Wound Repair Regen. 2002;10(3):161–8.

[91] Rioux LE. Université Laval; 2010.

[92] Ayoub A, Pereira JM, Rioux LE, Turgeon SL, Beaulieu M, Moulin VJ. Role of seaweed laminaran from Saccharina longicruris on matrix deposition during dermal tissue-engineered production. Int J Biol Macromol. 2015;75:13–20.

[93] Lehtovaara BC, Gu FX. Pharmacological, structural, and drug delivery properties and applications of 1,3-beta-glucans. J Agric Food Chem. 2011;59(13):6813–28.

[94] Pageon H, Zucchi H, Rousset F, Monnier VM, Asselineau D. Skin aging by glycation: lessons from the reconstructed skin model. Clin Chem Lab Med. 2014;52(1):169–74.

[95] Pageon H. Reaction of glycation and human skin: the effects on the skin and its components, reconstructed skin as a model. Pathol Biol. 2010;58(3):226–31.

[96] Bjork JW, Meier LA, Johnson SL, Syedain ZH, Tranquillo RT. Hypoxic culture and insulin yield improvements to fibrin-based engineered tissue. Tissue Eng Part A. 2012;18(7–8):785–95.

[97] Aijaz A, Faulknor R, Berthiaume F, Olabisi RM. Hydrogel microencapsulated insulin-secreting cells increase keratinocyte migration, epidermal thickness, collagen fiber density, and wound closure in a diabetic mouse model of wound healing. Tissue Eng Part A. 2015;21(21–22):2723–32.

[98] Dhall S, Silva JP, Liu Y, Hrynyk M, Garcia M, Chan A, et al. Release of insulin from PLGA-alginate dressing stimulates regenerative healing of burn wounds in rats. Clin Sci. 2015;129(12):1115–29.

[99] Pallet N, Hebert MJ. The apoptotic program promotes tissue remodeling and fibrosis. Kidney Int. 2011;80(10):1108; author reply.

[100] Buechler C, Krautbauer S, Eisinger K. Adipose tissue fibrosis. World J Diab. 2015;6(4): 548–53.

[101] Morissette A, Imbeault A, Cattan V, Bernard G, Taillon G, Chabaud S, Bolduc S. Strategies to reconstruct a functional urethral substitute by self-assembly method. Procedia Eng. 2013;59:8.

[102] Chabaud S, BC Boiroux B, Saba I, Leclerc M, Rousseau A, Bouhout S, Bolduc S. Urothelial cell expansion and urothelium maturation are improved by exposure to hypoxia. J Tissue Eng Regen Med. 2016; in revision.

[103] Montesinos MC, Desai A, Chen JF, Yee H, Schwarzschild MA, Fink JS, et al. Adenosine promotes wound healing and mediates angiogenesis in response to tissue injury via occupancy of A(2A) receptors. Am J Pathol. 2002;160(6):2009–18.

[104] Chan ES, Fernandez P, Merchant AA, Montesinos MC, Trzaska S, Desai A, et al. Adenosine A2A receptors in diffuse dermal fibrosis: pathogenic role in human dermal fibroblasts and in a murine model of scleroderma. Arthritis Rheum. 2006;54(8):2632–42.

[105] Chen Y, Epperson S, Makhsudova L, Ito B, Suarez J, Dillmann W, et al. Functional effects of enhancing or silencing adenosine A2b receptors in cardiac fibroblasts. Am J Physiol Heart Circ Physiol. 2004;287(6):H2478-86.

[106] Mills GB, Moolenaar WH. The emerging role of lysophosphatidic acid in cancer. Nat Rev Cancer. 2003;3(8):582–91.

[107] Pradere JP, Gonzalez J, Klein J, Valet P, Gres S, Salant D, et al. Lysophosphatidic acid and renal fibrosis. Biochim Biophys Acta. 2008;1781(9):582–7.

[108] Chabaud S, Marcoux TL, Deschenes-Rompre MP, Rousseau A, Morissette A, Bouhout S, et al. Lysophosphatidic acid enhances collagen deposition and matrix thickening in engineered tissue. J Tissue Eng Regen Med. 2015;9(11):E65-75.

[109] Carrier P, Deschambeault A, Audet C, Talbot M, Gauvin R, Giasson CJ, et al. Impact of cell source on human cornea reconstructed by tissue engineering. Invest Ophthalmol Vis Sci. 2009;50(6):2645–52.

[110] Bouhout S, Goulet F, Bolduc S. A novel and faster method to obtain a differentiated 3-dimensional tissue engineered bladder. J Urol. 2015;194(3):834–41.

[111] Bouhout S, Chabaud S, Bolduc S. Organ-specific matrix self-assembled by mesenchymal cells improves the normal urothelial differentiation in vitro. World J Urol. 2016;34(1):121–30.

[112] Baldwin J, Antille M, Bonda U, De-Juan-Pardo EM, Khosrotehrani K, Ivanovski S, et al. In vitro pre-vascularisation of tissue-engineered constructs A co-culture perspective. Vasc Cell. 2014;6:13.

[113] Blinder YJ, Freiman A, Raindel N, Mooney DJ, Levenberg S. Vasculogenic dynamics in 3D engineered tissue constructs. Sci Rep. 2015;5:17840.

[114] Tremblay PL, Hudon V, Berthod F, Germain L, Auger FA. Inosculation of tissue-engineered capillaries with the host's vasculature in a reconstructed skin transplanted on mice. Am J Transplant. 2005;5(5):1002–10.

[115] Ayata RE, Chabaud S, Auger M, Pouliot R. Behaviour of endothelial cells in a tridimensional in vitro environment. Biomed Res Int. 2015;2015:630461.

[116] Tung JJ, Tattersall IW, Kitajewski J. Tips, stalks, tubes: notch-mediated cell fate determination and mechanisms of tubulogenesis during angiogenesis. Cold Spring Harb Perspect Med. 2012;2(2):a006601.

[117] Szoke K, Reinisch A, Ostrup E, Reinholt FP, Brinchmann JE. Autologous cell sources in therapeutic vasculogenesis: in vitro and in vivo comparison of endothelial colony-forming cells from peripheral blood and endothelial cells isolated from adipose tissue. Cytotherapy. 2016;18(2):242–52.

[118] Klar AS, Guven S, Zimoch J, Zapiorkowska NA, Biedermann T, Bottcher-Haberzeth S, et al. Characterization of vasculogenic potential of human adipose-derived endothelial cells in a three-dimensional vascularized skin substitute. Pediatr Surg Int. 2016;32(1):17–27.

[119] Black AF, Berthod F, L'Heureux N, Germain L, Auger FA. In vitro reconstruction of a human capillary-like network in a tissue-engineered skin equivalent. FASEB J. 1998;12(13):1331–40.

[120] Gibot L, Galbraith T, Huot J, Auger FA. A preexisting microvascular network benefits in vivo revascularization of a microvascularized tissue-engineered skin substitute. Tissue Eng Part A. 2010;16(10):3199–206.

[121] Rochon MH, Fradette J, Fortin V, Tomasetig F, Roberge CJ, Baker K, et al. Normal human epithelial cells regulate the size and morphology of tissue-engineered capillaries. Tissue Eng Part A. 2010;16(5):1457–68.